[1]SOLDIER TRAINING PUBLICATION
No. 8-68W13-SM-TG

HEADQUARTERS
DEPARTMENT OF THE ARMY
Washington, DC, 15 April 2009

SOLDIER'S MANUAL
AND
TRAINER'S GUIDE
MOS 68W
Health Care Specialist
Skill Levels 1, 2 AND 3

TABLE OF CONTENTS

PAGE

[1]*This publication supersedes STP 8-91W15-SM-TG, 10 October 2001.

Subject Area 4: Airway Management

Subject Area 5: Venipuncture and IV Therapy

Subject Area 6: Primary Care

Subject Area 7: Musculosketeal

PREFACE

This publication is for skill level 1, 2 and 3 Soldiers holding military occupational specialty (MOS) 68W and for trainers and first-line supervisors. It contains standardized training objectives, in the form of task summaries, to train and evaluate Soldiers on critical tasks that support unit missions during wartime. Trainers and first-line supervisors should ensure Soldiers holding MOS/SL 68W1/2/3/ have access to this publication. This STP is available for download from the Reimer Digital Library (RDL).

This publication applies to the Active Army, the Army National guard (ARNGO/Army National Guard of the Unites States (ARNGUS), and the U.S. Army Reserve (USAR) unless otherwise stated.

The proponent of this publication is United States Army Training and Doctrine Command. Send comments and recommendations on DA Form 2028 (Recommended Changes to Publications and Blank Forms) directly to Academy of Health Sciences, ATTN: MCCS-HTI, 1750 Greely Rd, STE 135, Fort Sam Houston, TX 78234-5078.

Unless this manual states otherwise, masculine pronouns do not refer exclusively to men.

CHAPTER 1

Introduction

1-1. General. This Soldier training publication (STP) identifies the individual military occupational specialty (MOS) training requirements for Soldiers in MOS 68W. Another source of STP task data is the General Dennis J. Reimer Training and Doctrine Digital Library (RDL) at the Soldier's Training Homepage. Commanders, trainers, and Soldiers should use the STP to plan, conduct, and evaluate individual training in units. The STP is the primary MOS reference to support the self-development and training of every Soldier in the unit. It is used with the Soldier's Manual of Common Tasks, collective training products, and FM 7-0, Training for Full Spectrum Operation, to establish effective training plans and programs that integrate Soldier, leader, and collective tasks. This chapter explains how to use the STP in establishing an effective individual training program. It includes doctrinal principles and implications outlined in FM 7-0. Based on these guidelines, commanders and unit trainers must tailor the information to meet the requirements for their specific unit.

1-2. Training Requirement. Every Soldier, noncommissioned officer (NCO), warrant officer, and officer has one primary mission — to be trained and ready to fight and win our nation's wars. Success in battle does not happen by accident; it is a direct result of tough, realistic, and challenging training.

    a. Operational Environment.

        (1) Commanders and leaders at all levels must conduct training with respect to a wide variety of operational missions across the full spectrum of operations. These operations may include combined arms, joint, multinational, and interagency considerations, and span the entire breadth of terrain and environmental possibilities. Commanders must strive to set the daily training conditions as closely as possible to those expected for actual operations.

        (2) The operational missions of the Army include not only war, but also military operations other than war (MOOTW). Operations may be conducted as major combat operations, a small-scale contingency, or a peacetime military engagement. Offensive and defensive operations normally dominate military operations in war along with some small-scale contingencies. Stability operations and support operations dominate in MOOTW. Commanders at all echelons may combine different types of operations simultaneously and sequentially to accomplish missions in war and MOOTW. These missions require training since future conflict will likely involve a mix of combat and MOOTW, often concurrently. The range of possible missions complicates training. Army forces cannot train for every possible mission; they train for war and prepare for specific missions as time and circumstances permit.

        (3) One type of MOOTW is the chemical, biological, radiological, nuclear, and high-yield explosives (CBRNE) event. To assist commanders and leaders in training their units, CBRNE-related information is being included in AMEDD collective training. Even though most collective tasks may support a CBRNE event, the ones that will most directly be impacted are clearly indicated with a statement in the CONDITION that reads: "THIS TASK MAY BE USED TO SUPPORT A CBRNE EVENT." These collective tasks and any supporting individual tasks in this Soldier's manual should be considered for training emphasis.

        (4) Our forces today use a train-alert-deploy sequence. We cannot count on the

time or opportunity to correct or make up training deficiencies after deployment. Maintaining forces that are ready now, places increased emphasis on training and the priority of training. This concept is a key link between operational and training doctrine.

(5) Units train to be ready for war based on the requirements of a precise and specific mission. In the process they develop a foundation of combat skills that can be refined based on the requirements of the assigned mission. Upon alert, commanders assess and refine from this foundation of skills. In the train-alert-deploy process, commanders use whatever time the alert cycle provides to continue refinement of mission-focused training. Training continues during time available between alert notification and deployment, between deployment and employment, and even during employment as units adapt to the specific battlefield environment and assimilate combat replacements.

b.   How the Army Trains the Army.

(1) Training is a team effort and the entire Army — Department of the Army, major Army commands (MACOMs), the institutional training base, units, the combat training centers (CTCs), each individual Soldier, and the civilian workforce — has a role that contributes to force readiness. Department of the Army and MACOMs are responsible for resourcing the Army to train. The institutional Army, including schools, training centers, and NCO academies, for example, train Soldiers and leaders to take their place in units in the Army by teaching the doctrine and tactics, techniques, and procedures (TTP). Units, leaders, and individuals train to standard on their assigned critical individual tasks. The unit trains first as an organic unit and then as an integrated component of a team. Before the unit can be trained to function as a team, each Soldier must be trained to perform their individual supporting tasks to standard. Operational deployments and major training opportunities, such as major training exercises, CTCs, and collective evaluations provide rigorous, realistic, and stressful training and operational experience under actual or simulated combat and operational conditions to enhance unit readiness and produce bold, innovative leaders. The result of this Army-wide team effort is a training and leader development system that is unrivaled in the world. Effective training produces the force — Soldiers, leaders, and units — that can successfully execute any assigned mission.

(2) The Army Training and Leader Development Model (Figure 1-1) centers on developing trained and ready units led by competent and confident leaders. The model depicts an important dynamic that creates a lifelong learning process. The three core domains that shape the critical learning experiences throughout a Soldier's and leader's time span are the operational, institutional, and self-development domains. Together, these domains interact using feedback and assessment from various sources and methods to maximize warfighting readiness. Each domain has specific, measurable actions that must occur to develop our leaders.

- The operational domain includes home station training, CTC rotations, and joint training exercises and deployments that satisfy national objectives. Each of these actions provides foundational experiences for Soldier, leader, and unit development.

- The institutional domain focuses on educating and training Soldiers and leaders on the key knowledge, skills, and attributes required to operate in any environment. It includes individual, unit and joint schools, and advanced education.

- The self-development domain, both structured and informal, focuses on taking those actions necessary to reduce or eliminate the gap between operational and institutional experiences.

Figure 1-1. Army Training and Leader Development Model

(3) Throughout this lifelong learning and experience process, there is formal and informal assessment and feedback of performance to prepare leaders and Soldiers for their next level of responsibility. Assessment is the method used to determine the proficiency and potential of leaders against a known standard. Feedback must be clear, formative guidance directly related to the outcome of training events measured against standards.

c. Leader Training and Leader Development.

(1) Competent and confident leaders are a prerequisite to the successful training of units. It is important to understand that leader training and leader development are integral parts of unit readiness. Leaders are inherently Soldiers first and should be technically and tactically proficient in basic Soldier skills. They are also adaptive, capable of sensing their environment, adjusting the plan when appropriate, and properly applying the proficiency acquired through training.

(2) Leader training is an expansion of these skills that qualifies them to lead other Soldiers. As such, doctrine and principles of training require the same level of attention of senior commanders. Leader training occurs in the institutional Army, the unit, the CTCs, and

through self-development. Leader training is just one portion of leader development.

(3)　Leader development is the deliberate, continuous, sequential, and progressive process, grounded in Army values, that grows Soldiers and civilians into competent and confident leaders capable of decisive action. Leader development is achieved through the life-long synthesis of the knowledge, skills, and experiences gained through institutional training and education, organizational training, operational experience, and self-development. Commanders play the key roll in leader development that ideally produces tactically and technically competent, confident, and adaptive leaders who act with boldness and initiative in dynamic, complex situations to execute mission-type orders achieving the commander's intent.

(4)　A life cycle management diagram for MOS 68W Soldiers is on page 1-5. You can find more information at http://www.cs.amedd.army.mil/details.aspx?dt=127 (scroll down to Life Cycle Training Charts, select the Enlisted link, and find the appropriate tab along the bottom). This information, combined with the MOS Training Plan in Chapter 2, forms the career development model for the MOS.

d.　Training Responsibility. Soldier and leader training and development continue in the unit. Using the institutional foundation, training in organizations and units focuses and hones individual and team skills and knowledge.

(1)　Commander Responsibility.

(a)　The unit commander is responsible for the wartime readiness of all elements in the formation. The commander is, therefore, the primary trainer of the organization and is responsible for ensuring that all training is conducted in accordance with the STP to the Army standard.

(b)　Commanders ensure STP standards are met during all training. If a Soldier fails to meet established standards for identified MOS tasks, the Soldier must retrain until the tasks are performed to standard. Training to standard on MOS tasks is more important than completion of a unit training event such as collective task evaluation. The objective is to focus on sustaining MOS proficiency — this is the critical factor commanders must adhere to when training individual Soldiers in units.

(2)　NCO Responsibility.

(a)　A great strength of the US Army is its professional NCO Corps who takes pride in being responsible for the individual training of Soldiers, crews, and small teams. The NCO support channel parallels and complements the chain of command. It is a channel of communication and supervision from the command sergeant major (CSM) to the first sergeants (1SGs) and then to other NCOs and enlisted personnel. NCOs train Soldiers to the non-negotiable standards published in STPs. Commanders delegate authority to NCOs in the support channel as the primary trainers of individual, crew, and small team training. Commanders hold NCOs responsible for conducting standards-based, performance-oriented, battle-focused training and providing feedback on individual, crew, and team proficiency. Commanders define responsibilities and authority of their NCOs to their staffs and subordinates.

MOS 68W
Health Care Specialist
CAREER/TRAINING LIFE CYCLE

RANK	AMEDD Course NR	TRAINING	LENGTH	LOCATION	ATTENDANCE REQUIREMENT			Self-Development Course NR	SELF-DEVELOPMENT	LENGTH	LOCATION	ATTENDANCE REQUIREMENT
E1 - E6									PROFESSIONAL, POST-GRADUATE SHORT COURSE PROGRAM			
		Basic Combat Training Course	9 weeks	Ft. LW./Ft. Sill./Ft. Jackson./Ft. Benning	JBT				Surgical Support NCO Short Course	4 days	SA, TX	Sustainment
		Health Care Specialist	16 weeks	FSH, TX	AIT/MOS				MEDCOM CSM/SGM BR NCO	4 days	SA, TX	Sustainment
		SEE ASI's H3, H8, P1, P2, P3, M6, M3, Y6	4 weeks 2 days	Multiple sites	Leadership				CSM/SGM BR NCO Course	4 days	Landstuhl, Germany	Sustainment
		Warrior Leader Course	2 Weeks	Fort Rucker, AL	Voluntary SQI-F				Joint Services Emergency Medicine Symposium	5 days	SA, TX	Sustainment
		Flight Medic	26 Weeks	Fort Bragg, NC	Voluntary ASI W1				Annual Field Medicine Short Course	3 Days	FSH, TX	Sustainment
		Special Operations Combat Medic Course (SOCM)	2 weeks 2 days	Multiple sites	DA Selection Leadership				Baenwals, Plans Operations, Intelligence, Training	5 days	FSH, TX	Sustainment
		NCO Basic (Common Core) PH I	4 weeks, 2 days	FSH, TX	DA Selection Leadership				DISTRIBUTED LEARNING (DL) / CORRESPONDENCE COURSES			Initial (3 year) sustainment requirement
		68W TRANSITION COURSE	2 weeks	Multiple sites	DA Selection Leadership				CMFINC Operator / Respirator Course		Onsite	
		TATS AMEDD BNCOC (68W72) (DL)	4 weeks 2 days	FSH, TX	DA Selection Leadership				See page 5			
		NCO Basic (Common Core) Ph. 1 (DLTR)	2 weeks 2 days	Multiple sites	DA Selection Leadership				SPECIALTY COURSES			
		TATS AMEDD BNCOC (68W90)	1 week	Multiple sites	DA Selection Leadership				Orientation to the AMEDD Training Process	2 days	Ft Sam. Houston	JIT
E6 - E7		Instructor Training Course	2 weeks	AMS	JIT/SI (BI)				Air Assault School	2 weeks	Ft Campbell various sites	Graduation assignment for Air Assault position / volunteer
		Small Group Instructor Training Course (SGITC)	1 week	AMS	JIT				Airborne Training	3 weeks	Ft Benning	One course assignment to Airborne Div / Volunteer
		Battle Staff NCO	2 weeks 2 days	USASMA (Ft. Bliss)	JIT ASI V5				EMT-Intermediate		Various locations	
		Recruiter	7 weeks 2 days	USAREC	JIT SQI-4				EMT-Paramedic		Various locations	
		Drill Sgt School	9 weeks	Multiple Sites	JIT SQI-X							
		Inspector General NCO											
		First Sergeant NCO											
		Systems Automation Advisor											
		Observer/Controller NCO											
		AMEDD NCO Advanced (NCOES)	DL		DA Selection Leadership								
E7 - E8		AMEDD NCO Advanced (NCOES)	2 weeks days	FSH, TX	DA Selection Semi-Annual								
	TC-8-800	Combat Medical Skills Validation Test		Unit Training	Re-certification (every two years)								
		Emergency Medical Technician-Basic (EMT-B)		Multiple sites									
E8		First Sergeant Course (DL)	Ph 1-DL, Ph 2-4 weeks	USASMA	JIT SQI-M								
E9		Sergeants Major Non-Resident Course	39 weeks, 2 days	USASMA (Ft. Bliss)	DA Selection								
		Sergeants Major Resident Course	1 week	USASMA	Leadership								

NOTE: Genuine to R1 at 68W

(b) NCOs continue the soldierization process of newly assigned enlisted Soldiers, and begin their professional development. NCOs are responsible for conducting standards-based, performance-oriented, battle-focused training. They identify specific individual, crew, and small team tasks that support the unit's collective mission essential tasks; plan, prepare, rehearse, and execute training; and evaluate training and conduct after action reviews (AARs) to provide feedback to the commander on individual, crew, and small team proficiency. Senior NCOs coach junior NCOs to master a wide range of individual tasks.

(3) Soldier Responsibility. Each Soldier is responsible for performing individual tasks identified by the first-line supervisor based on the unit's mission essential task list (METL). Soldiers must perform tasks to the standards included in the task summary. If Soldiers have questions about tasks or which tasks in this manual they must perform, they are responsible for asking their first-line supervisor for clarification, assistance, and guidance. First-line supervisors know how to perform each task or can direct Soldiers to appropriate training materials, including current field manuals, technical manuals, and Army regulations. Soldiers are responsible for using these materials to maintain performance. They are also responsible for maintaining standard performance levels of all Soldier's Manual of Common Tasks at their current skill level and below. Periodically, Soldiers should ask their supervisor or another Soldier to check their performance to ensure that they can perform the tasks.

1-3. Battle-Focused Training. Battle focus is a concept used to derive peacetime training requirements from assigned and anticipated missions. The priority of training in units is to train to standard on the wartime mission. Battle focus guides the planning, preparation, execution, and assessment of each organization's training program to ensure its members train as they are going to fight. Battle focus is critical throughout the entire training process and is used by commanders to allocate resources for training based on wartime and operational mission requirements. Battle focus enables commanders and staffs at all echelons to structure a training program that copes with non-mission-related requirements while focusing on mission essential training activities. It is recognized that a unit cannot attain proficiency to standard on every task whether due to time or other resource constraints. However, unit commanders can achieve a successful training program by consciously focusing on a reduced number of METL tasks that are essential to mission accomplishment.

a. Linkage Between METL and STP. A critical aspect of the battle focus concept is to understand the responsibility for and the linkage between the collective mission essential tasks and the individual tasks that support them. For example, the commander and the CSM/1SG must jointly coordinate the collective mission essential tasks and supporting individual tasks on which the unit will concentrate its efforts during a given period. This task hierarchy is provided in the task database at the Reimer Digital Library. The CSM/1SG must select the specific individual tasks that support each collective task to be trained. Although NCOs have the primary role in training and sustaining individual Soldier skills, officers at every echelon remain responsible for training to established standards during both individual and collective training. Battle focus is applied to all missions across the full spectrum of operations.

b. Relationship of STPs to Battle-focused Training. The two key components of any STP are the Soldier's manual (SM) and trainer's guide (TG). Each gives leaders important information to help implement the battle-focused training process. The trainer's guide relates Soldier and leader tasks in the MOS and skill level to duty positions and equipment. It states where the task is trained, how often training should occur to sustain proficiency, and who in the unit should be trained. As leaders assess and plan training, they should rely on the trainer's guide to help identify training needs.

      (1)   Leaders conduct and evaluate training based on Army-wide training objectives and on the task standards published in the Soldier's manual task summaries or in the Reimer Digital Library. The task summaries ensure that --

- Trainers in every unit and location define task standards the same way
- Trainers evaluate all Soldiers to the same standards

      (2)   Figure 1-2 shows how battle-focused training relates to the trainer's guide and Soldier's manual:

- The left column shows the steps involved in training Soldiers.
- The right column shows how the STP supports each of these steps.

BATTLE-FOCUS PROCESS	STP SUPPORT PROCESS
Select supporting Soldier tasks	Use TG to relate tasks to METL
Conduct training assessment	Use TG to define what Soldier tasks to assess
Determine training objectives	Use TG to set objectives
Determine strategy; plan for training	Use TG to relate Soldier tasks to strategy
Conduct pre-execution checks	Use SM task summary as source for task performance
Execute training; conduct after action review	Use SM task summary as source for task performance
Evaluate training against established standards	Use SM task summary as standard for evaluation

Figure 1-2. Relationship of Battle-focused Training and STP

1-4. Task Summary Format. Task summaries outline the wartime performance requirements of each critical task in the SM. They provide the Soldier and the trainer with the information necessary to prepare, conduct, and evaluate critical task training. As a minimum, task summaries include information the Soldier must know and the skills that he must perform to standards for each task. The format of the task summaries included in this SM is as follows:

     a.   Task Title. The task title identifies the action to be performed.

     b.   Task Number. A 10-digit number identifies each task or skill. This task number, along with the task title, must be included in any correspondence pertaining to the task.

     c.   Conditions. The task conditions identify all the equipment, tools, references, job aids, and supporting personnel that the Soldier needs to use to perform the task in wartime. This section identifies any environmental conditions that can alter task performance, such as visibility, temperature, or wind. This section also identifies any specific cues or events that trigger task performance, such as a chemical attack or identification of a threat vehicle.

     d.   Standards. The task standards describe how well and to what level the task must be performed under wartime conditions. Standards are typically described in terms of accuracy, completeness, and speed.

e.   Performance Steps. This section includes a detailed outline of information on how to perform the task. Additionally, some task summaries include safety statements and notes. Safety statements (danger, warning, and caution) alert users to the possibility of immediate death, personal injury, or damage to equipment. Notes provide a small, extra supportive explanation or hint relative to the performance steps.

f.   Evaluation Preparation (when used). This subsection indicates necessary modifications to task performance in order to train and evaluate a task that cannot be trained to the wartime standard under wartime conditions. It may also include special training and evaluation preparation instructions to accommodate these modifications and any instructions that should be given to the Soldier before evaluation.

g.   Performance Measures. This evaluation guide identifies the specific actions that the Soldier must do to successfully complete the task. These actions are listed in a GO/NO-GO format for easy evaluation. Each evaluation guide contains an evaluation guidance statement that indicates the requirements for receiving a GO on the evaluation.

h.   References. This section identifies references that provide more detailed and thorough explanations of task performance requirements than those given in the task summary description.

1-5. Training Execution. All good training, regardless of the specific collective, leader, and individual tasks being executed, must comply with certain common requirements. These include adequate preparation, effective presentation and practice, and thorough evaluation. The execution of training includes preparation for training, conduct of training, and recovery from training.

a.   Preparation for Training. Formal near-term planning for training culminates with the publication of the unit training schedule. Informal planning, detailed coordination, and preparation for executing the training continue until the training is performed. Commanders and other trainers use training meetings to assign responsibility for preparation of all scheduled training. Preparation for training includes selecting tasks to be trained, planning the conduct of the training, training the trainers, reconnaissance of the site, issuing the training execution plan, and conducting rehearsals and pre-execution checks. Pre-execution checks are preliminary actions commanders and trainers use to identify responsibility for these and other training support tasks. They are used to monitor preparation activities and to follow up to ensure planned training is conducted to standard. Pre-execution checks are a critical portion of any training meeting. During preparation for training, battalion and company commanders identify and eliminate potential training distracters that develop within their own organizations. They also stress personnel accountability to ensure maximum attendance at training.

(1)   Subordinate leaders, as a result of the bottom-up feed from internal training meetings, identify and select the individual tasks necessary to support the identified training objectives. Commanders develop the tentative plan to include requirements for preparatory training, concurrent training, and training resources. As a minimum, the training plan should include confirmation of training areas and locations, training ammunition allocations, training simulations and simulators availability, transportation requirements, Soldier support items, a risk management analysis, assignment of responsibility for the training, designation of trainers responsible for approved training, and final coordination. The time and other necessary resources for retraining must also be an integral part of the original training plan.

(2)   Leaders, trainers, and evaluators are identified, trained to standard, and rehearsed prior to the conduct of the training. Leaders and trainers are coached on how to train, given time to prepare, and rehearsed so that training will be challenging and doctrinally correct. Commanders ensure that trainers and evaluators are not only tactically and technically competent on their training tasks, but also understand how the training relates to the organization's METL. Properly prepared trainers, evaluators, and leaders project confidence and enthusiasm to those being trained. Trainer and leader training is a critical event in the preparation phase of training. These individuals must demonstrate proficiency on the selected tasks prior to the conduct of training.

(3)   Commanders, with their subordinate leaders and trainers, conduct site reconnaissance, identify additional training support requirements, and refine and issue the training execution plan. The training plan should identify all those elements necessary to ensure the conduct of training to standard. Rehearsals are essential to the execution of good training. Realistic, standards-based, performance-oriented training requires rehearsals for trainers, support personnel, and evaluators. Preparing for training in Reserve Component (RC) organizations can require complex pre-execution checks. RC trainers must often conduct detailed coordination to obtain equipment, training support system products, and ammunition from distant locations. In addition, RC pre-execution checks may be required to coordinate Active Component assistance from the numbered CONUSA, training support divisions, and directed training affiliations.

b.   Conduct of Training. Ideally, training is executed using the crawl-walk-run approach. This allows and promotes an objective, standards-based approach to training. Training starts at the basic level. Crawl events are relatively simple to conduct and require minimum support from the unit. After the crawl stage, training becomes incrementally more difficult, requiring more resources from the unit and home station, and increasing the level of realism. At the run stage, the level of difficulty for the training event intensifies. Run stage training requires optimum resources and ideally approaches the level of realism expected in combat. Progression from the walk to the run stage for a particular task may occur during a one-day training exercise or may require a succession of training periods over time. Achievement of the Army standard determines progression between stages.

(1)   In crawl-walk-run training, the tasks and the standards remain the same; however, the conditions under which they are trained change. Commanders may change the conditions, for example, by increasing the difficulty of the conditions under which the task is being performed, increasing the tempo of the task training, increasing the number of tasks being trained, or by increasing the number of personnel involved in the training. Whichever approach is used, it is important that all leaders and Soldiers involved understand in which stage they are currently training and understand the Army standard.

(2)   An AAR is immediately conducted and may result in the need for additional training. Any task that was not conducted to standard should be retrained. Retraining should be conducted at the earliest opportunity. Commanders should program time and other resources for retraining as an integral part of their training plan. Training is incomplete until the task is trained to standard. Soldiers will remember the standard enforced, not the one discussed.

c.   Recovery from Training. The recovery process is an extension of training, and once completed, it signifies the end of the training event. At a minimum, recovery includes conduct of maintenance training, turn-in of training support items, and the conduct of AARs that review the overall effectiveness of the training just completed.

(1)   Maintenance training is the conduct of post-operations preventive maintenance checks and services, accountability of organizational and individual equipment, and final inspections. Class IV, Class V, TADSS, and other support items are maintained, accounted for, and turned-in, and training sites and facilities are closed out.

(2)   AARs conducted during recovery focus on collective, leader, and individual task performance, and on the planning, preparation, and conduct of the training just completed. Unit AARs focus on individual and collective task performance, and identify shortcomings and the training required to correct deficiencies. AARs with leaders focus on tactical judgment. These AARs contribute to leader learning and provide opportunities for leader development. AARs with trainers and evaluators provide additional opportunities for leader development.

1-6. Training Assessment. Assessment is the commander's responsibility. It is the commander's judgment of the organization's ability to accomplish its wartime operational mission. Assessment is a continuous process that includes evaluating individual training, conducting an organizational assessment, and preparing a training assessment. The commander uses his experience, feedback from training evaluations, and other evaluations and reports to arrive at his assessment. Assessment is both the end and the beginning of the training management process. Training assessment is more than just training evaluation, and encompasses a wide variety of inputs. Assessments include such diverse systems as training, force integration, logistics, and personnel, and provide the link between the unit's performance and the Army standard. Evaluation of training is, however, a major component of assessment. Training evaluations provide the commander with feedback on the demonstrated training proficiency of Soldiers, leaders, battle staffs, and units. Commanders cannot personally observe all training in their organization and, therefore, gather feedback from their senior staff officers and NCOs.

a.   Evaluation of Training. Training evaluations are a critical component of any training assessment. Evaluation measures the demonstrated ability of Soldiers, commanders, leaders, battle staffs, and units against the Army standard. Evaluation of training is integral to standards-based training and is the cornerstone of leader training and leader development. STPs describe standards that must be met for each Soldier task.

(1)   All training must be evaluated to measure performance levels against the established Army standard. The evaluation can be as fundamental as an informal, internal evaluation performed by the leader conducting the training. Evaluation is conducted specifically to enable the individual undergoing the training to know whether the training standard has been achieved. Commanders must establish a climate that encourages candid and accurate feedback for the purpose of developing leaders and trained Soldiers.

(2)   Evaluation of training is not a test; it is not used to find reasons to punish leaders and Soldiers. Evaluation tells Soldiers whether or not they achieved the Army standard and, therefore, assists them in determining the overall effectiveness of their training plans. Evaluation produces disciplined Soldiers, leaders, and units. Training without evaluation is a waste of time and resources.

(3)   Evaluations are used by leaders as an opportunity to coach and mentor Soldiers. A key element in developing leaders is immediate, positive feedback that coaches

and leads subordinate leaders to achieve the Army standard. This is a tested and proven path to develop competent, confident adaptive leaders.

    b.   Evaluators. Commanders must plan for formal evaluation and must ensure the evaluators are trained. These evaluators must also be trained as facilitators to conduct AARs that elicit maximum participation from those being trained. External evaluators will be certified in the tasks they are evaluating and normally will not be dual-hatted as a participant in the training being executed.

    c.   Role of Commanders and Leaders. Commanders ensure that evaluations take place at each echelon in the organization. Commanders use this feedback to teach, coach, and mentor their subordinates. They ensure that every training event is evaluated as part of training execution and that every trainer conducts evaluations. Commanders use evaluations to focus command attention by requiring evaluation of specific mission essential and battle tasks. They also take advantage of evaluation information to develop appropriate lessons learned for distribution throughout their commands.

    d.   After Action Review. The AAR, whether formal or informal, provides feedback for all training. It is a structured review process that allows participating Soldiers, leaders, and units to discover for themselves what happened during the training, why it happened, and how it can be done better. The AAR is a professional discussion that requires the active participation of those being trained. FM 7-1 provides detailed instructions for conducting an AAR and detailed guidance on coaching and critiquing during training.

1-7. Training Support. This manual includes the following information which provides additional training support information.

    a.   Glossary. The glossary, which follows the last appendix, is a single comprehensive list of acronyms, abbreviations, definitions, and letter symbols.

    b.   References. This section contains two lists of references, required and related, which support training of all tasks in this SM. Required references are listed in the conditions statement and are required for the Soldier to do the task. Related references are materials that provide more detailed information and a more thorough explanation of task performance.

    c.   Appendix. A Drug Dosage Calculations.

This page intentionally left blank.

CHAPTER 2

Training Guide

2-1. General. The MOS Training Plan (MTP) identifies the essential components of a unit training plan for individual training. Units have different training needs and requirements based on differences in environment, location, equipment, dispersion, and similar factors. Therefore, the MTP should be used as a guide for conducting unit training and not a rigid standard. The MTP consists of two parts. Each part is designed to assist the commander in preparing a unit training plan which satisfies integration, cross training, training up, and sustainment training requirements for Soldiers in this MOS.

Part One of the MTP shows the relationship of an MOS skill level between duty position and critical tasks. These critical tasks are grouped by task commonality into subject areas.

Section I lists subject area numbers and titles used throughout the MTP. These subject areas are used to define the training requirements for each duty position within an MOS.

Section II identifies the total training requirement for each duty position within an MOS and provides a recommendation for cross training and train-up/merger training.

- **Duty Position Column.** This column lists the duty positions of the MOS, by skill level, which have different training requirements.

- **Subject Area Column.** This column lists, by numerical key (see Section I), the subject areas a Soldier must be proficient in to perform in that duty position.

- **Cross-Train Column.** This column lists the recommended duty position for which Soldiers should be cross-trained.

- **Train-Up/Merger Column.** This column lists the corresponding duty position for the next higher skill level or MOSC the Soldier will merge into on promotion.

Part Two lists, by general subject areas, the critical tasks to be trained in an MOS and the type of training required (resident, integration, or sustainment).

- **Subject Area Column.** This column lists the subject area number and title in the same order as Section I, Part One of the MTP.

- **Task Number Column.** This column lists the task numbers for all tasks included in the subject area.

- **Title Column.** This column lists the task title for each task in the subject area.

- **Training Location Column.** This column identifies the training location where the task is first trained to Soldier training publications standards. If the task is first trained to standard in the unit, the word "Unit" will be in this column. If the task is first trained to standard in the training base, it will identify, by brevity code (ALC, SLC, etc.), the resident course where the task was taught. Figure 2-1 contains a list of training locations and their corresponding brevity codes.

UNIT	Trained in the Unit
AIT	Advanced Individual Training
INSTITUT	Institution
SLC	Senior Leader Course
ALC	Advanced Leader Course
ASI/SD	Additional Skill Identifier/Special Duty
IET	Initial Entry Training
PLDC	Primary Leadership Development Course
BCT	Basic Combat Course

Figure 2-1. Training Locations

- **Sustainment Training Frequency Column.** This column indicates the recommended frequency at which the tasks should be trained to ensure Soldiers maintain task proficiency. Figure 2-2 identifies the frequency codes used in this column.

BA	- Biannually
AN	- Annually
SA	- Semiannually
QT	- Quarterly
MO	- Monthly
BW	- Biweekly
WK	- Weekly

Figure 2-2. Sustainment Training Frequency Codes

- **Sustainment Training Skill Level Column.** This column lists the skill levels of the MOS for which Soldiers must receive sustainment training to ensure they maintain proficiency to Soldier's manual standards.

2-2. Part One, Section I. Subject Area Codes.

Skill Level 1
1   Vital Signs
2   Medical Treatment
3   Trauma Treatment
4   Airway Management
5   Venipuncture and IV Therapy
6   Primary Care
7   Musculosketeal
8   Chemical Agent Injuries
9   Triage and Evacuation
10   Medication Administration
11   Force Protection

Skill Level 2
12   Advanced Procedures (SL 2)

Skill Level 3
13   Advanced Procedures (SL 3)

2-3. Part One, Section II. Duty Position Training Requirements.

	DUTY POSITION	SUBJECT AREAS	CROSS TRAIN	TRAIN-UP/ MERGER
SL 1	Health Care Specialist	1-11	NA	68W1 Health Care Specialist
SL 2	Health Care Sergeant	1-12	NA	68W2 Health Care NCO
SL 3	Health Care NCO	1-13	NA	68W3 Health Care Specialist NCO

2-4. Part Two. Critical Tasks List.

MOS TRAINING PLAN
68W13

CRITICAL TASKS

Task Number	Title	Training Location	Sust Tng Freq	Sust Tng SL
	Skill Level 1			
Subject Area 1. Vital Signs				
081-831-0010	MEASURE A PATIENT'S RESPIRATIONS	AIT	SA	1-3
081-831-0011	MEASURE A PATIENT'S PULSE	AIT	SA	1-3
081-831-0012	MEASURE A PATIENT'S BLOOD PRESSURE	AIT	SA	1-3
081-831-0013	MEASURE A PATIENT'S TEMPERATURE	AIT	SA	1-3
081-833-0164	MEASURE A PATIENT'S PULSE OXYGEN SATURATION	AIT	SA	1-3
Subject Area 2. Medical Treatment				
081-831-0033	INITIATE A FIELD MEDICAL CARD	AIT	SA	1-3
081-831-0035	MANAGE A CONVULSIVE AND/OR SEIZING PATIENT	AIT	AN	1-3
081-831-0046	ADMINISTER EXTERNAL CHEST COMPRESSIONS	AIT	SA	1-3
081-833-0006	MEASURE A PATIENT'S INTAKE AND OUTPUT	UNIT	AN	1-3
081-833-0031	TREAT A CASUALTY FOR ANAPHYLACTIC SHOCK	AIT	SA	1-3
081-833-0116	ASSIST IN VAGINAL DELIVERY	AIT	AN	1-3
081-833-0143	TREAT A POISONED CASUALTY	AIT	AN	1-3
081-833-0144	TREAT A DIABETIC EMERGENCY	AIT	AN	1-3
081-833-0156	PERFORM A MEDICAL PATIENT ASSESSMENT	AIT	SA	1-3
081-833-0159	TREAT A CARDIAC EMERGENCY	AIT	AN	1-3
081-833-0160	TREAT A RESPIRATORY EMERGENCY	AIT	AN	1-3
081-833-0187	INSERT AN OROGASTRIC TUBE	UNIT	AN	1-3
081-833-0201	TREAT A NEAR DROWNING VICTIM	AIT	AN	1-3
081-833-0224	TREAT A PATIENT WITH AN ALLERGIC REACTION	AIT	AN	1-3
081-833-3027	CARDIAC ARREST USING AN AUTOMATED EXTERNAL DEFIBRILLATOR	AIT	SA	1-3
081-835-3005	PERFORM A GASTRIC LAVAGE	UNIT	AN	1-3
081-835-3007	OBTAIN AN ELECTROCARDIOGRAM	UNIT	SA	1-3
Subject Area 3. Trauma Treatment				
081-833-0045	TREAT A CASUALTY WITH AN OPEN ABDOMINAL WOUND	AIT	SA	1-3
081-833-0046	TREAT A CASUALTY WITH AN IMPALEMENT	AIT	SA	1-3
081-833-0047	INITIATE TREATMENT FOR HYPOVOLEMIC SHOCK	AIT	SA	1-3
081-833-0049	TREAT A CASUALTY WITH A CHEST INJURY	AIT	SA	1-3
081-833-0052	TREAT A CASUALTY WITH AN OPEN OR CLOSED HEAD INJURY	AIT	SA	1-3
081-833-0056	TREAT FOREIGN BODIES OF THE EYE	AIT	AN	1-3

CRITICAL TASKS

Task Number	Title	Training Location	Sust Tng Freq	Sust Tng SL
081-833-0057	TREAT LACERATIONS, CONTUSIONS, AND EXTRUSIONS OF THE EYE	AIT	AN	1-3
081-833-0058	TREAT BURNS OF THE EYE	AIT	AN	1-3
081-833-0070	ADMINISTER INITIAL TREATMENT FOR BURNS	AIT	SA	1-3
081-833-0155	PERFORM A TRAUMA CASUALTY ASSESSMENT	AIT	SA	1-3
081-833-0157	PROVIDE BASIC EMERGENCY MEDICAL CARE FOR AN AMPUTATION	AIT	SA	1-3
081-833-0161	CONTROL BLEEDING	AIT	SA	1-3
081-833-0194	PREPARE AN AID BAG	UNIT	AN	1-3
081-833-0209	TREAT A CASUALTY FOR CONTUSIONS OR ABRASIONS	AIT	AN	1-3
081-833-0210	APPLY A TOURNIQUET TO CONTROL BLEEDING	AIT	SA	1-3
081-833-0211	APPLY A HEMOSTATIC DRESSING	AIT	SA	1-3
081-833-0212	APPLY A PRESSURE DRESSING TO AN OPEN WOUND	AIT	SA	1-3
081-833-0213	PERFORM A TACTICAL CASUALTY ASSESSMENT	AIT	SA	1-3
081-833-0229	APPLY KERLIX TO AN OPEN WOUND	AIT	SA	1-3
Subject Area 4. Airway Management				
081-831-0018	OPEN THE AIRWAY	AIT	SA	1-3
081-831-0019	CLEAR AN UPPER AIRWAY OBSTRUCTION	AIT	SA	1-3
081-831-0048	PERFORM RESCUE BREATHING	AIT	SA	1-3
081-833-0016	INSERT AN OROPHARYNGEAL AIRWAY (J TUBE)	AIT	SA	1-3
081-833-0017	VENTILATE A PATIENT WITH A BAG-VALVE-MASK SYSTEM	AIT	SA	1-3
081-833-0018	SET UP A D-SIZED OXYGEN TANK	AIT	SA	1-3
081-833-0021	PERFORM ORAL AND NASOPHARYNGEAL SUCTIONING OF A PATIENT	AIT	AN	1-3
081-833-0142	INSERT A NASOPHARYNGEAL AIRWAY	AIT	SA	1-3
081-833-0158	ADMINISTER OXYGEN	AIT	SA	1-3
081-833-0169	INSERT A COMBITUBE	AIT	SA	1-3
081-833-0170	PERFORM ENDOTRACHEAL SUCTIONING OF A PATIENT	UNIT	AN	1-3
081-833-0230	INSERT A KING LT	AIT	SA	1-3
081-833-3005	PERFORM A SURGICAL CRICOTHYROIDOTOMY	AIT	SA	1-3
081-833-3007	PERFORM A NEEDLE CHEST DECOMPRESSION	AIT	SA	1-3
Subject Area 5. Venipuncture and IV Therapy				
081-833-0032	OBTAIN A BLOOD SPECIMEN USING A VACUTAINER	AIT	AN	1-3
081-833-0033	INITIATE AN INTRAVENOUS INFUSION	AIT	SA	1-3
081-833-0034	MANAGE AN INTRAVENOUS INFUSION	AIT	SA	1-3
081-833-0185	INITIATE A FAST 1	AIT	SA	1-3
081-835-3025	INITIATE A SALINE LOCK	AIT	SA	1-3
Subject Area 6. Primary Care				

CRITICAL TASKS

Task Number	Title	Training Location	Sust Tng Freq	Sust Tng SL
081-831-0007	PERFORM A PATIENT CARE HANDWASH	AIT	AN	1-3
081-833-0054	IRRIGATE EYES	AIT	AN	1-3
081-833-0059	IRRIGATE AN OBSTRUCTED EAR	UNIT	AN	1-3
081-833-0125	TREAT SKIN DISORDERS	AIT	AN	1-3
081-833-0139	TREAT ABDOMINAL DISORDERS	AIT	AN	1-3
081-833-0145	DOCUMENT PATIENT CARE USING SUBJECTIVE, OBJECTIVE, ASSESSMENT, PLAN (SOAP) NOTE FORMAT	AIT	SA	1-3
081-833-0165	PERFORM PATIENT HYGIENE	UNIT	AN	1-3
081-833-0193	PERFORM VISUAL ACUITY TESTING	AIT	AN	1-3
081-833-0195	REMOVE A PATIENT'S RING	AIT	AN	1-3
081-833-0203	TREAT COMMON EYE, EAR, NOSE, AND THROAT (EENT) DISORDERS	AIT	AN	1-3
081-833-0208	TREAT SUBUNGUAL HEMATOMA	UNIT	AN	1-3
081-833-0220	MAINTAIN A HEALTH RECORD	UNIT	AN	1-3
081-833-0222	TREAT COMMON MUSCULOSKELETAL DISORDERS	AIT	AN	1-3
081-833-0223	TREAT COMMON RESPIRATORY DISORDERS	AIT	AN	1-3
Subject Area 7. Musculosketeal				
081-831-1052	APPLY A SAM SPLINT	AIT	SA	1-3
081-833-0060	APPLY AN ELASTIC BANDAGE	AIT	SA	1-3
081-833-0062	IMMOBILIZE A SUSPECTED FRACTURE OF THE ARM OR DISLOCATED SHOULDER	AIT	SA	1-3
081-833-0064	IMMOBILIZE THE HIP	AIT	SA	1-3
081-833-0141	APPLY A TRACTION SPLINT	AIT	SA	1-3
081-833-0154	PROVIDE BASIC EMERGENCY TREATMENT FOR A PAINFUL, SWOLLEN, DEFORMED EXTREMITY	AIT	SA	1-3
081-833-0176	TREAT A CASUALTY WITH A SUSPECTED SPINAL INJURY	AIT	AN	1-3
081-833-0177	APPLY A CERVICAL COLLAR	AIT	SA	1-3
081-833-0178	APPLY A KENDRICK EXTRICATION DEVICE	AIT	AN	1-3
081-833-0180	APPLY A KENDRICK TRACTION DEVICE	AIT	AN	1-3
081-833-0181	APPLY A LONG SPINE BOARD	AIT	AN	1-3
081-833-0182	APPLY A REEL SPLINT	AIT	SA	1-3
Subject Area 8. Chemical Agent Injuries				
081-833-0083	TREAT A NERVE AGENT CASUALTY IN THE FIELD	AIT	AN	1-3
081-833-0084	TREAT A BLOOD AGENT (HYDROGEN CYANIDE) CASUALTY IN THE FIELD	UNIT	AN	1-3
081-833-0085	TREAT A CHOKING AGENT CASUALTY IN THE FIELD	UNIT	AN	1-3
081-833-0086	TREAT A BLISTER AGENT CASUALTY (MUSTARD, LEWISITE, PHOSGENE OXIME) IN THE FIELD	UNIT	AN	1-3
081-833-0095	DECONTAMINATE A CASUALTY	AIT	AN	1-3
081-833-0137	TREAT A BIOLOGICAL EXPOSED CASUALTY	UNIT	AN	1-3

CRITICAL TASKS

Task Number	Title	Training Location	Sust Tng Freq	Sust Tng SL
081-833-0202	TREAT A RADIATION CASUALTY	UNIT	AN	1-3
Subject Area 9. Triage and Evacuation				
071-334-4001	GUIDE A HELICOPTER TO A LANDING POINT	UNIT	AN	1-3
071-334-4002	ESTABLISH A HELICOPTER LANDING POINT	UNIT	AN	1-3
081-831-0101	REQUEST MEDICAL EVACUATION	AIT	AN	1-3
081-833-0080	TRIAGE CASUALTIES ON A CONVENTIONAL BATTLEFIELD	AIT	AN	1-3
081-833-0082	TRIAGE CASUALTIES ON AN INTEGRATED BATTLEFIELD	AIT	AN	1-3
081-833-0151	LOAD CASUALTIES ONTO GROUND EVACUATION PLATFORMS	AIT	AN	1-3
081-833-0152	ESTABLISH A CASUALTY COLLECTION POINT	AIT	AN	1-3
081-833-0171	LOAD CASUALTIES ONTO NONSTANDARD VEHICLES, 1 1/4 TON, 4X4, M998	UNIT	AN	1-3
081-833-0172	LOAD CASUALTIES ONTO NONSTANDARD VEHICLES, 2 1/2 TON, 6X6 OR 5 TON, 6X6, CARGO TRUCK	UNIT	AN	1-3
081-833-0173	LOAD CASUALTIES ONTO NONSTANDARD VEHICLES, 5 TON M-1085, M-1093, 2 1/2 TON M-1081	UNIT	AN	1-3
081-833-0214	LOAD CASUALTIES ONTO A UH-60 SERIES HELICOPTER	AIT	AN	1-3
081-833-0226	LOAD CASUALTIES ONTO A STRYKER ARMORED AMBULANCE	UNIT	AN	1-3
081-833-0227	COORDINATE CASUALTY TREATMENT AND EVACUATION	AIT	AN	1-3
Subject Area 10. Medication Administration				
081-833-0088	PREPARE AN INJECTION FOR ADMINISTRATION	AIT	AN	1-3
081-833-0089	ADMINISTER AN INJECTION (INTRAMUSCULAR, SUBCUTANEOUS, INTRADERMAL)	AIT	AN	1-3
081-833-0174	ADMINISTER MORPHINE	AIT	AN	1-3
081-833-0179	ADMINISTER MEDICATIONS	AIT	AN	1-3
Subject Area 11. Force Protection				
081-831-0037	DISINFECT WATER FOR DRINKING	AIT	AN	1-3
081-831-0038	TREAT A CASUALTY FOR A HEAT INJURY	AIT	SA	1-3
081-831-0039	TREAT A CASUALTY FOR A COLD INJURY	AIT	SA	1-3
081-831-9018	IMPLEMENT SUICIDE PREVENTION MEASURES	AIT	SA	1-3
081-833-0072	TREAT A CASUALTY FOR INSECT BITES OR STINGS	AIT	SA	1-3
081-833-0073	TREAT A CASUALTY FOR SNAKEBITE	AIT	SA	1-3
081-833-0076	APPLY RESTRAINING DEVICES TO PATIENTS	UNIT	AN	1-3
Skill Level 2				
Subject Area 12. Advanced Procedures (SL 2)				
081-833-0167	PLACE A PATIENT ON A CARDIAC MONITOR	UNIT	AN	2-3
081-833-0189	MAINTAIN IMMUNIZATION PROGRAM	UNIT	AN	2-3

CRITICAL TASKS

Task Number	Title	Training Location	Sust Tng Freq	Sust Tng SL
081-833-0196	REMOVE A TOENAIL	UNIT	AN	2-3
081-833-0197	REMOVE A URINARY CATHETER	UNIT	AN	2-3
081-833-0207	TREAT PARONYCHIA	UNIT	AN	2-3
081-833-3017	INSERT A URINARY CATHETER	UNIT	AN	2-3
081-835-3010	MAINTAIN AN INDWELLING URINARY CATHETER	UNIT	AN	
Skill Level 3				
Subject Area 13. Advanced Procedures (SL 3)				
081-825-0001	PROCESS ITEMS FOR STERILIZATION	ALC	AN	3
081-830-3016	INTUBATE A PATIENT	ALC	SA	3
081-831-0008	PUT ON STERILE GLOVES	ALC	AN	3
081-833-0007	ESTABLISH A STERILE FIELD	ALC	AN	3
081-833-0010	CHANGE A STERILE DRESSING	ALC	AN	3
081-833-0012	PERFORM A WOUND IRRIGATION	ALC	AN	3
081-833-0093	SET UP A CASUALTY DECONTAMINATION STATION	UNIT	AN	3
081-833-0168	INSERT A CHEST TUBE	ALC	AN	3
081-833-0184	ESTABLISH AN AMBULANCE EXCHANGE POINT	UNIT	AN	3
081-833-0192	PERFORM ABSCESS INCISION AND DRAINAGE	ALC	AN	3
081-833-0206	TREAT HIGH ALTITUDE ILLNESS	ALC	AN	3
081-833-0215	CONDUCT CRITICAL INCIDENT STRESS DEBRIEF	UNIT	AN	3
081-833-0219	MAINTAIN A HUMAN PATIENT SIMULATOR	UNIT	AN	3
081-833-0234	PROVIDE INITIAL SCREENING FOR TRAUMATIC BRAIN INJURY	ALC	AN	3
081-833-3014	PERFORM A NEUROLOGICAL EXAMINATION ON A PATIENT WITH SUSPECTED CENTRAL NERVOUS SYSTEM (CNS) INJURIES	UNIT	AN	3
081-833-3208	SUTURE A MINOR LACERATION	ALC	AN	3
081-835-3000	ADMINISTER BLOOD	ALC	AN	3
081-835-3002	ADMINISTER MEDICATIONS BY IV PIGGYBACK	UNIT	AN	3
081-835-3054	ADMINISTER BLOOD PRODUCTS	ALC	AN	3
081-850-0049	INSPECT BASE CAMPS	UNIT	AN	3

CHAPTER 3

MOS/Skill Level Tasks

Skill Level 1

Subject Area 1: Vital Signs

MEASURE A PATIENT'S RESPIRATIONS
081-831-0010

Conditions: You have a patient requiring medical assessment. You will need a watch and SF 600 (Medical Record-Chronological Record of Medical Care). You are not in a CBRNE environment.

Standards: Count a patient's respirations for 30 seconds and multiply by two.

Performance Steps

NOTE: The patient should not be aware that his respirations are being counted. The conscious patient that is aware their respirations are being counted will often alter respiratory rate by breathing slower and deeper.

CAUTION: All body fluids should be considered potentially infectious. Always observe body substance isolation (BSI) precautions by wearing gloves and eye protection as a minimal standard of protection.

1. Count the number of times the chest rises (inspiration) and returns to its normal position (expiration). Each respiratory cycle (inspiration/expiration) counts as one respiration. Normal ranges for respirations for each age group are as follows:
   a. Adults: 12-20 breaths/min.
   b. Children: 15-30 breaths/min.
   c. Infants: 25-50 breaths/min.

2. Evaluate the respirations.
   a. Depth.
      (1) Normal: deep, even movement of the chest.
      (2) Shallow: minimal rise and fall of the chest and abdomen.
      (3) Labored: increased effort to breathe, with possible gasping.
   b. Quality (character).
      (1) Normal: effortless, automatic, regular rate, even depth, noiseless, and free of discomfort.
      (2) Dyspnea: difficult or labored breathing.
      (3) Tachypnea: rapid respiratory rate; usually a rate exceeding 24 breaths/min (adult).
      (4) Noisy: snoring, rattling, wheezing (whistling), or grunting.
      (5) Apnea: temporary absence of breathing.

3. Observe for physical characteristics of abnormal respirations.
   a. Appearance: the patient may appear restless, anxious, pale, ashen, or cyanotic (blue skin color).

Performance Steps

  b. Position: the patient may alter his position by leaning forward with his hands on his legs (tripod position) or may be unable to breathe while lying down.

4. Record the rate of respirations and any observations noted (depth and quality) on the appropriate medical forms, according to local protocols and SOP.

5. Report any abnormal respirations to the supervisor immediately.

Evaluation Preparation:

Setup: You must count the rate with the Soldier being evaluated. If you are using a simulated patient, you may evaluate step 2 by having the patient purposely exhibit abnormal breathing characteristics. A tolerance of ± 2 counts will be allowed during the evaluation.

Brief Soldier: Tell the Soldier to count and record a patient's respirations.

Performance Measures	GO	NO-GO
1. Counted the number of times the chest rose and fell in 30 seconds and multiplied by two.	——	——
2. Evaluated the depth and quality of the respirations.	——	——
3. Observed for the physical characteristics of abnormal respirations.	——	——
4. Recorded the rate of respirations and any observations noted on the appropriate medical forms.	——	——
5. Reported any abnormal respirations to the supervisor immediately.	——	——

Evaluation Guidance: Score each Soldier according to the performance measures. Unless otherwise stated in the task summary, the Soldier must pass all performance measures to be scored GO. If the Soldier fails any steps, show what was done wrong and how to do it correctly.

References

  Required
  None

        Related
        None

MEASURE A PATIENT'S PULSE
081-831-0011

Conditions: You have a patient requiring medical assessment. You will need a watch and SF 600 (Medical Record-Chronological Record of Medical Care). You are not in a CBRNE environment.

Standards: Count a patient's pulse for a minimum of 30 seconds, a full minute if irregularities were detected. Identify any demonstrated abnormalities in the pulse rate, rhythm, and strength.

Performance Steps
CAUTION: All body fluids should be considered potentially infectious. Always observe body substance isolation (BSI) precautions by wearing gloves and eye protection as a minimal standard of protection.

1. Position the patient so the pulse site is accessible.

2. Palpate (feel) the pulse site.
    a. Place the tips of your index and middle fingers on the pulse site. Do not use your thumb to palpate a pulse as your thumb has its own pulse.
    b. Apply moderate pressure with your fingers to palpate the pulse.
NOTE: In responsive patients, older than 1 year, you should palpate the radial pulse at the wrist. In unresponsive patients, older than 1 year, you should palpate the carotid pulse at the neck. In patients less than 1 year of age, palpate the brachial pulse.

3. Count the number of pulses felt in a 30 second period and multiply times two. A pulse that is weak, difficult to palpate, or irregular should be palpitated and counted for a full minute.
    a. The normal pulse rates (at rest) are as follows:
        (1) Adults: 60-100 beats/min.
        (2) Children: 70-150 beats/min.
        (3) Infants: 100-160 beats/min.
    b. Pulse rates in an adult patient that exceed 100 beats/min are described as tachycardia.
    c. Pulse rates less than 60 beats/min are described as bradycardia.
NOTE: Pulse rates can vary from patient to patient. In well-conditioned athletes or patients taking certain heart medications, the pulse rate may be considerably lower. In these adult patients, bradycardia may be considered normal.

4. Evaluate the pulse rhythm (regularity).
    a. Regular rhythm.
        (1) Usually easy to find.
        (2) Has a regular rate and rhythm.
        (3) Varies with the individual.
    b. Irregular rhythm (any change from a regular beating pattern).
NOTE: If the pulse is irregular or intermittent, you should palpate a second pulse at the carotid or femoral site. All patients presenting with an irregular pulse rhythm should be referred to a medical officer.

5. Evaluate the pulse strength.
    a. Strong (full) pulse.
        (1) Usually easy to find.
        (2) Beats evenly and forcefully.

Performance Steps

      b. Bounding (stronger than normal) pulse.
         (1) Easy to find.
         (2) Exceptionally strong heartbeats that make the arteries difficult to compress.
      c. Weak (thready) pulse.
         (1) Usually difficult to find.
         (2) Weak and thin.

6. Record the rate, rhythm, strength, and any significant deviations from normal on the appropriate medical forms, according to local protocols and SOP.

7. Report any significant pulse abnormalities to the supervisor immediately.

Evaluation Preparation:

Setup: You must count the pulse rate with the Soldier being evaluated by palpating a corresponding site. Specify which site the Soldier is to palpate. A tolerance of ± 2 beats/min will be allowed during the evaluation.

Brief Soldier: Tell the Soldier to count and record the patient's pulse.

Performance Measures	GO	NO-GO
1. Positioned the patient so the pulse site was accessible.	——	——
2. Palpated the pulse site.	——	——
3. Counted and evaluated pulse rate.	——	——
4. Evaluated pulse rhythm	——	——
5. Evaluated pulse strength.	——	——
6. Recorded the rate, rhythm, strength, and any significant deviations from normal on the appropriate medical forms.	——	——
7. Reported any significant pulse abnormalities to the supervisor immediately.	——	——

Evaluation Guidance: Score each Soldier according to the performance measures. Unless otherwise stated in the task summary, the Soldier must pass all performance measures to be scored GO. If the Soldier fails any steps, show what was done wrong and how to do it correctly.

References
    Required   **Related**
    SF 600   EMERG CARE AND TRANS 9

MEASURE A PATIENT'S BLOOD PRESSURE
081-831-0012

Conditions: You have a patient requiring medical assessment. You will need a sphygmomanometer, clean stethoscope, and SF 600 (Medical Record-Chronological Record of Medical Care). You are not in a CBRNE environment.

Standards: Measure a patient's blood pressure.

Performance Steps
CAUTION: All body fluids should be considered potentially infectious. Always observe body substance isolation (BSI) precautions by wearing gloves and eye protection as a minimal standard of protection.

1. Explain the procedure to the patient, if necessary.
    a. The length of time the procedure will take.
    b. The site to be used.
    c. The physical sensations the patient will feel.

2. Select the proper size of sphygmomanometer cuff.
NOTE: The cuff width should wrap around the arm 1-1.5 times and take up two-thirds of the upper arm length, if using the brachial artery, and two-thirds of the upper leg length if using the popliteal artery.

NOTE: A cuff that is too small may result in falsely high readings; a cuff that is too large may result in falsely low readings.

3. Check the equipment.
    a. Ensure the cuff is completely deflated and fully retighten the one-way valve thumbscrew.
    b. Ensure the sphygmomanometer pressure gauge is reading zero.
NOTE: The following procedures describe the procedure for obtaining a blood pressure reading using the upper arm (brachial artery).

4. Position the patient and cuff.
    a. Place the patient in a relaxed and comfortable sitting, standing, or lying position.
NOTE: Measuring the blood pressure of a standing patient will result in a slightly higher reading.
    b. With the patient's arm extended, at approximately heart level and with the palm up, place the cuff over the brachial artery. Ensure the lower edge of the cuff is 1-2 inches above the elbow and the bladder portion is over the artery.
    c. Wrap the cuff just tightly enough to prevent slippage.
    d. Support the arm so it is in a relaxed state.

5. Palpate the brachial artery to determine where to place the stethoscope.

6. Place the diaphragm of the stethoscope over the pulse site and hold it firmly pressed against the artery with the fingers of your nondominant hand.

7. With the valve closed tightly, inflate the cuff using the ball-pump until the cuff reads at least 140 mm HG or until you no longer hear the pulse sounds. Continue pumping to increase the cuff's pressure by an additional 20 mm Hg.
CAUTION: The cuff should not remain inflated for more than 2 minutes.

Performance Steps

8. Determine the blood pressure reading.
   a. If a stethoscope is used, complete the following steps:
      (1) Rotate the thumbscrew in a clockwise motion, allowing the cuff to deflate slowly at about 3 mm Hg per second.
      (2) Watch the gauge and listen carefully. Note the patient's systolic blood pressure as the first distinct "taps" or "thumps" of the pulse waves that can be heard clearly.
      (3) Continue to watch the gauge and note the reading where the sound changes again or becomes muffled or disappears. This will be the diastolic blood pressure.
      (4) As soon as the pulse sounds cease, open the valve by rotating the thumbscrew and release the remaining air rapidly.
   b. If a stethoscope is not used, complete the following steps:
   NOTE: If in a very noisy environment where hearing the pulse waves is difficult or impossible, the palpation method may be used.
      (1) With your nondominant hand, palpate the radial pulse (at the wrist) on the same arm as the cuff.
      (2) While palpating the radial pulse, rapidly inflate the cuff to 200 mm Hg. As the cuff inflates, you will no longer feel the pulse under your fingertips.
      (3) Rotate the thumbscrew in a clockwise motion, allowing the cuff to deflate slowly at about 3 mm Hg per second.
      (4) Watch the gauge, when you again feel the radial pulse return, note the reading on the gauge (systolic blood pressure).
   NOTE: The diastolic pressure cannot be determined using this method. If the procedure must be repeated, wait at least 1 minute before repeating the procedure.
      (5) As soon as you note the systolic reading, open the valve by rotating the thumbscrew and release the remaining air rapidly.

9. Record the blood pressure on the appropriate medical forms.
   a. Record the systolic reading over the diastolic reading, for example 120/80.
   b. If obtaining the blood pressure without a stethoscope (by palpation), record the systolic reading alone alongside the letter "P", for example 120/P.
   c. Record all readings in even numbers.
   NOTE: Record the blood pressure readings with the time it was taken.

10. Evaluate the blood pressure readings by noting the normal ranges for the blood pressure.
    a. Adults: 90-140 mm Hg (systolic); 60-90 mm Hg (diastolic).
    b. Children: 80-110 mm Hg (systolic).
    c. Infants: 60 mm Hg (systolic).
    NOTE: Blood pressure readings vary with age and gender. A patient has hypotension when the blood pressure is lower than the normal range; the patient has hypertension when the blood pressure is higher than the normal range.

11. Report any abnormal blood pressure findings to the supervisor immediately.

Evaluation Preparation:
Setup: A double stethoscope should be used if available. A tolerance of ± 4 mm Hg will be allowed. If other methods are used, such as independent measurements on different sites or at different times, the evaluator must apply discretion in applying the ± 4 mm Hg standard. You will allow the Soldier to retake the blood pressure at least once if the Soldier feels that it is

necessary to obtain an accurate reading. You will use discretion in allowing additional repetitions based upon the difficulty of obtaining a reading on the patient during the evaluation.

Brief Soldier: Tell the Soldier to take a patient's blood pressure. Tell the Soldier that the blood pressure may be retaken, if necessary, to obtain an accurate reading.

Performance Measures	GO	NO-GO
1. Explained the procedure to the patient, if necessary.	——	——
2. Selected the proper size of sphygmomanometer cuff.	——	——
3. Checked the equipment.	——	——
4. Positioned the patient and cuff.	——	——
5. Palpated brachial artery.	——	——
6. Positioned the stethoscope, if used.	——	——
7. Inflated the cuff until the gauge read at least 140 mm Hg or until pulse sounds could no longer be heard.	——	——
8. Determined the blood pressure.	——	——
9. Recorded the blood pressure on the appropriate medical forms.	——	——
10. Evaluated the blood pressure.	——	——
11. Reported any abnormal readings to the supervisor immediately.	——	——

Evaluation Guidance: Score each Soldier according to the performance measures. Unless otherwise stated in the task summary, the Soldier must pass all performance measures to be scored GO. If the Soldier fails any steps, show what was done wrong and how to do it correctly.

References
  Required   **Related**
  SF 600   EMERG CARE AND TRANS 9

MEASURE A PATIENT'S TEMPERATURE
081-831-0013

Conditions: You have a patient requiring medical assessment. You have performed a patient care hand-wash. You will need properly disinfected oral and rectal thermometers or an electronic thermometer, canisters labeled "used," water soluble lubricant, gauze pads, a watch, and SF 600 (Medical Record-Chronological Record of Medical Care). You are not in a CBRNE environment.

Standards: Measure the patient's temperature to the nearest 0.2° F.

Performance Steps
CAUTION: All body fluids should be considered potentially infectious so always observe body substance isolation (BSI) precautions by wearing gloves and eye protection as a minimal standard of protection.

   1. Determine which site to use.
      a. Take an oral temperature if the patient is a conscious adult or child who can follow directions and can breathe normally through their nose.
CAUTIONS: Do not take an oral temperature when the patient-- 1. Has had recent facial or oral surgery. 2. Is unable to follow directions (confused, disturbed, or heavily sedated). 3. Is being administered oxygen by mouth or nose. 4. Is likely to bite down on the thermometer. 5. Has smoked, chewed gum or has eaten or drank anything hot or cold within the last 15 to 30 minutes.
      b. The tympanic method may be used with conscious or unconscious patients and is the preferred method if the patient has recently had anything to eat or drink.
CAUTION: Do not attempt to take a tympanic temperature if the patient has had recent facial or ear surgery or has cerumen (ear wax) impaction.
      c. Obtain the patient's temperature by the rectal method if the oral or tympanic methods are ruled out by the patient's condition.
CAUTION: Do not attempt to take a rectal temperature if the patient has had recent rectal surgery, unless directed to by a medical officer. Do not attempt to take a rectal temperature on an infant unless directed by a medical officer.
      d. Obtain the patient's temperature by the axillary (least preferred) method if the patient's condition rules out using the other methods.

   2. Select the appropriate thermometer.
      a. Tympanic thermometer.
      b. Oral thermometer: has a blue tip and may be labeled "Oral."
      c. Rectal thermometer: has a red tip and may be labeled "Rectal."
      d. Axillary temperatures may be obtained using an oral thermometer.

   3. Explain the procedure and position the patient appropriately.
      a. Tympanic method. Position the patient with their head turned to make the ear canal easily accessible.
      b. Oral method. Position the patient seated or lying down.
      c. Rectal method. Position the patient lying on either side with the top knee flexed.
      d. Axillary method. Position the patient either seated or lying face up with the armpit exposed.

   4. Measure the temperature.
      a. Shake the thermometer several times to ensure it is reading below 94° F.

Performance Steps
    b. Place the thermometer at the proper site.
        (1) Oral method. Place the thermometer underneath the tongue. Instruct the patient
            to close their lips around the instrument firmly but not to bite down.
        (2) Rectal method. In an adult, insert the thermometer 1 to 2 inches into their rectum.
CAUTION: The rectal thermometer must be lubricated with a water soluble lubricant prior to
insertion. Once inserted, do not let go of the thermometer.
        (3) Tympanic method. Pull the ear pinna upward and rearward; insert the
            thermometer speculum into the ear canal snugly to create a seal, pointing toward
            the nose.
        (4) Axillary method. Pat the armpit dry and place the bulb end of an oral
            thermometer in the center of the armpit with the glass tip protruding to the front of
            the patient's body. Place the patient's arm across his chest.
    c. Leave the thermometer in place for the required time.
        (1) Oral method: must remain in place for at least 3 minutes.
NOTE: If using a digital oral thermometer, leave in place until testing is complete. The digital
unit will normally have an audible tone.
        (2) Rectal method: must be held in place for at least 2 minutes.
        (3) Tympanic method: must remain in place until an audible signal occurs and the
            patient's temperature appears on the digital display.
        (4) Axillary method: must remain in place for at least 10 minutes.

  5. Remove the thermometer and wipe it down with a gauze pad or discharge the protective
     plastic sheath as appropriate.

  6. Read the temperature scale or digital display.

  7. Evaluate the temperature reading. The normal temperature ranges are as follows:
     a. Oral method: 97.0° to 99.0° F.
     b. Rectal method: 98.0° to 100.0° F.
     c. Tympanic method: 97.0° to 99.0° F.
     d. Axillary method: 96.0° to 98.0° F.

  8. Place the thermometer in the proper "used" canister or dispose of the plastic sheath as
     appropriate.

  9. Record the patient's temperature to the nearest 0.2° F on the appropriate medical forms.
     Record a rectal temperature with an "R" on the patient's record; with an axillary reading,
     use an "A" on the patient's record.

  10. Report any significant temperature abnormalities to the supervisor immediately.

Evaluation Preparation:
Setup: To test step 1 for evaluation purposes, create a scenario in which the patient's condition
will dictate which site the Soldier must choose. A tolerance of ± 0.2° F will be allowed during the
evaluation.

Brief Soldier: Tell the Soldier to measure, evaluate, and record a patient's temperature.

Performance Measures	GO	NO-GO
1. Determined an appropriate site to use.	——	——
2. Selected the appropriate thermometer.	——	——
3. Explained the procedure and positioned the patient.	——	——
4. Inserted the thermometer properly and left in place for the appropriate time.	——	——
5. Removed the thermometer and wiped it down with a gauze pad or discharged the plastic sheath appropriately.	——	——
6. Read the temperature scale or digital display.	——	——
7. Evaluated the temperature.	——	——
8. Placed the thermometer in the proper "used" canister or disposed of the plastic sheath as appropriate.	——	——
9. Recorded the temperature to the nearest 0.2° F correctly on the appropriate medical forms.	——	——
10. Reported any significant temperature abnormalities to the supervisor immediately.	——	——

Evaluation Guidance: Score each Soldier according to the performance measures. Unless otherwise stated in the task summary, the Soldier must pass all performance measures to be scored GO. If the Soldier fails any step, show what was done wrong and how to do it correctly.

References

Required	Related
SF 600	BASIC NURSING 7

MEASURE A PATIENT'S PULSE OXYGEN SATURATION
081-833-0164

Conditions: You need to measure a patient's pulse oxygen saturation. You will need the pulse oximetry monitor, sensing probe, and alcohol swabs. You are not in a CBRNE environment.

Standards: Measure a patient's pulse oxygen saturation.

Performance Steps
CAUTION: All body fluids should be considered potentially infectious. Always observe body substance isolation (BSI) precautions by wearing gloves and eye protection as a minimal standard of protection.

1. Select the appropriate sensing probe location for the patient.
    a. For adults, sensing probes can be placed on the index, middle, or ring finger.
    b. Sensing probes can also be placed on the toe unless the patient has decreased circulation to the lower extremities.
    c. Earlobe clips and neonate sensing probes for the foot are available for infants and newborns.

2. Wipe the selected site with alcohol to ensure it is clean and dry.

3. Apply the sensor so that the emitting light is directly opposite to the detector.

4. Attach the sensor cable to the machine and turn the power on.

5. Notify the medical officer if the digital readout is below the prescribed parameters.
NOTE: Normally, pulse oximetry values will be greater than 95% in room air, with the majority being between 98% and 100%. Factors that may provide falsely high readings include carbon monoxide poisoning, hypovolemia, and certain types of toxins.

6. Document the oximeter reading, the location of the device, the time taken, and the amount of oxygen being delivered (if applicable).

7. Take appropriate measures for continuous monitoring, if applicable.
    a. Ensure the alarms are on before leaving the patient.
NOTE: Monitors come with preset limits. These limits can be changed per medical officer's order.
    b. Move sensing probe locations every 2 hours; move adhesive sensors every 4 hours.
CAUTION: The pulse oximeter is just a tool; do not rely on it solely for indications of the patient's condition. Treat the patient, not the machine.

Evaluation Preparation:
Setup: Have all materials present for the evaluation. Have another Soldier act as the patient. Tell the patient not to assist the Soldier in any way.

Brief the Soldier: Tell the Soldier to measure the patient's pulse oxygen saturation.

Performance Measures   <u>GO</u>   <u>NO-</u>
                                                                                    <u>GO</u>

1. Selected the appropriate location to attach the sensing probe.   ____   ____

2. Wiped the selected site and ensured it was clean and dried.   ____   ____

3. Applied the sensing probe.   ____   ____

4. Attached the sensing probe cable to the monitor and turned the power on. ____   ____

5. Notified the medical officer of abnormal readings.   ____   ____

6. Documented the oximeter reading, the location of the device, the time   ____   ____
   taken, and the amount of oxygen being delivered (if applicable).

7. Took appropriate measures for continuous monitoring, if applicable.   ____   ____

Evaluation Guidance: Score each Soldier according to the performance measures in the evaluation guide. Unless otherwise stated in the task summary, the Soldier must pass all performance measures to be scored GO. If the Soldier fails any step, show what was done wrong and how to do it correctly.

References
  Required
  None

  Related
  BASIC NURSING 7
  ISBN 0-7637-3901-4

Subject Area 2: Medical Treatment

INITIATE A FIELD MEDICAL CARD
081-831-0033

Conditions: You have treated a casualty for injury or illness and must record the treatment given. You will need DD Form 1380 (U.S. Field Medical Card) and a pen or pencil. You are not in a CBRNE environment.

Standards: Complete the minimum required blocks.

Performance Steps
CAUTION: All body fluids should be considered potentially infectious. Always observe body substance isolation (BSI) precautions by wearing gloves and eye protection as a minimal standard of protection.

1. Remove the protective sheet from the carbon copy.
NOTE: FMCs are issued as a pad of 20 cards, each containing an original card, a carbon protective sheet, and a duplicate sheet.

2. Complete the minimum required blocks.
 a. Block 1. Enter the casualty's name, rank, and complete social security number (SSN). If the casualty is a foreign military person (including prisoners of war), enter their military service number. Enter the casualty's military occupational specialty (MOS) or area of concentration for specialty code. Enter the casualty's religion and sex.
 b. Block 3. Use the figures in the block to show the location of the injury or injuries. Check the appropriate box(es) to describe the casualty's injury or injuries.
NOTES: 1. Use only authorized abbreviations. Except for those listed below, however, abbreviations may not be used for diagnostic terminology.
Abr W--Abraded wound; Cont W--Contused wound; FC--Fracture (compound) open; FCC--Fracture (compound) open comminuted; FS--Fracture (simple) closed; LW--Lacerated wound; MW--Multiple wounds; Pen W--Penetrating wound; Perf W--Perforating wound; SL—Slight; SV--Severe.
2. When more space is needed, attach another DD Form 1380 to the original. Label the second card in the upper right corner "DD Form 1380 #2." It will show the casualty's name, grade, and SSN.
 c. Block 4. Check the appropriate box.
 d. Block 7. Check the yes or no box. Write the dose administered and the date and time that it was administered.
 e. Block 9. Write the information requested. If you need additional space, use Block 14.
 f. Block 11. Initial the far right side of the block.

3. Complete the other blocks as time permits. Most blocks are self-explanatory. The following specifics are noted:
 a. Block 2. Enter the casualty's unit of assignment and the country of whose armed forces they are a member. Check the armed service of the casualty, that is, A/T = Army, AF/A = Air Force, N/M = Navy, and MC/M Marine.
 b. Block 5. Write in the casualty's pulse rate and the time the pulse was measured.
 c. Block 6. Check the yes or no box. If a tourniquet is applied, you should write in the time and date it was applied.

Performance Steps

    d. Block 8. Write in the time, date, and type of IV solution given. If you need additional space, use Block 9.

    e. Block 10. Check the appropriate box. Write the date and time of disposition.

    f. Block 12. Write the time and date of the casualty's arrival. Record the casualty's blood pressure, pulse, and respirations in the space provided.

    g. Block 13. Document the appropriate comments by the date and time of observation.

    h. Block 14. Document the provider's orders by date and time. Record the dose of tetanus administered and the time it was administered. Record the type and dose of antibiotic administered and the time it was administered.

    i. Block 15. The signature of the provider or medical officer is written in this block.

    j. Block 16. Check the appropriate box and enter the date and time.

    k. Block 17. This block will be completed by the United Ministry Team. Check the appropriate box of the service provided. The signature of the chaplain providing the service is written in this block.

NOTE: As the FMC is the first, and sometimes only, record of treatment of combat casualties, accuracy and thoroughness of information provided is of the utmost importance.

    4. Attach the completed FMC to the casualty's uniform by twisting the wire after threading it through the top buttonhole of the uniform. Keep the FMC in plain view.

NOTE: Do not attach the FMC to the casualty's body armor as this equipment will be separated from the casualty once they arrive at the medical treatment facility (MTF).

Evaluation Preparation:

Setup: For training and evaluation construct a combat casualty scenario. Have another Soldier act as a casualty and have him respond to the Soldier's questions with personal data according to the scenario provided. Ensure the Soldier acting as the casualty has read the scenario thoroughly.

Brief Soldier: Tell the Soldier to complete the FMC by asking appropriate questions of the casualty. Tell the casualty to respond to the Soldier's questions with necessary information according to the scenario provided. To test step 2, you may either have the Soldier complete the required blocks, or you may require the completion of all blocks. After step 2, ask the Soldier what must be done with each copy of the FMC.

Performance Measures	GO	NO-GO
1. Removed the protective sheet from the carbon copy.	——	——
2. Completed the minimum required blocks (1, 3, 4, 7, 9, and 11).	——	——
3. Completed the other blocks as time permitted.	——	——
4. Attached the FMC to the top buttonhole of the casualty's uniform.	——	——

Evaluation Guidance: Score each Soldier according to the performance measures. Unless otherwise stated in the task summary, the Soldier must pass all performance measures to be scored GO. If the Soldier fails any steps, show what was done wrong and how to do it correctly.

References

Required	Related
DD FORM 1380	AR 40-66

MANAGE A CONVULSIVE AND/OR SEIZING PATIENT
081-831-0035

Conditions: You need to manage a convulsive and/or seizing patient. You have already taken the appropriate body substance isolation precautions. You will need padding materials, oxygen, suctioning equipment, and non-rebreather or bag-valve-masks. You are not in a CBRNE environment.

Standards: Complete all steps to manage a convulsive and/or seizing patient without allowing or causing unnecessary injury to the patient.

Performance Steps

1. Identify the type of convulsions and/or seizures based upon the following characteristic signs and symptoms:
   a. Petit mal.
      (1) Brief loss of concentration or awareness without loss of motor tone.
      (2) Other signs are lip smacking and eye blinking.
      (3) Occurs mainly in children and is rarely an emergency.
   b. Grand mal (generalized).
      (1) May be preceded by an aura.
      (2) Has two phases.
         (a) Tonic/Clonic phase--characterized by rigidity and stiffening of the body, drooling and occasional cyanosis around the face and lips.
         (b) Postictal phase--begins when convulsions stop. The patient may regain consciousness and enter a state of drowsiness and confusion or remain unconscious for several hours.
      (3) May involve incontinence, biting of the tongue (rare), cyanosis, or mental confusion.
   c. Status epilepticus.
      (1) Two or more seizures without an intervening period of consciousness or a seizure lasting more than 30 minutes.
      (2) A medical emergency, if untreated it may lead to--
         (a) Aspiration of secretions.
         (b) Cerebral or tissue hypoxia.
         (c) Brain damage or death.
         (d) Fractures of long bones.
         (e) Head trauma.
         (f) Injured tongue from biting.
NOTE: Mentally note the aspects of seizure activity for recording after the seizure.

2. Maintain the airway of a patient exhibiting tonic-clonic movement.
CAUTION: Never place anything in the mouth of a seizing patient.

3. Place the patient on his side, if possible.
   a. Observe the patient to prevent aspiration and suffocation.
   b. Place patient on high-flow oxygen at 15L/min via non-rebreathing mask, if available.
CAUTIONS: 1. Do not elevate the patient's head. 2. Do not restrain the patient's limbs during seizures.

4. Prevent injury to tissue and bones by padding or removing objects on which the patient may injure himself.

Performance Steps

5. Manage the patient after the convulsive state has ended.
   a. Place the patient on his side, if necessary.
   b. Continue to maintain the patient's airway.
NOTE: A patient who has just had a grand mal seizure will sometimes drool and will usually be drowsy so you must be prepared to suction, if equipment is available.
   c. Administer supplemental oxygen, if available, via non-rebreather mask or bag-valve-mask if not available earlier.
   d. If possible, place the patient in a quiet, reassuring atmosphere.
CAUTION: Sudden, loud noises may cause another seizure.

6. Record the seizure activity.
   a. Duration of the seizure.
   b. Presence of cyanosis, breathing difficulty, or apnea.
   c. Level of consciousness before, during, and after the seizure.
   d. Whether preceded by aura (ask the patient).
   e. Muscles involved.
   f. Type of motor activity.
   g. Incontinence.
   h. Eye movement.
   i. Previous history of seizures, head trauma, and/or drug or alcohol abuse.

7. Evacuate the patient on his side.

Evaluation Preparation:
Setup: For training and evaluation, have another Soldier act as a patient.

Brief Soldier: Tell the Soldier to manage the patient.

Performance Measures	GO	NO-GO
1. Identified the type of convulsions and/or seizures.	——	——
2. Maintained the airway of a patient exhibiting tonic-clonic movement.	——	——
3. Placed the patient on his side, if possible.	——	——
4. Prevented injury to tissue and bones by padding or removing objects on which the patient may injure himself.	——	——
5. Managed the patient after the convulsive state ended.	——	——
6. Recorded the seizure activity.	——	——
7. Evacuated the patient.	——	——
8. Did not cause further injury to the patient.	——	——

Evaluation Guidance: Score each Soldier according to the performance measures. Unless otherwise stated in the task summary, the Soldier must pass all performance measures to be scored GO. If the Soldier fails any steps, show what was done wrong and how to do it correctly.

References

Required   **Related**

None   EMERG CARE AND TRANS 9

ADMINISTER EXTERNAL CHEST COMPRESSIONS
081-831-0046

Conditions: You need to administer external chest compressions. Another Soldier who is cardiopulmonary resuscitation (CPR) qualified may be available to assist or may arrive while you are performing one-rescuer CPR. You are not in a CBRNE environment.

Standards: Continue CPR until the pulse is restored or until you are relieved by other competent person(s), are too exhausted to continue, the casualty is pronounced dead by an authorized person, or enemy fire prevents you from continuing until the casualty can be moved behind cover.

Performance Steps
CAUTION: All body fluids should be considered potentially infectious. Always observe body substance isolation (BSI) precautions by wearing gloves and eye protection as a minimal standard of protection.

1. Establish unresponsiveness (gently shake the casualty, asking, "Are you OK?")
    a. If the casualty is unresponsive, continue with step 2.
    b. If responsive, continue evaluating the casualty.

2. Open the airway. (See task 081-831-0018.)

3. Check for breathing. (See task 081-831-0048.) Look for rise and fall of the chest, listen for breath sounds from the casualty's mouth, and/or feel for breath coming from the casualty's nose or mouth. If the casualty is not breathing adequately, position a barrier device or pocket mask over the casualty's face. Give two slow rescue breaths, each over 1 full second, causing the chest to rise.

4. Check for signs of circulation; attempt to palpate the casualty's carotid pulse (do not take more than 10 seconds).
    a. If the casualty has a carotid pulse but is not breathing, perform rescue breathing. (See task 081-831-0048.)
    b. If casualty does not have a carotid pulse, begin chest compressions.

5. Ensure that the casualty is positioned on a hard, flat surface, in a supine position. Kneel next to the casualty.

6. Position your hands for external chest compressions.
    a. Place the heel of one hand on the sternum (center of chest) between the casualty's nipples (lower half of the sternum).
    b. Place the heel of your other hand over the first hand on the sternum so both of your hands are parallel to each other.
NOTE: You may either extend or interlace your fingers but keep your fingers off the casualty's chest.

7. Position your body.
    a. Lock your elbows with the arms straight.
    b. Position your shoulders directly over your hands.

Performance Steps

8. Give 30 compressions.
    a. Press straight down to depress the sternum 1½ to 2 inches.
    b. Come straight up and completely release pressure on the sternum to allow the chest to return to its normal position. The time allowed for release should equal the time required for compression.
CAUTION: Do not remove the heel of your hand from the casualty's chest or reposition your hand between compressions.
    c. Give 30 compressions at a rate of 100 per minute. (Each set of 30 compressions should take about 20 seconds.)

9. Give two full rescue breaths.
    a. Move quickly to the casualty's head and lean over his mouth.
    b. Open the casualty's airway. (See task 081-831-0018.)
    c. Give two full rescue breaths (each lasting 1 second).

10. Repeat steps 6 through 9 four more times.

11. Reassess the casualty.
    a. Check for the return of the carotid pulse for 5 seconds (no longer than 10 seconds).
        (1) If the pulse is present, continue with step 11b.
        (2) If the pulse is absent, continue with step 12.
    b. Check breathing for 3 to 5 seconds.
        (1) If breathing is present, monitor breathing and pulse closely.
        (2) If breathing is absent, perform rescue breathing only. (See task 081-831-0048.)

12. Resume CPR with compressions.

13. Recheck for pulse every few minutes.

14. Continue to alternate chest compressions and rescue breathing until--
    a. The casualty is revived.
    b. You are too exhausted to continue.
    c. You are relieved by another health care provider.
    d. The casualty is pronounced dead by an authorized person.
    e. A second rescuer states, "I know CPR," and joins you in performing two-rescuer CPR.
NOTE: A qualified second rescuer joins the first rescuer at the end of the compression-ventilation cycle after a check for pulse by the first rescuer. The new cycle starts with the second rescuer continuing with 30 compressions after an absent pulse is verified by the first rescuer. Two-rescuer CPR is then initiated.

15. Perform two-rescuer CPR, if applicable.
    a. Compressor: Give 30 chest compressions at the rate of 100 per minute.
        Ventilator: Maintain an open airway and monitor the carotid pulse occasionally for adequacy of chest compressions.
    b. Compressor: Pause.
        Ventilator: Give two full rescue breaths (each breath lasts 1 second).
    c. Compressor: Continue to give chest compressions until a change in positions is initiated.
        Ventilator: Continue to give ventilations until the compressor indicates that a change is to be made.
NOTE: When performing two-rescuer CPR, the rescuers must change positions every 2 minutes to avoid fatigue and increase the effectiveness of compressions.

Performance Steps

    d. Compressor: Give a clear signal to change positions.
       Ventilator: Remain in the rescue breathing position.
    e. Compressor: Give the 30th compression.
       Ventilator: Give two breaths following the 30th compression.
    f. Compressor and ventilator simultaneously switch positions.
    g. New Ventilator: Check the casualty's carotid pulse for 5 seconds.
       * If present state, "There is a pulse," and perform rescue breathing.
       * If not present state, "No pulse." Tell the new compressor to give chest compressions.
       New compressor: Position the hands to begin chest compressions as directed by the ventilator
    h. Compressor: Continue to give chest compressions at the rate of 100 per minute.
       Ventilator: Continue to give two breaths on each 30th upstroke of chest compressions and ensure that the chest rises.

NOTE: If signs of gastric distension are noted, do the following: 1. Recheck and reposition the airway. 2. Watch for rise and fall of the chest. 3. Ventilate the casualty only enough to cause the chest to rise.

CAUTIONS: 1. Do not push on the abdomen. 2. If the casualty vomits, turn the casualty on his side, clear the airway, and then continue CPR.

NOTE: If the casualty is intubated, the ratio of breaths to compressions becomes asynchronous. Give 100 compressions per minute with a ventilation rate of approximately 10 to 12 per minute.

  16. Continue to perform CPR as stated in the task standard.
NOTE: The rescuer doing rescue breathing should recheck the carotid pulse every 3 to 5 minutes.

  17. When the pulse and breathing are restored, continue to evaluate the casualty. If the casualty's condition permits, place him in the recovery position. (See task 081-831-0018.)
CAUTION: During evacuation, CPR or rescue breathing should be continued en route if necessary.

Evaluation Preparation:
Setup: For training and evaluation a CPR mannequin must be used. Place the mannequin face up on the floor (in the supine position). One-rescuer CPR, two-rescuer CPR, or a combination of both (see NOTE after step 14e) can be evaluated. If two Soldiers are involved, they will be designated as "rescuer #1" and "rescuer #2." Rescuer #1 will start in the chest compression position and will be the only one scored during performance of the task. The evaluator will ensure that all aspects of the task are evaluated by indicating whether the pulse is present and when the rescuers should change positions.

Brief Soldier: If two Soldiers are involved, tell them about their roles as rescuer #1 and #2. Ask rescuer #1 on what kind of surface the casualty should be positioned. Then, tell the Soldier(s) to perform one-rescuer or two-rescuer CPR, as appropriate.

Performance Measures	GO	NO-GO
1. Established unresponsiveness.	——	——
2. Opened the airway.	——	——
3. Checked for breathing.	——	——
4. Checked for signs of circulation.	——	——
5. Positioned the casualty on a hard flat surface.	——	——
6. Properly positioned the hands during chest compressions.	——	——
7. Positioned your body.	——	——
8. Gave 30 chest compressions at the rate of 100 per minute.	——	——
9. Gave 2 full rescue breaths.	——	——
10. Repeated steps 6-9.	——	——
11. Reassessed the casualty.	——	——
12. Resumed CPR with compressions.	——	——
13. Recheck for pulse every few minutes.	——	——
14. Continued to alternate chest compressions and rescue breathing.	——	——
15. Performed two-rescuer CPR, if applicable.	——	——
16. Continued to perform CPR as stated in the task standard.	——	——
17. Continued to evaluate the casualty when the pulse and breathing was restored.	——	——

Evaluation Guidance: Score each Soldier according to the performance measures. Unless otherwise stated in the task summary, the Soldier must pass all performance measures to be scored GO. If the Soldier fails any steps, show what was done wrong and how to do it correctly.

References

Required	Related
None	ISBN 0-7637-3901-4

MEASURE A PATIENT'S INTAKE AND OUTPUT
081-833-0006

Conditions: You need to measure a patient's intake and output. You have a medical officer's order and have performed a patient care hand-wash. You will need a DD Form 792 (Nursing Service - Twenty-Four Hour Patient Intake and Output Worksheet), SF 511 (Vital Signs Record), or other appropriate forms, calibrated graduated container, gloves, common serving items, urinal, bedpan, urinary drainage bag, emesis basin, and nasogastric drainage container. You are not in a CBRNE environment.

Standards: Accurately measure and record the patient's fluid intake and output on appropriate forms.

Performance Steps

1. Explain the procedure to the patient.
    a. Inform the patient of the length of time during which the intake and output will be measured and the purpose of taking the measurements.
    b. Tell the patient of any medical orders on fluid intake, such as forcing fluids or restricting the amount of intake.

2. Tell the patient what types of items require intake and/or output measurement.
    a. Intake measurement.
        (1) Items that are naturally fluid at room temperature such as Jell-O, ice cream, ice, and infant cereals.
        (2) Fluids consumed with and between meals, such as water, coffee, tea, broth, juice, milk, milk shakes, and carbonated beverages.
        (3) Intravenous (IV) infusion fluids and blood.
        (4) Oral liquid medications.
        (5) Irrigating solutions that are not returned.
    b. Output measurement.
        (1) Urine.
        (2) Liquid stool.
        (3) Vomitus.
        (4) Drainage from wounds and suction devices.

3. Tell the patient to use specified containers, such as a bedpan or urinal, to save all fluid output.

4. Measure the intake.
    a. Calculate the oral fluid intake.
NOTE: Check the water pitcher at the beginning and end of each shift. Check the meal tray for the amount of liquids consumed before removing it from the room.
        (1) Note the type and size of the oral fluid containers.
        (2) Check the container to find the fluid capacity.
        (3) Check the "Equivalents Table" on DD Form 792.
NOTE: If an unmarked container is not listed on DD Form 792, fill it with water and pour its contents into a graduate to check its capacity.
    b. Calculate the amount of IV solution or blood given.
    c. Calculate the amount of any irrigating solutions that are not returned, if applicable.
        (1) Subtract the amount of solution returned from the known amount used for the irrigating procedure.

Performance Steps
    (2) Record the difference as intake.

  5. Record, in cubic centimeters (cc), the fluid intake under the appropriate heading on DD
     Form 792.
NOTE: To convert ounces to cc, multiply the number of fluid ounces by 30. Example: 12 fluid
ounces multiplied by 30 equals 360 cc. One milliliter (ml) is approximately equal to one cc.

  6. Measure the output.
    a. Put on gloves.
    b. Record the level of output (urine, liquid stool, or emesis) in a graduated container.
NOTE: If it is not possible to weigh or measure liquid stool, estimate the amount IAW local
SOP. Estimate the amount of solid stool IAW local SOP.
    c. Estimate the amount of wound drainage, if present, IAW local SOP.
    d. Estimate any output not in a container, such as on the floor, skin, or sheets, IAW local
      SOP.
    e. Observe characteristics of the output.
      (1) Color and odor of urine.
      (2) Color, odor, and consistency of stool.
      (3) Color and consistency of nasogastric drainage.

  7. Remove gloves and perform a patient care hand-wash.

  8. Record in cc, the amount and characteristics of output under the appropriate headings on
     DD Form 792.
NOTE: If no output was available to measure, enter this information in the "Remarks" section of
DD Form 792.

  9. Compute accumulated intake and output totals at the end of the 24-hour period and record
     on the appropriate forms IAW local SOP.

Evaluation Preparation:
Setup: If the performance of this task must be simulated for training and evaluation,
premeasure at least two fluid items into common serving utensils. The Soldier will use them as
the remains of a patient's simulated intake. You may also partially empty a bag or bottle of IV
solution and have the Soldier calculate the amount of intravenous intake. Have at least two
premeasured containers of simulated waste fluid to use for simulated output. Have the Soldier
explain steps 1-3 to you.

Brief Soldier: Tell the Soldier to measure and record the intake and output of a specified
patient.

Performance Measures	GO	NO-GO
1. Explained procedures to the patient.	——	——
2. Told the patient what types of items require intake and/or output measurement.	——	——
3. Told the patient to use specific containers, such as a bedpan or urinal, to save all fluid.	——	——
4. Measured the intake.	——	——

Performance Measures	GO	NO-GO

5. Recorded, in cubic centimeters (cc), the fluid intake. ____ ____

6. Put on gloves and measured the output. ____ ____

7. Removed gloves and performed a patient care hand-wash. ____ ____

8. Recorded, in cc, the amount and characteristics of output under the appropriate headings on DD Form 792. ____ ____

9. Computed accumulated intake and output totals at the end of the 24-hour period and recorded on the appropriate forms IAW local SOP. ____ ____

Evaluation Guidance: Score each Soldier according to the performance measures. Unless otherwise stated in the task summary, the Soldier must pass all performance measures to be scored GO. If the Soldier fails any steps, show what was done wrong and how to do it correctly.

References

Required	**Related**
DD FORM 792	BASIC NURSING 7
SF 511	

TREAT A CASUALTY FOR ANAPHYLACTIC SHOCK
081-833-0031

Conditions: You need to treat a casualty for an anaphylactic shock. You will need a needle, syringe, epinephrine (1:1000 solution), a constricting band, stethoscope, sphygmomanometer, bag-valve-mask system, intravenous (IV) infusion, oxygen equipment, and DD Form 1380 (U.S. Field Medical Card). You are not in a CBRNE environment.

Standards: Initiate treatment for anaphylactic shock, stabilize the casualty, and minimize the effects of anaphylaxis without causing further injury to the casualty.

Performance Steps
NOTE: Anaphylactic reactions occur within minutes or even seconds after contact with the substance to which the casualty is allergic. Reactions occur in the skin, respiratory system, and circulatory system.

1. Check the casualty for signs and symptoms of anaphylactic shock.
    a. Skin.
        (1) Flushed or ashen.
        (2) Burning or itching.
        (3) Edema (swelling), especially in the face, tongue, or airway.
        (4) Urticaria (hives) spreading over the body.
        (5) Marked swelling of the lips and cyanosis about the lips.
    b. Respiratory.
        (1) Tightness or pain in the chest.
        (2) Sneezing and coughing.
        (3) Wheezing, stridor, or difficulty in breathing (dyspnea).
        (4) Sputum (may be blood tinged).
        (5) Respiratory failure.
    c. Circulatory.
        (1) Weak, rapid pulse.
        (2) Falling blood pressure.
        (3) Hypotension.
        (4) Dizziness or fainting.
        (5) Coma.

2. Open the airway, if necessary. (See task # 081-833-0018.)
NOTE: In cases of airway obstruction from severe glottic edema, a cricothyroidotomy may be necessary. (See task 081-833-3006.)

3. Administer high concentration oxygen. (See task 081-833-0158.)

4. Administer epinephrine.
    a. Administer 0.3 - 0.5 ml of epinephrine, 1:1000 solution, subcutaneously (SQ) or intramuscularly (IM).
NOTE: Annotate the time of injection on the FMC.
    b. Additional epinephrine may be required if anaphylaxis progresses. Additional doses may be administered every 5 to 10 minutes if needed.

5. Initiate an intravenous (IV) infusion. (See task 081-833-0033.)

Performance Steps

6. Provide supportive measures for the treatment of shock, respiratory failure, circulatory collapse, or cardiac arrest.
   a. Infuse additional IV fluid if blood pressure continues to drop.
   b. Position the casualty in the supine position with legs elevated if injuries permit.
   c. Perform rescue breathing, if necessary. (See task 081-831-0048.)
   d. Administer external chest compressions, if necessary. (See task 081-831-0046.)

7. Check the casualty's vital signs every 3 to 5 minutes until the casualty is stable.

8. Record the treatment given on the FMC.

9. Evacuate the casualty, providing supportive measures en route.

Performance Measures	**GO**	**NO-GO**
1. Checked the casualty for signs and symptoms of anaphylactic shock.	⎯⎯	⎯⎯
2. Opened the airway, if necessary.	⎯⎯	⎯⎯
3. Administered oxygen.	⎯⎯	⎯⎯
4. Administered epinephrine.	⎯⎯	⎯⎯
5. Initiated an IV.	⎯⎯	⎯⎯
6. Provided supportive measures for the treatment of shock, respiratory failure, circulatory collapse, or cardiac arrest.	⎯⎯	⎯⎯
7. Checked the casualty's vital signs every 3 to 5 minutes until the casualty was stable.	⎯⎯	⎯⎯
8. Recorded the treatment given on the FMC.	⎯⎯	⎯⎯
9. Evacuated the casualty and provided supportive measures en route.	⎯⎯	⎯⎯

Evaluation Guidance: Score each Soldier according to the performance measures. Unless otherwise stated in the task summary, the Soldier must pass all performance measures to be scored GO. If the Soldier fails any steps, show what was done wrong and how to do it correctly.

References

Required	**Related**
DD FORM 1380	EMERG CARE AND TRANS 9
	ISBN 0-07-065351-9

ASSIST IN VAGINAL DELIVERY
081-833-0116

Conditions: You encounter a woman who is in labor. You will need a sterile obstetric kit (if kit is not available, you will need clean sheets and towels, heavy flat twine or new shoelaces, plastic bag, scissors, sterile pad, warm blanket, and clean, unused examination gloves). You are not in a CBRNE environment.

Standards: Assist with vaginal delivery without causing further injury to the patient.

Performance Steps

1. Assist with the first stage of labor.
NOTES: 1. Scene size-up, initial assessment, focused history, examination, detailed physical examination, ongoing assessment, and evacuate assessment steps must be taken to ensure that injury(ies) or illness is/are not over looked resulting in further injury to the patient. 2. Evacuate an expecting mother unless delivery is expected within a few minutes.
    a. Interview the pregnant woman. Request health history.
        (1) Present pregnancy history. Is this your first pregnancy? Have there been complications during your pregnancy?
        (2) Medical history. Is there a history of diabetes, hypertension, or chronic diseases?
        (3) Obstetric history. How many times have you been pregnant?
    b. Assess general appearance and behavior.
    c. Check vital signs between contractions. If hypotension occurs, place the patient on her left side, administer oxygen (if available), and notify the health care provider immediately.
    d. Assess the labor pattern status.
        (1) Contractions--initial onset, frequency, and duration.
        (2) Discomfort or pain.
    e. Assess amniotic membranes status. Inquire if the patient has experienced constant leakage or rupture of vaginal fluid.

2. Assist with the second stage of labor.
    a. Assist with delivery of the infant as directed by health care provider.
NOTE: If the medic is in an isolated environment and is unable to evacuate the patient, the medic will deliver the infant.
    b. Determine if the umbilical cord is around the infant's neck as the infant is being born. Place two fingers under the cord at the back of the baby's neck. Bring the cord forward, over the baby's upper shoulder and head. If you cannot loosen or slip the cord over the baby's head, clamp the cord in two places and, with extreme care, cut the cord between the two clamps and unwrap the ends of the cord from around the baby's neck and proceed with the delivery.
    c. Support the head after the infant's head is born.
    d. Suction the mouth two or three times and the nostrils. Avoid contact with the back of the mouth.
    e. Support the infant with both hands as the torso and full body are born.
    f. Wipe blood and mucus from the mouth and nose with sterile gauze. Suction the mouth and nose again.
    g. Clamp, tie, and cut the umbilical cord (between the clamps) as pulsations cease approximately four finger widths from the infant.

Performance Steps

    h. Wrap the infant in a warm blanket and place on its side, head slightly lower than the trunk.

3. Assist with the third stage of labor.
    a. Observe for delivery of the placenta while preparing the mother and infant for evacuation.
    b. Place a sterile pad over the vaginal opening and lower the patient's legs.
    c. Record the time of delivery and evacuate the mother, infant, and placenta to the hospital.

4. Provide initial care for the newborn.
    a. Position, dry, wipe, and wrap the newborn in a blanket and cover the head.
    b. Perform appearance, pulse, grimace, activity, and respirations (APGAR) testing at 1 and 5 minutes after birth.
        (1) Appearance (color)--no central (trunk) cyanosis.
        (2) Pulse--greater than 100/min.
        (3) Grimace--vigorous and crying.
        (4) Activity--good motion in extremities.
        (5) Respirations, breathing effort--normal, crying.

Performance Measures	GO	NO-GO
1. Assisted with the first stage of labor.	___	___
2. Assisted with the second stage of labor.	___	___
3. Assisted with the third stage of labor.	___	___
4. Provided initial care for the newborn.	___	___

Evaluation Guidance: Score each Soldier according to the performance measures. Unless otherwise stated in the task summary, the Soldier must pass all performance measures to be scored GO. If the Soldier fails any steps, show what was done wrong and how to do it correctly.

References

    Required   **Related**
    None   EMERG CARE AND TRANS 9

TREAT A POISONED CASUALTY
081-833-0143

Conditions: You have a casualty that has been poisoned. All other more serious injuries have been assessed and treated. You have taken body substance isolation (BSI) precautions and have performed an initial assessment. You will need activated charcoal, airway adjuncts, oxygen, intravenous (IV) catheter, IV tubing, 0.9 percent normal saline, water source, suction equipment and the casualty's medical record. You are not in a CBRNE environment.

Standards: Determine the type of poisoning and provide treatment, minimizing the effects of the poisoning, without causing further injury to the casualty.

Performance Steps

1. Determine the type of poisoning.
CAUTION: If determination cannot be made to the type of poisoning, the casualty should be treated by the symptoms presented.
    a. Ingested poisons.
        (1) Altered mental status.
        (2) Nausea/vomiting.
        (3) Abdominal pain.
        (4) Diarrhea.
        (5) Chemical burns around the mouth.
        (6) Unusual breath odors.
    b. Inhaled poisons.
        (1) Carbon monoxide.
            (a) Headache.
            (b) Dizziness.
            (c) Dyspnea.
            (d) Nausea/vomiting.
            (e) Cyanosis.
            (f) Coughing.
        (2) Smoke Inhalation.
            (a) Dyspnea.
            (b) Coughing.
            (c) Breath that has a smoky smell or the odor of chemicals involved at the scene.
            (d) Black residue in any sputum coughed up by the casualty.
            (e) Nose-hairs singed from super-heated air.
    c. Injected poisons.
        (1) Sympathomimetics (Uppers- example: cocaine).
            (a) Excitement.
            (b) Tachycardia.
            (c) Tachypnea.
            (d) Dilated pupils.
            (e) Sweating.
        (2) Sedative-Hypnotics (Downers- example; Valium, Xanax).
            (a) Sluggish.
            (b) Sleepy typical coordination of body and speech.
            (c) Pulse and breathing rates are low, often to the point of a true emergency.

Performance Steps
        (3) Hallucinogens.
           (a) Tachycardia.
           (b) Dilated pupils.
           (c) Flushed face.
           (d) Often sees or hears things, has very little concept of time.
        (4) Narcotics.
           (a) Reduced rate of breathing.
           (b) Dyspnea.
           (c) Low skin temperature.
           (d) Muscles relaxed.
           (e) Pinpoint pupils.
           (f) Very sleepy.
    d. Absorbed poisons.
        (1) Liquid or powder on the casualty's skin.
        (2) Burns.
        (3) Itching.
        (4) Irritation.
        (5) Redness.

  2. Administer emergency care.
    a. Ingested poisons.
        (1) Maintain the airway.
        (2) Gather all information about the type of ingested poisoning.
        (3) Initiate IV therapy.
        (4) Administer activated charcoal.
CAUTION: Activated charcoal is contraindicated for casualties that have an altered mental status, that you suspect have swallowed acids or alkalis, or that are unable to swallow.

NOTE: Be prepared to provide oral suctioning if the casualty starts to vomit. All vomitus must be saved.
           (a) Adults and children: 1 gram of activated charcoal/kg of body weight.
           (b) Usual adult dose: 25 - 50 grams.
           (c) Usual pediatric dose: 12.5 - 25 grams.
        (5) Give supplemental oxygen.
        (6) Record the name, dose, and time of administration of medication.
        (7) Transport to the nearest medical treatment facility.
    b. Inhaled poisons.
        (1) Remove the casualty from the unsafe environment.
           (a) Maintain the airway.
           (b) Administer high concentrations of oxygen.
NOTE: This is the most important treatment for inhalation poisoning.
           (c) Transport to the nearest medical treatment facility.
           (d) Document interventions.
    c. Absorbed poisons.
        (1) Remove the casualty from the source.
        (2) Remove contaminated clothing.
        (3) Brush off any powders from the casualty's skin.
        (4) Flush the skin with large amounts of water for at least 20 minutes.

Performance Steps
    d. Injected poisons.
        (1) Maintain the airway and be prepared to provide assisted ventilations.
        (2) Give supplemental oxygen.
        (3) Initiate IV therapy.
        (4) Look for gross soft tissue damage ("tracks").
        (5) Protect the casualty from harming self and others.
NOTE: Be prepared to use restraints.
        (6) Transport to the nearest medical treatment facility.

  3. Document procedures. (See tasks 081-831-0033 and 081-833-0145.)

Evaluation Preparation:
Setup: For training and evaluation, have another Soldier act as the casualty.

Brief Soldier: Tell the Soldier that the casualty has an ingested or inhaled poison. Have the Soldier state what actions should be taken when an IV infusion is initiated.

Performance Measures	<u>GO</u>	<u>NO-</u> <u>GO</u>
1. Determined the type of poisoning.	——	——
2. Administered emergency care.	——	——
3. Documented the procedure.	——	——

Evaluation Guidance: Score each Soldier according to the performance measures. Unless otherwise stated in the task summary, the Soldier must pass all performance measures to be scored GO. If the Soldier fails any steps, show what was done wrong and how to do it correctly.

References
    Required   **Related**
    None   ISBN 0-7637-3901-4

TREAT A DIABETIC EMERGENCY
081-833-0144

Conditions: You have a patient with a diabetic emergency. You have taken body substance isolation precautions and have performed an initial assessment, focused history, and physical exam. You will need oral glucose, tongue depressors, intravenous (IV) infusion set, oxygen, and the patient's medical record. You are not in a CBRNE environment.

Standards: Initiate treatment for hypoglycemia or hyperglycemia, stabilize the patient, and minimize the effects without causing further injury to the patient.

Performance Steps

    1. Identify the signs and symptoms of a diabetic emergency.
        a. Hypoglycemia. (Low blood sugar)
NOTE: Hypoglycemia is the most common of all diabetic emergencies.
            (1) Rapid onset of altered mental status.
NOTE: This is especially so after missing a meal, vomiting, or an unusual amount of physical exertion.
            (2) Intoxicated appearance, staggering, slurred speech, or unconsciousness.
            (3) Elevated heart rate.
            (4) Cold, clammy skin.
            (5) Hunger.
            (6) Seizures.
            (7) Uncharacteristic behavior.
            (8) Anxiety.
            (9) Combativeness.
        b. Hyperglycemia. (High blood sugar)
            (1) Slow onset.
            (2) Warm, red, dry skin.
            (3) Sweet, fruity breath odor (acetone).
            (4) Deep, rapid breathing.
            (5) Dry mouth.
            (6) Intense thirst.
            (7) Abdominal pain.
            (8) Nausea and vomiting.

    2. Administer the appropriate treatment.
NOTE: If you are unsure whether the patient has hyperglycemia or hypoglycemia, it is safer to treat the patient for hypoglycemia.
        a. Hypoglycemia.
            (1) If conscious, administer oral glucose IAW local protocol.
NOTE: Give it only if the patient has a history of diabetes, the patient has an altered mental status, and the patient is awake enough to swallow.
                (a) Apply glucose to a tongue depressor and place it in the patient's mouth between the cheek and gum.
                (b) Or if the patient is able, let the patient squeeze the glucose from the tube directly into his mouth.
            (2) Monitor the patient for complications.
            (3) Assess vital signs.
            (4) If unconscious--

Performance Steps
        (a) Secure the airway and administer oxygen.
        (b) Assess vital signs.
        (c) Start an intravenous IV at to keep open (TKO) rate.
        (d) Place the patient in the recovery position.
        (e) Transport to the nearest medical treatment facility.
    b. Hyperglycemia.
        (1) Maintain an open airway and administer oxygen.
        (2) Assess vital signs.
        (3) Start an IV at TKO rate.
        (4) Place the patient on a cardiac monitor, if available.
        (5) Transport to the nearest medical treatment facility.

  3. Document all treatment given.
NOTE: Document the patient's mental status using the alert, verbal, painful, unresponsive (AVPU) scale and vital signs every 5 minutes. A change in mental status may indicate an alteration in the patient's blood sugar level.

Evaluation Preparation:
Setup: For training and evaluation, have another Soldier act as the patient and exhibit signs and symptoms of hyperglycemia or hypoglycemia.

Brief Soldier: Tell the Soldier to state the signs and symptoms of hypoglycemia or hyperglycemia, and then treat the patient.

Performance Measures	GO	NO-GO
1. Identified the type of diabetic emergency (hypoglycemia or hyperglycemia).	——	——

2. Administered appropriate treatment for
   a. Hypoglycemia. (Conscious or unconscious)
   b. Hyperglycemia.
	——	——

3. Documented all treatment given.	——	——

Evaluation Guidance: Score each Soldier according to the performance measures. Unless otherwise stated in the task summary, the Soldier must pass all performance measures to be scored GO. If the Soldier fails any steps, show what was done wrong and how to do it correctly.

References
  Required   **Related**
  None   ISBN 0-7637-3901-4

PERFORM A MEDICAL PATIENT ASSESSMENT
081-833-0156

Conditions: You have a patient with a complaint that is medical in nature and no significant mechanism of injury. You will need a sphygmomanometer, stethoscope, thermometer, and airway adjuncts. You are not in a CBRNE environment.

Standards: Perform a medical patient assessment without causing further injury.

Performance Steps

1. Take body substance isolation precautions.

2. Perform scene size-up.
   a. Determine the safest route to access the patient.
   b. Determine the mechanism of injury/nature of illness.
   c. Determine the number of patients.
   d. Request additional help if necessary.
   e. Consider stabilization of the spine.

3. Perform an Initial Assessment.
   a. Form a general impression of the patient and the patient's environment.
   b. Assess the patient's mental status using the Alert, Verbal, Pain, Unresponsive (AVPU) scale.
      (1) A - Alert and oriented.
      (2) V - Responsive to verbal stimuli.
      (3) P - Responsive to painful stimuli.
      (4) U - Unresponsive.
   c. Determine the chief complaint/apparent life-threatening condition.
   d. Assess the airway.
      (1) Perform an appropriate maneuver to open and maintain the airway if necessary. (See task 081-831-0018.)
      (2) Insert an appropriate airway adjunct, if necessary. (See tasks 081-833-0016, 081-833-0142, and 081-833-0169. Also if skill level 30, See task 081-830-3016.)
   e. Assess breathing.
      (1) Determine the rate, rhythm, and quality of breathing.
      (2) Administer oxygen if necessary using the appropriate delivery device. (See tasks 081-833-0158 and 081-831-0048.)
   f. Assess circulation.
      (1) Check skin color and temperature.
      (2) Assess the pulse for rhythm and force.
         (a) Check the radial pulse in adults.
         (b) Check the radial pulse and capillary refill in children under 6 years old.
         (c) Check the brachial pulse and capillary refill in infants.
      (3) Check for major bleeding.
      (4) Control major bleeding. (See tasks 081-833-0161 and 081-833-0046.)
      (5) Treat for shock. (See task 081-833-0047.)
   g. Identify priority patients and make a transport decision (load and go or stay and play).

Performance Steps
NOTE: High priority conditions that require immediate transport include poor general impression, unresponsive, responsive but not following commands, difficulty breathing, shock, complicated childbirth, chest pain with systolic blood pressure less than 100, uncontrolled bleeding, and severe pain.

4. Conduct a rapid physical exam if the patient is unconscious. Inspect each of the following areas for deformities, contusions, abrasions, punctures or penetration, burns, tenderness, lacerations, swelling (DCAP-BTLS).
    a. Assess the head.
    b. Assess the neck.
    c. Assess the chest.
    d. Assess the abdomen.
    e. Assess the pelvis.
    f. Assess the extremities.
    g. Assess the posterior.

5. Gather a SAMPLE history from the patient.
NOTE: If the patient is unable to give you this information, gather as much information about the SAMPLE history as you can from the patient's family and/or bystanders.
    a. Signs and symptoms. Gather history of the present illness (OPQRST) from the patient.
        (1) Respiratory.
            (a) Onset - When did it begin?
            (b) Provocation - What were you doing when this came on?
            (c) Quality - Can you describe the feeling you have?
            (d) Radiation - Does the feeling seem to spread to any other part of your body? Do you have pain or discomfort anywhere else in your body?
            (e) Severity - On a scale of 1 to 10, how bad is your breathing trouble (10 is worst, 1 is best)?
            (f) Time - How long have you had this feeling?
            (g) Interventions - Have you taken any medication to help you breathe? Did it help?
        (2) Cardic.
            (a) Onset - When did it begin?
            (b) Provocation - What were you doing when this came on?
            (c) Quality - Can you describe the feeling you have?
            (d) Radiates - Does the feeling seem to spread to any other part of your body? Do you have pain or discomfort anywhere else in your body?
            (e) Severity - On a scale of 1 to 10, how bad is your breathing trouble (10 is worst, 1 is best)?
            (f) Time - How long have you had this feeling?
            (g) Interventions - Have you taken any medication to help you? Did it help?
        (3) Altered mental status.
            (a) Description of the episode - Can you tell me what happened? How did the episode occur?
            (b) Onset - How long ago did it occur?
            (c) Duration - How long did it last?
            (d) Associated symptoms - Was the patient sick or complaining of not feeling well before this happened?
            (e) Evidence of trauma - Was the patient involved in falls or accidents recently?

Performance Steps

      (f) Interventions - Has the patient taken anything to help with this problem? Did it help?

      (g) Seizures - Did the patient have a seizure?

      (h) Fever - Did the patient have a fever? What was the patient's temperature?

    (4) Allergic reaction.

      (a) History of allergies - Do you have any allergies?

      (b) What were you exposed to - Is there any chance that you were exposed to something that you may be allergic to?

      (c) How were you exposed - How did you come into contact with _____ (whatever the patient is allergic to)?

      (d) Effects - What kind of symptoms are you having? How long after you were exposed did the symptoms start?

      (e) Progression - How long after you were exposed did the symptoms start? Are they worse now than they were before?

      (f) Interventions - Have you taken anything to help? Did it help?

    (5) Poisoning/overdose.

      (a) Substance - What substance was involved?

      (b) When did you ingest/become exposed - When did the exposure/ingestion occur?

      (c) How much did you ingest - How much did the patient ingest?

      (d) Over what time period - Over how long a period did the ingestion occur?

      (e) Interventions - What interventions did the family or bystanders take?

      (f) Estimated weight - What is the patient's estimated weight?

    (6) Environmental emergency.

      (a) Source - What caused the injury?

      (b) Environment - Where did the injury occur?

      (c) Duration - How long were you exposed?

      (d) Loss of consciousness - Did you lose consciousness at any time?

      (e) Effects (general or local) - What signs and symptoms are you having? What effect did being exposed have on the patient?

    (7) Obstetrics.

      (a) Are you pregnant?

      (b) How long have you been pregnant?

      (c) Are you having pain or contractions?

      (d) Are you bleeding? Are you having any discharge?

      (e) Do you feel the need to push?

      (f) When was your last menstrual period?

    (8) Behavioral.

      (a) How do you feel?

      (b) Determine suicidal tendencies - Do you have a plan to hurt yourself or anyone else? .

      (c) Is the patient a threat to self or others?

      (d) Is there a medical problem?

      (e) Interventions?

  b. Allergies.

  c. Medications.

  d. Past pertinent history.

  e. Last oral intake.

  f. Event(s) leading to present illness.

Performance Steps

6. Perform a focused physical examination on the affected body part/system.

7. Obtain baseline vital signs (See tasks 081-831-0013, 081-831-0011, 081-831-0010, and 081-831-0012).

8. Provide medication, interventions, and treatment as needed. (See tasks 081-831-0035, 081-833-0103, 081-833-0116, 081-833-0143, 081-833-0144, 081-833-0159, 081-833-0160, 081-833-0163, 081-833-0166, 081-833-0054, 081-831-0038, 081-831-0039, 081-833-0031, 081-833-0073, and 081-833-3206.)

9. Reevaluate the transport decision.

10. Consider completing a detailed physical examination.

11. Perform Ongoing Assessment.
    a. Repeat the initial assessment.
    b. Repeat vital signs.
    c. Repeat the focused assessment regarding the patient's complaint or injuries.

Performance Measures	**GO**	**NO-GO**
1. Took BSI precautions.	——	——
2. Performed a scene size-up.	——	——
3. Performed an Initial Assessment.	——	——
4. Conducted a rapid physical exam if the patient was unconscious.	——	——
5. Gathered a SAMPLE history from the patient.	——	——
6. Performed a focused physical examination on the affected body part/system.	——	——
7. Obtained baseline vital signs.	——	——
8. Provided medication, interventions and treatment.	——	——
9. Reevaluated the transport decision.	——	——
10. Considered completing a detailed physical examination.	——	——
11. Performed Ongoing Assessment.	——	——

Evaluation Guidance: Score each Soldier according to the performance measures in the evaluation guide. Unless otherwise stated in the task summary, the Soldier must pass all performance measures to be scored GO. If the Soldier fails any step, show what was done wrong and how to do it correctly.

References

Required	**Related**
None	EMERG CARE AND TRANS 9

TREAT A CARDIAC EMERGENCY
081-833-0159

Conditions: You have a conscious patient who is complaining of chest pain. You have already taken the appropriate body substance isolation (BSI) precautions. You have already done the initial patient assessment, focused history, and physical. You will need a sphygmomanometer, stethoscope, oxygen tank setup, non-rebreather mask, nitroglycerin, patient's medical record, and intravenous (IV) materials. You are not in a CBRNE environment.

Standards: Complete all necessary steps to manage a patient with a cardiac emergency, without causing any further injury.

Performance Steps

   1. Identify the signs and symptoms of cardiac emergency or compromise.
      a. Pain, pressure, or discomfort in the chest or upper abdomen (epigastrium).
      b. Dyspnea.
      c. Palpitations.
      d. Sudden onset of sweating with nausea or vomiting.
      e. Anxiety (feeling of impending doom or irritability).
      f. Abnormal pulse.
        (1) Bradycardia (less than 60 beats per minute).
        (2) Tachycardia (greater than 100 beats per minute).
      g. Abnormal blood pressure.
        (1) Hypotensive (systolic pressure less than 90).
        (2) Hypertensive (systolic pressure greater than 140).
      h. Pulmonary edema.
        (1) Shortness of breath.
        (2) Dyspnea.
        (3) Rales upon auscultation.
        (4) Blood tinged sputum.
      i. Pedal edema.

   2. Administer the appropriate treatment.
      a. Place the patient in a position of comfort.
NOTE: This is usually in the Fowler's position.
      b. Apply a high concentration of oxygen via a non-rebreather mask.
      c. Assist the patient in taking nitroglycerin, if available.
NOTE: Administer the nitroglycerin only if ALL of the following conditions are met:
1. Patient complains of chest pain.
2. Patient has a history of cardiac problems.
3. Patient has a current prescription for nitroglycerin.
4. Patient has the nitroglycerin with him.
5. Patient's systolic blood pressure is greater than 100.
        (1) Check the five rights.
        (2) Remove the oxygen mask.
        (3) Ask the patient to open his mouth and lift his tongue.
        (4) Place the tablet or spray (if using mist) under the tongue with a gloved hand.
CAUTION: Avoid contacting the nitroglycerin tablet or mist with bare skin. The vasodilation affects could cause unconsciousness.
        (5) Have the patient close his mouth and hold the tablet under the tongue.

Performance Steps
      (6) Replace the oxygen mask.
      (7) Recheck the blood pressure within 2 minutes.
NOTE: If the blood pressure falls below 100, treat the patient for shock and transport immediately.
      d. If the patient experiences no relief, repeat step 2c every 5 minutes until the patient has taken a total of three tablets.
      e. If the patient experiences no relief after three nitroglycerin tablets or his condition worsens, initiate an IV at to keep open (TKO) rate. (See task 081-833-0033.)

3. Transport promptly to the nearest medical treatment facility.

4. Perform an ongoing assessment while en route.

5. Document all interventions.

Evaluation Preparation:
Setup: Have one Soldier be the patient while the Soldier being tested administers treatment. Tell the Soldier who is acting as the patient the signs and symptoms he should exhibit and how to answer the questions asked by the Soldier being tested.

Brief Soldier: Tell the Soldier to treat the patient for a cardiac emergency.

Performance Measures	GO	NO-GO
1. Identified the signs and symptoms of cardiac emergency or compromise.	——	——
2. Administered the appropriate treatment.	——	——
3. Transported promptly to the nearest medical treatment facility.	——	——
4. Performed an ongoing assessment while en route.	——	——
5. Documented all interventions.	——	——

Evaluation Guidance: Score each Soldier according to the performance measures in the evaluation guide. Unless otherwise stated in the task summary, the Soldier must pass all performance measures to be scored GO. If the Soldier fails any step, show what was done wrong and how to do it correctly.

References
    Required   **Related**
    None   EMERG CARE AND TRANS 9

TREAT A RESPIRATORY EMERGENCY
081-833-0160

Conditions: You have a conscious patient with a respiratory emergency. You will need a stethoscope, pulse oximeter, oxygen tank, nasal cannula, oxygen mask and tubing, hand held metered dose inhaler (MDI) with spacer, nebulizer set up, medicated solution, normal saline for inhalation therapy, and the patient's medical records. You are not in a CBRNE environment.

Standards: Correctly identify and treat a respiratory emergency without causing further harm to the patient.

Performance Steps

1. Examine the patient.
    a. Assess the airway and open it, if necessary. (See task 081-831-0018.)
CAUTION: A patient experiencing respiratory distress can rapidly progress to full arrest. Always be prepared to utilize advanced airway procedures.
        (1) Ask the patient a question requiring more than a yes or no answer.
        (2) Note whether or not the patient can speak in full sentences.
        (3) Look for the presence of drooling that may indicate a partial or complete airway obstruction.
    b. Assist with artificial ventilations if respiratory effort and rate are inadequate.
        (1) Look for the rise and fall of the chest during inspiration and expiration.
        (2) Listen for the presence of noisy respirations (e.g., stridor, wheezing).
    c. Apply supplemental oxygen by mask or nasal cannula.
NOTE: Any casualty complaining of difficulty breathing should receive supplemental oxygen.
    d. Place the patient in the position of comfort.
NOTE: Most patients experiencing difficulty breathing prefer to remain in a seated position.
    e. Obtain a complete set of vital signs to include pulse oximetry, if available.

2. Perform a focused physical examination.
    a. Listen to the anterior and posterior lung fields with the stethoscope.
    b. Look at the chest and abdomen and note the presence of any retractions.
    c. Check the skin for the presence of cyanosis.
    d. Check the lower extremities for the presence of edema.

3. Obtain a focused history.
    a. Ask the patient if there is an existing condition such as asthma.
    b. Ask the patient if he is taking any medications.
    c. Question the patient about allergies to medications.
    d. Ask the patient if difficulty breathing was of sudden or gradual onset.

4. Assist the patient in using a metered dose inhaler.
NOTE: This step may only be performed if the casualty has an inhaler prescribed to him.
    a. Perform the five rights of medication usage.
    b. Have the patient exhale deeply.
    c. Have the patient place his lips around the opening and press the inhaler to activate the spray as he inhales deeply.
    d. Instruct the patient to hold his breath as long as possible before exhaling.
    e. Repeat steps 4b through 4d.

5. Administer a nebulizer treatment.

Performance Steps
NOTE: This step may only be performed with a medical officer's order for nebulization.
  a. Set up the nebulizer per manufacturer's guidelines.
  b. Instill the appropriate medicine IAW local SOP.
  c. Connect the nebulizer to an oxygen source.
NOTE: Compressed air can be used but it doesn't supply the casualty with supplemental oxygen.
  d. Turn on the flow of oxygen and check for the formation of mist (smoke).
  e. Have the patient place his lips on the mouth piece and slowly inhale and exhale the mist.
  f. Monitor the patient's vital signs every 5 minutes. If available, attach the casualty to a pulse oximeter.

 6. Document the procedure.

 7. Transport the patient.

Performance Measures	**GO**	**NO-GO**
1. Examined the patient.	——	——
2. Performed a focused physical examination.	——	——
3. Obtained a focused history.	——	——
4. Assisted the patient in using a metered dose inhaler.	——	——
5. Administered a nebulizer treatment.	——	——
6. Documented the procedure.	——	——
7. Transported the patient.	——	——

Evaluation Guidance: Score each Soldier according to the performance measures in the evaluation guide. Unless otherwise stated in the task summary, the Soldier must pass all performance measures to be scored a GO. If the Soldier fails any step, show what was done wrong and how to do it correctly.

References
 Required
 None

      Related
      ISBN 0-7637-3901-4

INSERT AN OROGASTRIC TUBE
081-833-0187

Conditions: You need to insert an orogastric tube. A patient care hand-wash has been performed. You will need gloves, an orogastric tube, surgical lubricant, scissors, adhesive tape, stethoscope, large (50 or 60 cc) syringe, small (10 to 20 cc) syringe, container for contaminated waste, water, small cup, a drinking straw, and the patient's medical records. You are not in a CBRNE environment.

Standards: Insert an orogastric tube without causing further harm.

Performance Steps

   1. Determine if the indicators for performing orogastric intubation are present.
      a. Burns.
      b. Intestinal obstruction.
      c. Preoperative and postoperative care.
      d. Patient requires a gastrointestinal lavage.
        (1) Overdose.
        (2) Gastrointestinal bleeding.
      e. Patient requires a gavage.
        (1) Comatose patients.
        (2) Debilitated patients.
      f. Need to analyze stomach contents.
        (1) Patient may be bleeding internally.
        (2) Overdose.

   2. Put on gloves.
WARNING Wear gloves for self-protection against transmission of contaminants whenever handling body fluids.

   3. Explain the procedure to the patient.
      a. Tell the patient that a tube will be inserted along the oral passage and that he may feel some discomfort.
      b. Tell the patient that breathing through the nose/mouth, panting, and swallowing can help in passing the tube.
      c. Ask the patient about any history of oral or esophageal injury.
      d. Tell the patient that the tube must be placed about 20 inches down the oral passageway.
      e. Tell the patient that the procedure may cause him to gag.

   4. Examine the patient's oral cavity. Ensure that there are no obstructions in the oral cavity.

   5. Position the patient.
      a. Position the responsive, awake, and alert patient in the Fowler's position. Elevate the head of the bed to about 30 to 45 degrees.
      b. Place a comatose or unconscious patient in the lateral position (turn the patient onto his or side).

Performance Steps

6. Prepare the equipment.
    a. Select the correct type and size of orogastric tube to use.
    b. Cut four or five pieces of tape 3 to 4 inches long and attach one end of each where they will be easily accessible.
    c. Unwrap the tube from the plastic wrapper.
    d. Measure the tube for insertion.
    e. Lubricate 5 to 6 centimeters of the distal end of the tube with water-soluble lubricant.

7. Insert the lubricated tip of the tube into the oropharynx.

8. Advance the tube into the esophagus.

9. Continue advancing the tube until the tape marker touches the lips.

10. Check the placement of the tube, and take corrective action when the tube is not correctly placed.

11. Using tape, secure the tube to the patient's cheek.

12. Connect the tube to the suction apparatus when required.

13. Remove the gloves and wash your hands.

14. Record the procedure on the appropriate form.

Evaluation Preparation: This task is best evaluated by performance of the steps. Give the Soldier a scenario in which he must perform orogastric intubation.

Performance Measures	GO	NO-GO
1. Determined if the indicators for performing orogastric intubation were present.	——	——
2. Put on gloves.	——	——
3. Explained the procedure to the patient.	——	——
4. Examined the patient's oral cavity.	——	——
5. Positioned the patient.	——	——
6. Prepared the equipment.	——	——

    a. Selected the correct type and size of orogastric tube to use.
    b. Cut four or five pieces of tape 3 to 4 inches long and attached one end of each where they would be easily accessible.
    c. Unwrapped the tube from the plastic wrapper.
    d. Measured the tube for insertion.
    e. Lubricated 5 to 6 centimeters of the distal end of the tube with water-soluble lubricant.

7. Inserted the lubricated tip of the tube into the oropharynx.	——	——
8. Advanced the tube into the esophagus.	——	——

Performance Measures	<u>GO</u>	<u>NO-</u> <u>GO</u>
9. Continued advancing the tube until the tape marker touched the lips.	——	——
10. Checked the placement of the tube, and took corrective action when the tube was not correctly placed.	——	——
11. Using tape, secured the tube to the patient's cheek.	——	——
12. Connected the tube to the suction apparatus when required.	——	——
13. Removed gloves and washed hands.	——	——
14. Recorded the procedure on the appropriate form.	——	——

Evaluation Guidance: Score each Soldier according to the performance measures. Unless otherwise stated in the task summary, the Soldier must pass all performance measures to be scored GO. If the Soldier fails any steps, show what was done wrong and how to do it correctly.

References
    Required
    None

    Related
    BASIC NURSING 7

TREAT A NEAR DROWNING VICTIM
081-833-0201

Conditions: You come upon a casualty who appears to be a near drowning victim. You will need a medical aid bag, buoyant backboard, automated external defibrillator (AED), oxygen, oxygen administration equipment, a blanket, and a DD Form 1380 (U.S. Field Medical Card). You are not in a CBRNE environment.

Standards: Solicit a casualty history, perform a physical examination, and administer supportive care, without causing further injury to the casualty.

Performance Steps

1. Recognize the signs and symptoms of near drowning.
   a. Change in level of consciousness.
   b. Restlessness.
   c. Chest pain.
   d. Rales, rhonchi, or wheezing.
   e. Vomiting.
   f. Cyanosis.
   g. Signs of shock (common in near-drowning). When shock is present, try to determine if shock is hypovolemic, hypoxic, or neurogenic spinal injury.
   h. Pink froth from nose and mouth.

2. Ensure the safety of all rescuers, including yourself, before any water rescue can begin.

3. Perform prehospital management for near drowning and aspiration.
   a. Raise the casualty to the surface and remove him from the water as soon as possible.
NOTE: Cervical or spinal injuries are always a primary concern. You must assume that the casualty has a spinal injury and treat accordingly. This means that initial resuscitation and spine immobilization must occur while the casualty is still in the water. (See task 081-833-0176.)
   b. Immediately perform rescue breathing (in the water, if possible), and then cardiopulmonary resuscitation (CPR), if needed.
   c. Float a buoyant backboard under the casualty as ventilation is continued.
   d. Secure the trunk and neck to the backboard to eliminate spine motion. Do not remove the casualty from the water until this is done.
   e. Remove casualty from water.
   f. Place the casualty in the lateral recumbent position, with the backboard in place.
   g. Cover the casualty with a blanket.
   h. Administer oxygen by mask.

4. Perform a trauma casualty assessment (See task 081-833-0155.)
NOTE: All victims should be hospitalized for at least 24 hours for observation. Common complications of near drowning are respiratory failure and circulatory collapse.

5. Record all treatment on the FMC.

6. Evacuate the casualty.

Performance Measures	GO	NO-GO
1. Recognized the signs and symptoms of near drowning.	——	——
2. Ensured the safety of all rescuers, including yourself, before any water rescue began.	——	——
3. Performed prehospital management for near drowning or aspiration.	——	——
4. Performed a trauma casualty assessment.	——	——
5. Recorded all treatment on the FMC.	——	——
6. Evacuated the casualty.	——	——

Evaluation Guidance: Score each Soldier according to the performance measures. Unless otherwise stated in the task summary, the Soldier must pass all performance measures to be scored GO. If the Soldier fails any steps, show what was done wrong and how to do it correctly.

References

Required	**Related**
DD FORM 1380	ISBN 0-7637-4406-9

TREAT A PATIENT WITH AN ALLERGIC REACTION
081-833-0224

Conditions: You have a patient demonstrating signs and symptoms of an allergic reaction. You will need a combat medic aid bag, and an epinephrine autoinjector. You are not in a CBRNE environment.

Standards: Treat a patient with an allergic reaction without causing further harm.

Performance Steps

1. Recognize the causes of allergic reactions.
    a. Drugs (penicillin).
    b. Insect bites (bee stings).
    c. Pollen.
    d. Food (peanuts).

2. Recognize the early manifestations of an allergic or anaphylactic reaction.
    a. Skin.
        (1) Flushing.
        (2) Urticaria (hives).
        (3) Swelling of face (especially eyes and lips), hands, feet, neck.
        (4) Swelling of mouth, tongue, airway (angioedema).
    b. Respiratory.
        (1) Tightness in throat and chest.
        (2) Cough.
        (3) Rapid, labored noisy breathing.
        (4) Stridor (harsh, high pitched sound during inspiration).
        (5) Wheezing (may be audible without a stethoscope).
    c. Cardiac.
        (1) Increased heart rate.
        (2) Decreased blood pressure.
    d. Generalized feelings.
        (1) Itchy, watery eyes.
        (2) Headache.
        (3) Runny nose.
        (4) Sense of impending doom.

3. Recognize the signs of anaphylactic shock.
    a. May have any of the above, but must have signs of respiratory distress or shock.
    b. Altered mental status.
    c. Signs of respiratory distress.
    d. Signs of shock.

4. Treat allergic reactions.
    a. Perform initial assessment ABCs (treat any life-threatening conditions).
    b. Perform a focused history and physical exam.
    c. Assess baseline vital signs and SAMPLE history.
    d. Manage the patient's airway and breathing. If the patient has an epinephrine autoinjector and has symptoms of anaphylaxis, assist in the epinephrine administration.

Performance Steps

5. Evacuate the patient to the nearest medical treatment facility (MTF).

Evaluation Preparation:
Setup: This task is best evaluated by verbalization of the steps.

Brief Soldier: Give the Soldier a scenario in which he must manage allergic reactions.

Performance Measures	GO	NO-GO
1. Recognized the causes of allergic reactions.	——	——
2. Recognized the early manifestations of an allergic or anaphylactic reaction.	——	——
3. Recognized the signs of anaphylactic shock.	——	——
4. Treated the allergic reaction.	——	——
5. Evacuated the patient to the nearest MTF.	——	——

Evaluation Guidance: Score each Soldier according to the performance measures. Unless otherwise stated in the task summary, the Soldier must pass all performance measures to be scored GO. If the Soldier fails any steps, show what was done wrong and how to do it correctly.

References

    Required   **Related**
    None   ISBN 0-7637-4406-9

MANAGE CARDIAC ARREST USING AN AUTOMATED EXTERNAL DEFIBRILLATOR
081-833-3027

Conditions: You and an assistant arrive at a scene where an adult patient is in ventricular fibrillation or pulseless ventricular tachycardia and is receiving basic cardiac life support from a rescuer. You have already taken the necessary body substance isolation. You will need an automated external defibrillator (AED), oropharyngeal airway, bag-valve-mask, non-rebreather mask, and oxygen tank set up. You are not in a CBRNE environment.

Standards: Complete all the steps necessary to perform cardiac defibrillation with an AED.

Performance Steps

1. Briefly question the rescuer about the arrest event.
   a. How long has the patient been in arrest?
   b. How long has cardiopulmonary resuscitation (CPR) been in progress?
   c. Do you know two man CPR?

2. Direct the rescuer to stop CPR.
NOTE: Allow the rescuer to complete the current cycle.

3. Determine whether the patient is a candidate for an AED.
NOTE: If the patient has sustained trauma before collapse, do not attach the AED. Continue CPR and transport immediately.
   a. Unresponsive.
   b. Apneic.
   c. Pulseless.

4. Direct the rescuer to resume CPR.

5. Turn the AED on.

6. Attach the monitoring-defibrillation pads to the cables if the pads aren't attached.

7. Attach the AED to the patient.
   a. Place the top right pad below the right mid-clavicular.
   b. Place the lower pad over the lower left ribs.

8. Direct the rescuer to stop CPR.

9. Ensure all individuals are standing clear of the patient.
   a. Give the order, "ALL CLEAR."
   b. Visually check to ensure that no one is in contact with the patient.
   c. Visually check to ensure no one is in direct contact with any electrically conductive material touching the patient, such as intravenous (IV) lines, monitor wires, or the bed frame.

10. Initiate analysis of rhythm.
    a. Press the analysis button.
    b. Wait for the machine to analyze the rhythm.

11. If shock is indicated by the AED, ensure everyone is clear and deliver the shock.
    a. Give the order, "ALL CLEAR."
    b. Visually check to ensure that no one is in contact with the patient.

Performance Steps

    c. Visually check to ensure no one is in direct contact with any electrically conductive material touching the patient, such as IV lines, monitor wires, or the bed frame.

    d. Press the button to deliver the shock.

CAUTION: Do not defibrillate if anyone is touching the patient or the patient is wet (dry the patient), touching metal (move away from metal), or wearing a nitroglycerin patch (remove the patch with a gloved hand).

12. Reanalyze the rhythm (have everyone stand clear).

    a. If additional shock is indicated, proceed to step 13.

    b. If no shock is indicated--

NOTE: The patient may be in asystole or pulseless electrical activity (PEA), which are not shockable rhythms.

    (1) Immediately start CPR after shock is delivered.

    (2) Do not reassess pulse.

13. Insert an oropharyngeal or nasopharyngeal airway. (See tasks 081-833-0016 and 081-833-0142.)

14. Direct ventilation of the patient by beginning ventilations with a bag-valve-mask, if available.

15. Add supplemental oxygen at 15L/min, if available.

16. Reevaluate patient after 5 cycles of CPR.

17. Check the carotid pulse.

18. Press the analyze button on the AED and ensure everyone is clear of the patient.

19. Repeat step 12, if shock is advised.

20. Direct rescuers to continue CPR.

21. After 5 more cycles of CPR, reassess the patient.

22. Continue sequence of defibrillation and CPR, as needed.

23. Transport the patient.

Performance Measures	GO	NO-GO
1. Questioned the rescuer about the arrest event.	___	___
2. Directed the rescuer to stop CPR.	___	___
3. Determined whether the patient was a candidate for an AED.	___	___
4. Directed the rescuer to resume CPR.	___	___
5. Turned the AED on.	___	___
6. Attached the monitoring-defibrillation pads to the cables, if necessary.	___	___
7. Attached the AED to the patient.	___	___
8. Directed the rescuer to stop CPR.	___	___

Performance Measures	GO	NO-GO
9. Ensured all individuals were standing clear of the patient.	——	——
10. Initiated analysis of rhythm.	——	——
11. If shock was indicated by the AED, ensured everyone was clear and delivered the shock.	——	——
12. Reanalyzed the rhythm and took appropriate action.	——	——
13. Inserted an oropharyngeal or nasopharyngeal airway.	——	——
14. Directed ventilation of the patient by beginning ventilations with a bag-valve-mask, if available.	——	——
15. Added supplemental oxygen at 15L/min, if available.	——	——
16. Reevaluated patient after 5 cycles of CPR.	——	——
17. Checked the carotid pulse.	——	——
18. Pressed the analyze button on the AED and ensured everyone was clear of the patient.	——	——
19. Repeated step 12, if shock was advised.	——	——
20. Directed rescuers to continue CPR.	——	——
21. After 5 more cycles of CPR, reassessed the patient.	——	——
22. Continued sequence of defibrillation and CPR, as needed.	——	——
23. Transported the patient.	——	——

Evaluation Guidance: Score each Soldier according to the performance measures in the evaluation guide. Unless otherwise stated in the task summary, the Soldier must pass all performance measures to be scored GO. If the Soldier fails any step, show what was done wrong and how to do it correctly.

References
    Required
    None

                                    Related
                                      EMERG CARE AND TRANS 9

PERFORM A GASTRIC LAVAGE
081-835-3005

Conditions: You have verified a medical officer's orders requiring a gastric lavage. A patient care hand-wash been performed. You will need an Ewald tube, nasogastric tubes, water-soluble lubricant, 50 cc catheter-tip syringes, basins, protective pads, towels, sphygmomanometer, stethoscope, thermometer, graduated containers, ice, prescribed lavage solution, gloves, DD Form 792 (Nursing Service - Twenty-Four Hour Patient Intake and Output Worksheet), and the patient's clinical record. You are not in a CBRNE environment.

Standards: Perform the gastric lavage IAW the medical officer's orders and without causing further injury to the patient.

Performance Steps

1. Assemble the necessary equipment and set it up at the patient's bedside.
    a. Ensure that there is enough irrigating solution on hand.
NOTE: In most cases, the medical officer's order will be to lavage "until clear". Lavage will continue until the stomach contents return clear (nothing is returned but the irrigating solution itself). This requires preparation of at least 6 liters of the prescribed irrigating solution (usually normal saline).
    b. Ensure that ice or chilled solution is available when the medical officer orders "ice lavage".
NOTE: When lavage is done to control gastric bleeding, the order is usually for "ice lavage". Chilling the irrigating solution promotes vasoconstriction, thereby helping to control bleeding.

2. Explain the procedure to the patient.

3. Establish baseline vital signs.

4. Position the patient.
    a. A patient who is alert should be placed in the Fowler's or semi-Fowler's position.
    b. A patient who is not alert, or too weak to sit, should be positioned on the left side, with the head of the bed elevated 15 degrees.
NOTE: This left lateral recumbent position will allow the tip of the tube to lie in the greater curvature of the stomach.

5. Insert the appropriate tube if one is not already in place.
    a. For a stomach wash, the medical officer will specify insertion of large lumen nasogastric tube or the Ewald stomach tube.
NOTE: The Ewald stomach tube is normally inserted through the mouth rather than the nose, because it is a large bore tube.
    b. For control of gastric bleeding, the medical officer will specify insertion of a large lumen nasogastric tube.
    c. In the event of severe bleeding, as in the case of esophageal varices, the medical officer will specify insertion of a nasogastric tube that has gastric and esophageal balloons (Blakemore tube, for example).
NOTE: In any situation, a large lumen tube is indicated. Particles of food, mucous, or blood may occlude the lumen of a small tube.
CAUTION: Gloves should be worn for self-protection against transmission of contaminants whenever handling body fluids.

Performance Steps

6. Aspirate all stomach contents.
   a. Using a 50 cc syringe, aspirate stomach contents and place the aspirate in a measured container.
   b. Repeat until all stomach contents have been aspirated.
   c. Record the total amount as output on the I/O worksheet.
   d. Save the aspirate for disposition as directed by the medical officer.

7. Instill the irrigating solution, using the method specified by the medical officer's order or local SOP.
   a. Syringe method. Using a 50 cc catheter-tip syringe, instill 100 cc of the solution. (Instillation and withdrawal are repeated 100 cc at a time, until clear or IAW medical officer's order).
   b. Funnel method. Using a funnel (or syringe barrel), instill up to 500 cc of the solution by pouring it slowly into the funnel.

CAUTION: When using the funnel method, it is imperative that the patient be carefully assessed for abdominal distention. The size and tolerance of the patient will determine how much fluid can be instilled at one time.

8. Withdraw the irrigating solution.
   a. Syringe method.
      (1) Using a 50 cc catheter-tip syringe, withdraw all the irrigating solution and stomach contents.
      (2) Place the aspirate into a measured container.
      (3) Note the amount and character of the aspirate.

NOTE: If syringe aspiration is difficult, or no aspirate can be obtained, the gastric tube may be resting against the gastric mucosa. Reposition the patient and aspirate again. If aspiration is still difficult, reposition the tube by advancing or withdrawing it slightly.

   b. Funnel method.
      (1) Lower the funnel end of the tube below the level of the patient's stomach to facilitate gravity drainage.

NOTE: If the solution does not begin to drain by gravity, aspirate with a syringe (creating a siphon effect) to start the backflow of solution. If gravity drainage cannot be established, withdraw the solution by the syringe method.

      (2) Allow the irrigating solution and stomach contents to drain into a measured container.
      (3) Note the amount and character of the return solution.

9. Continue the lavage by repeating steps 7 and 8 IAW the medical officer's order. That is, continue until the stomach contents are clear, the prescribed amount of solution has been administered, or as otherwise directed.

10. Clamp the tubing.
    a. Clamp and secure the tube if it is to remain in place.
    b. Clamp and withdraw the tube if it is to be removed.

11. Remove all used equipment from the bedside.

12. Measure the lavage return.
    a. Measure and record the total lavage return.
    b. Estimate the amount of stomach contents by subtracting the known amount of irrigating solution used from the measured amount of total lavage return.
    c. Record the amount of stomach contents as output on the I/O worksheet.

Performance Steps

13. Dispose of the initial stomach aspirate and all lavage solution returned as directed by the medical officer's orders or local SOP. That is, hold it for examination by the medical officer, send fluid samples to the laboratory, or discard it into the appropriate receptacle.

14. Document the procedure and significant nursing observations on the appropriate forms IAW local SOP.
    a. Note the type and amount of lavage solution used.
    b. Note the color, odor, character, and amount of initial stomach contents aspirated.
    c. Note the color, odor, character, and amount of lavage return.
    d. Describe the patient's tolerance of the procedure.
    e. Note the disposition of any specimens.

Performance Measures	**GO**	**NO-GO**
1. Assembled the equipment.	——	——
2. Explained the procedure to the patient.	——	——
3. Established baseline vital signs.	——	——
4. Positioned the patient.	——	——
5. Inserted the tube if necessary.	——	——
6. Aspirated all the stomach contents.	——	——
7. Instilled the irrigating solution.	——	——
8. Withdrew the irrigating solution.	——	——
9. Continued the lavage as directed by the medical officer's order.	——	——
10. Clamped the tubing.	——	——
11. Removed the equipment.	——	——
12. Measured the lavage return.	——	——
13. Disposed of the initial aspirate and lavage return as directed by the medical officer's order.	——	——
14. Documented the procedure.	——	——

Evaluation Guidance: Score each Soldier according to the performance measures in the evaluation guide. Unless otherwise stated in the task summary, the Soldier must pass all performance measures to be scored GO. If the Soldier fails any step, show what was done wrong and how to do it correctly.

References

Required	**Related**
DD FORM 792	BASIC NURSING 7

OBTAIN AN ELECTROCARDIOGRAM
081-835-3007

Conditions: You have verified a medical officer's order to obtain an electrocardiogram (EKG) on a patient. You have identified the patient and explained the procedure. A patient care hand-wash has been performed. You will need an EKG machine, electrodes, alcohol prep pads, towels, tape, OF 520 (Electrocardiograph Record), and the patient's clinical record. You are not in a CBRNE environment.

Standards: Obtain an EKG in IAW with the medical officer's order.

Performance Steps

1. Prepare the equipment.
    a. Read the manufacturer's instructions for the proper use of the equipment on hand.
    b. Plug in the machine and turn it on.
    c. Allow it to perform self-checks, if computerized, or warm up for 5 minutes if not computerized.

2. Prepare the patient.
    a. Provide for the patient's privacy.
    b. Provide a female chaperone, if necessary, for female patients.
    c. Ask or assist the patient to remove wristwatch, shoes, socks or hose, and all clothing from the waist up.
    d. Provide a chest drape for female patients.
    e. Ask or assist the patient to lie supine on the bed or examination table.
    f. Ensure that the patient's body is not in contact with any metal objects, and that all limbs are firmly supported.
NOTE: Some metal objects, watches, or jewelry may interfere with the accurate recording of the electrical impulses.
    g. Instruct the patient to relax and breathe normally throughout the entire procedure.

3. Apply limb electrodes.
    a. Clean the site for electrode placement by wiping with an alcohol prep pad to remove dead skin and oils as needed.
NOTE: An area of broken down or irritated skin should not be used for the electrode connection.
    b. Position the electrode.
        (1) Secure the leg electrodes on the medial or lateral aspect of the calf.
        (2) Secure the arm electrodes on the arm or forearm, ensuring that the connections are not on, or adjacent to, an intravenous (IV) site.
        (3) Ensure that all the connections are made over a fleshy area, not over bone, as bone may interfere with conduction of the electrical impulse to the electrode.
NOTE: Make the usual electrode connection to a fleshy part of the stump if the patient is missing a limb. Secure the electrode with tape if necessary.

4. Apply the chest electrodes.
    a. Clean the sites for electrode placement by wiping with an alcohol prep pad to remove dead skin and oils as needed.
    b. Position the electrodes, being careful to place them over the intercostal spaces and not directly over the ribs. Refer to Figure 3-1.

Performance Steps

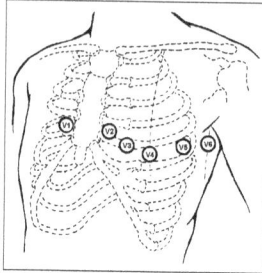

Figure 3-1. Position the electrodes

(1) V1: 4th intercostal space at the right sternal border.
(2) V2: 4th intercostal space at the left sternal border.
(3) V3: Halfway between V2 and V4.
(4) V4: 5th intercostal space at the left midclavicular line.
(5) V5: 5th intercostal space at the left anterior axillary line.
(6) V6: 5th intercostal space at the left midaxillary line.

NOTE: The standard EKG machine utilizes 12 "leads". These leads represent paths of electrical activity, and are designated as leads I, II, III, AVR, AVL, AVF, V1, V2, V3, V4, V5, and V6. Do not confuse these 12 leads with the 10 electrodes (sometimes referred to as "leads") that are attached to the patient.

5. Obtain the EKG tracing.
    a. Operate the equipment IAW the manufacturer's operating instructions.
    b. Ensure a complete and readable EKG tracing. Refer to Figure 3-2.

6. Observe and assess the EKG tracing as it is printed and take appropriate action.
    a. Observe the tracing for the presence of the normal waves in each heartbeat.

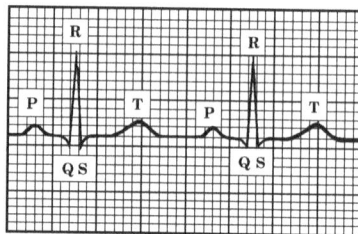

Figure 3-2. EKG tracing

Performance Steps
NOTE: Each heartbeat is normally represented as 5 major waves: P, Q, R, S, and T. The Q, R, and S waves all represent the same portion of the heartbeat and are referred to as a unit: QRS complex. Occasionally, a 6th wave will appear. It is referred to as the U wave. Although it does not always appear, its presence is perfectly normal.

    b. Observe for irregularities that are a result of artifact, interference, or equipment malfunction.
      (1) Check the patient's position.
      (2) Check the placement of the electrodes.
      (3) Obtain new equipment if necessary.
      (4) Repeat the EKG.
    c. Observe for irregularities of the heart's rhythm.
      (1) Notify the medical officer IMMEDIATELY if you note the presence of any of the life-threatening ventricular arrhythmias.
NOTE: Ventricular arrhythmias are characterized by an ectopic (out of place) focus in the wall of the ventricle which initiates ventricular contraction. A distorted and prolonged QRS complex occurs as a result of the aberrant conduction pathway.
        (a) Ventricular Fibrillation. V-Fib is an irregular and chaotic ventricular arrhythmia characterized by a rapid rate and disorganized conduction of impulses throughout the ventricular myocardium. Death will occur within minutes without immediate defibrillation or initiation of cardiopulmonary resuscitation (CPR). Refer to Figure 3-3.

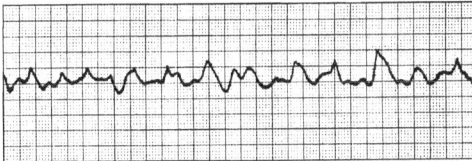

Figure 3-3. Ventricular fibrillation

        (b) Ventricular Tachycardia. V-Tach is a ventricular arrhythmia characterized by broad QRS complexes and a regular rate that falls between 100 to 200 beats per minute. Immediate correction is essential, as V-Tach may lead to V-Fib. Refer to Figure 3-4.

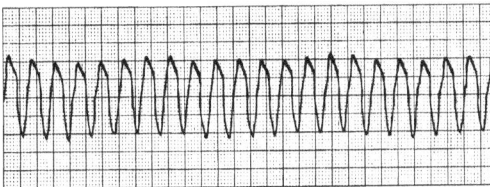

Figure 3-4. Ventricular tachycardia

Performance Steps

(c) Premature Ventricular Contractions (PVCs). PVCs occur when an ectopic focus in one of the ventricles initiates contraction of the ventricles. When this occurs, there will be no atrial contraction associated with that beat, and no P wave will be seen in front of the QRS complex. A PVC usually has a tall, broad QRS complex. PVCs that come from different focal points in the ventricle will have different shapes on the EKG. PVCs may be harmless, but they may also be the forerunners of V-Tach and V-Fib. For this reason, even occasional PVCs should be considered important. PVCs are considered life threatening when they are frequent (more than 6 per minute), when they occur in groups of two or more (back-to-back), when they are multi-focal, and when they occur in a pattern of every other beat or every third beat (bigeminy, trigeminy). Refer to Figure 3-5.

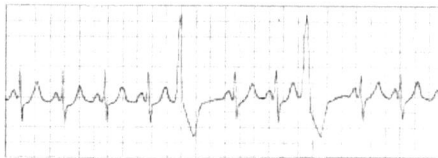

Figure 3-5. Premature ventricular contractions

(2) Notify the medical officer of any other irregularities of the heart's rhythm after you have completed the tracing, but before you remove the electrodes from the patient. (A second tracing may be ordered.)

7. Calculate the heart rate and report any abnormalities.
   a. Time is measured on the horizontal axis of the EKG graph paper. Refer to Figure 3-6.

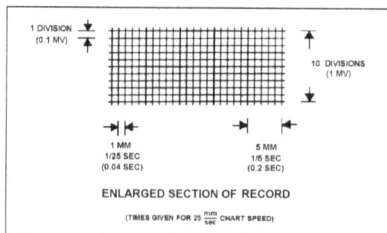

Figure 3-6. EKG graph paper

(1) Each small box = 0.04 seconds.
(2) Each large box = 5 small boxes = 0.20 seconds.
(3) 5 large boxes = 1.0 second = 1 inch of graph paper.
(4) 300 large boxes = 60 seconds = 1 minute.
   b. Calculate the heart rate using one of the following methods:

Performance Steps

      (1) Count the number of large boxes between any two R waves and divide that number into 300. Example: 300 divided by 5 large boxes = 60.

      (2) Count the number of R waves in a 6 second strip and multiply by 10. Example: 6 R waves X 10 = 60.

    c. Notify the medical officer of irregularities.

      (1) Bradycardia--less than 60 beats per minute.

      (2) Tachycardia--more than 100 beats per minute.

NOTE: Paper speed must be set on the normal (25 mm/sec) setting.

  8. Remove the electrodes.

    a. Remove all the chest and limb electrodes.

    b. Wipe the patient's skin with a damp towel to remove the excess electrode paste.

NOTE: Instruct the patient to wash with soap and water as soon as convenient to avoid skin irritation from the EKG paste.

  9. Ask or assist the patient to dress.

  10. Prepare the report.

    a. Remove the EKG tracing from the machine.

    b. Mark the EKG tracing printout with the patient's identification.

    c. Attach the completed OF 520 to the EKG tracing printout.

    d. Make proper distribution of the report as directed by the medical officer's order or IAW local SOP.

  11. Store the equipment.

    a. Dispose of used electrodes IAW local SOP.

    b. Restock the machine with EKG paper, electrodes, alcohol prep pads, towels, and drapes, as necessary.

    c. Store the machine in the area and manner directed by local policy.

  12. Document the procedure and significant nursing observations on the appropriate forms IAW local SOP.

Performance Measures	GO	NO-GO
1. Prepared the equipment.	——	——
2. Prepared the patient.	——	——
3. Applied the limb electrodes.	——	——
4. Applied the chest electrodes.	——	——
5. Obtained the EKG tracing.	——	——
6. Assessed the EKG tracing.	——	——
7. Calculated the heart rate.	——	——
8. Removed the electrodes.	——	——
9. Asked or assisted the patient to dress.	——	——
10. Prepared the report.	——	——

Performance Measures	GO	NO-GO
11. Stored the equipment.	____	____
12. Documented the procedure.	____	____

Evaluation Guidance: Score each Soldier according to the performance measures. Unless otherwise stated in the task summary, the Soldier must pass all performance measures to be scored GO. If the Soldier fails any steps, show what was done wrong and how to do it correctly.

References
    Required
    None

    Related
    BASIC NURSING 7

Subject Area 3: Trauma Treatment

TREAT A CASUALTY WITH AN OPEN ABDOMINAL WOUND
081-833-0045

Conditions: You have a casualty with an open abdominal wound. All other more serious injuries have been assessed and treated. You will need field dressings, sterile abdominal dressings, cravats, scissors, gauze, saline solution, intravenous (IV) equipment, and DD Form 1380 (U.S. Field Medical Card). You are not in a CBRNE environment.

Standards: Treat an open abdominal wound, minimize the effects of the injury, and stabilize the casualty without causing additional injury.

Performance Steps

1. Position the casualty.
   a. Place the casualty on his back (face up).
   b. Ensure the casualty has a patent airway.
   c. Flex the casualty's knees.
   d. Turn the casualty's head to the side and keep the airway clear if vomiting occurs.
2. Treat for shock. (See task 081-833-0047.) Initiate one large bore (18 gauge) IV if the casualty is exhibiting signs and symptoms of shock.

WARNING: The most important concern in the initial management of abdominal injuries is shock. Shock may be present initially or develop later. Neither the presence or absence of a wound, nor the size of the external wound are safe guidelines for judging the severity of the wound.

3. Expose the wound. Inspect for distention, contusions, penetration, eviscerations or obvious bleeding.

CAUTION: Do not attempt to replace protruding internal organs or remove any protruding foreign objects.

4. Stabilize any protruding objects. (See task 081-833-0046.)

5. Apply a sterile abdominal dressing.

NOTE: Protruding abdominal organs should be kept moist to prevent the tissue from drying out. A moist, sterile dressing should be applied if available.

   a. Using the sterile side of the dressing, or other clean material, place any protruding organs near the wound.
   b. Ensure that the dressing is large enough to cover the entire mass of protruding organs or area of the wound.
   c. If large enough to cover the affected area, place the sterile side of the plastic wrapper directly over the wound.
   d. Place the dressing directly on top of the wound or plastic wrapper, if used.
   e. Tie the dressing tails loosely at the casualty's side.

CAUTION: Do not apply pressure on the wound or expose internal parts.

   f. If two dressings are needed to cover a large wound, repeat steps 5a through 5e. Ensure that the ties of additional dressings are not tied over each other.
   g. If necessary, loosely cover the dressings with cravats. Tie them on the side of the casualty opposite that of the dressing ties.

Performance Steps

    6. Do not cause further injury to the casualty.
        a. Do not touch any exposed organs with bare hands.
        b. Do not try to push any exposed organs back into the body.
        c. Do not tie the dressing tails tightly or directly over the dressing.
        d. Do not give the casualty anything by mouth (NPO).
NOTE: Continue to assess the casualty.

    7. Prepare the casualty for evacuation.
        a. Place the casualty on his back (face up) with the knees flexed.
        b. If evacuation is delayed, check the casualty for signs of shock every 5 minutes.
        c. Consider pain management as necessary. (See task 081-833-0174.)

    8. Record the treatment given on the FMC.

Performance Measures	**GO**	**NO-GO**
1. Positioned the casualty.	____	____
2. Treated for shock.	____	____
3. Exposed the wound.	____	____
4. Stabilized any protruding objects.	____	____
5. Applied a sterile abdominal dressing.	____	____
6. Prepared the casualty for evacuation.	____	____
7. Recorded the treatment given on the FMC.	____	____
8. Did not cause further injury to the casualty.	____	____

Evaluation Guidance: Score each Soldier according to the performance measures. Unless otherwise stated in the task summary, the Soldier must pass all performance measures to be scored GO. If the Soldier fails any steps, show what was done wrong and how to do it correctly.

References
    Required   **Related**
    DD FORM 1380   EMERG CARE AND TRANS 9
                              PHTLS

TREAT A CASUALTY WITH AN IMPALEMENT
081-833-0046

Conditions: The casualty you are assessing has an impalement injury. All other more serious injuries have been assessed and treated. You will need field dressings, cravats, bandages, gauze, scissors, splinting equipment, sling, oxygen delivery device, and DD Form 1380 (U.S. Field Medical Card). You are not in a CBRNE environment.

Standards: Immobilize the impaled object and minimize the effect of the injury without causing further injury to the casualty.

Performance Steps
WARNING: Do not exert any force on or attempt to remove the impaled object unless the object is impaled in the cheek and both ends of the object can be seen or unless the object is blocking the airway. Severe bleeding or nerve and muscle damage may result.

1. Position the casualty.
    a. Tell the casualty to remain still and not to move the impaled object.
    b. Expose the injury by cutting away or removing clothing or equipment around the wound site.
    c. If the impalement injury is on an extremity, check the pulse distal to the injury site.
    d. If the impalement is found in the cheek and both ends of the object can be seen.
        (1) Remove the object in the direction it entered the cheek.
        (2) Position the casualty to allow for drainage and be prepared to suction the casualty.
    e. If both ends of the object in the cheek cannot be seen, go to step 2.

2. Immobilize the impaled object.
NOTE: If an assistant is available, one person should immobilize the object while the other applies the dressings and bandages.
    a. If necessary, apply direct pressure using gloved hands on either side of the object.
WARNING: Do not exert force on the impaled object.
    b. Place several layers of bulky dressing around the injury site so that the dressings surround the object.
    c. Use additional bulky materials or dressings to build up the area around the object.

3. Apply the support bandages.
    a. Apply the bandage over the bulky support material to hold it in place.
    b. Apply the bandage tightly but not so tight as to impair circulation or breathing.
WARNING: Do not anchor the bandage on or exert pressure on the impaled object.
    c. Check circulation after applying the support bandages.
NOTE: If a pulse was palpated in step 1c and it cannot be palpated after the bandage has been applied, the bandage must be loosened until a pulse can be palpated.

4. Immobilize the affected area with a splint or sling, if applicable.
WARNINGS: 1. Do not anchor a splint or sling to the impaled object. 2. Avoid undue motion of the impaled object when applying a splint.

5. Check for a pulse distal to the injury site.

6. Provide oxygen.

7. Treat for shock, if necessary.

Performance Steps

   8. Consider pain management as necessary. (See task 081-833-0174.)

   9. Record the treatment on the FMC.

  10. Evacuate the casualty.

Evaluation Preparation:
Setup: For training and evaluation, have another Soldier act as the casualty. Use a moulage set or similar materials to create a simulated impalement injury. You may also have another Soldier assist in immobilizing the object.

Brief Soldier: Tell the Soldier to treat the casualty for an impalement injury and to direct the actions of the assistant, if applicable. Tell the Soldier that they are not in a CBRNE environment.

Performance Measures
                                                                  GO   NO-GO

Performance Measure	GO	NO-GO
1. Positioned the casualty.	____	____
2. Immobilized the impaled object.	____	____
3. Applied the support bandages.	____	____
4. Immobilized the affected area with a splint or sling, if applicable.	____	____
5. Checked for a pulse distal to the injury site.	____	____
6. Provided oxygen.	____	____
7. Treated for shock, if necessary.	____	____
8. Considered pain management as necessary.	____	____
9. Recorded the treatment on the FMC.	____	____
10. Evacuated the casualty.	____	____
11. Did not cause further injury to the casualty.	____	____

Evaluation Guidance: Score each Soldier according to the performance measures. Unless otherwise stated in the task summary, the Soldier must pass all performance measures to be scored GO. If the Soldier fails any steps, show what was done wrong and how to do it correctly.

References
    Required   **Related**
    DD FORM 1380   EMERG CARE AND TRANS 9
                                      PHTLS

INITIATE TREATMENT FOR HYPOVOLEMIC SHOCK
081-833-0047

Conditions: You are in the field and are assessing a casualty who is suffering from significant blood loss. You will need an intravenous (IV) infusion set, IV fluids, splints, stethoscope, sphygmomanometer, a blanket or poncho, and a DD Form 1380 (U.S. Field Medical Card). You are not in a CBRNE environment.

Standards: Initiate treatment for hypovolemic shock, stabilize the casualty, minimize the effect of shock, and prepare for immediate evacuation without further injury to the casualty.

Performance Steps

1. Control bleeding. (See task 081-833-0161.)

2. Maintain the airway.
NOTE: Administer oxygen, if available. (See task 081-833-0019).

   3. Reassure the casualty to reduce anxiety.
NOTE: Anxiety increases the heart rate, which worsens the casualty's condition. Anyone who has just been shot or who has experienced detonation of explosives nearby will have tachycardia.

   4. Initiate one large bore (18 gauge) IV. (See task 081-833-0033).

   5. Maintain the IV flow with Hextend.
      a. Continue the flow until the systolic blood pressure stabilizes at greater than 80 mm Hg.
         (1) The usual amount is 500 ml; you can repeat the dose of 500 ml one time. A total of 1000 ml maximum amount of Hextend can be used for hypovolemia.
         (2) A palpable radial pulse usually indicates that the casualty has a systolic blood pressure of 80 mm Hg.

   6. Elevate the casualty's legs.
      a. Elevate the casualty's legs above chest level, without lowering the head below chest level.
NOTE: Splint leg or ankle fractures before elevating the legs, if necessary.
      b. If the casualty is on a litter, elevate the foot of the litter.

   7. Maintain normal body temperature. Aggressively treat for hypothermia in a trauma patient.

   8. Monitor the casualty.
NOTE: Give nothing by mouth. Moisten the casualty's lips with a wet cloth.
      a. Check vital signs every 5 minutes until they return to normal, and then check every 15 minutes.
      b. Check the casualty's level of consciousness.
NOTE: If the blood pressure is unstable or drops, the pneumatic anti-shock garment should be applied by qualified personnel.

   9. Record the procedure on the FMC.

   10. Evacuate the casualty.

Evaluation Preparation:
Setup: For training and evaluation, have another Soldier act as the casualty. For step 3, have the Soldier state what actions are taken when an IV infusion is initiated.

Brief Soldier: Tell the Soldier to initiate treatment for hypovolemic shock.

Performance Measures	GO	NO-GO
1. Controlled bleeding.	____	____
2. Maintained the airway.		
3. Reassured the casualty to reduce anxiety.	____	____
4. Initiated one large bore IV.	____	____
5. Maintained the IV flow.	____	____
6. Elevated the casualty's legs.	____	____
7. Maintained normal body temperature.	____	____
8. Monitored the casualty.	____	____
9. Recorded the procedure on the FMC.	____	____
10. Evacuated the casualty.	____	____

Evaluation Guidance: Score each Soldier according to the performance measures. Unless otherwise stated in the task summary, the Soldier must pass all performance measures to be scored GO. If the Soldier fails any steps, show what was done wrong and how to do it correctly.

References

Required	Related
DD Form 1380	EMERG CARE AND TRANS 9

TREAT A CASUALTY WITH A CHEST INJURY
081-833-0049

Conditions: You have a casualty with a chest injury. All other more serious injuries have been assessed and treated. You will need scissors, adhesive tape, field dressings, padding, ace wrap, cravats, field jacket, poncho, blanket, or similar material, oxygen and DD Form 1380 (U.S. Field Medical Card). You are not in a CBRNE environment.

Standards: Treat a chest wound without causing additional injury to the casualty.

Performance Steps

1. Perform an initial assessment of the casualty. (See task 081-833-0213.)

2. Check the casualty for signs and symptoms of chest injuries.
    a. Deformities, contusions, abrasions, punctures/penetrations (DCAP), bleeding, tenderness, lacerations, swelling (BTLS).
    b. Pleuritic pain that is increased by or occurs with respirations and is localized around the injury site.
    c. Labored or difficult breathing.
    d. Diminished or absent breath sounds.
    e. Cyanotic lips, fingertips, or fingernails.
    f. Rapid, weak pulse and low blood pressure.
    g. Coughing up blood or bloody sputum.
    h. Failure of one or both sides of the chest to expand normally upon inhalation.
    i. Paradoxical breathing - the motion of the injured segment of a flail chest, opposite to the normal motion of the chest wall.
    j. Enlarged neck veins.
    k. Coughing up blood or bloody sputum.
    l. Tracheal deviation - shift of the trachea from the midline toward the unaffected side due to pressure buildup on the injured side.

3. Check for an exit wound or injury.

4. Determine the type of injury.
    a. Open pneumothorax - air entering pleural space through defect in pleural wall.
        (1) Signs and symptoms.
            (a) Respiratory distress.
            (b) Anxiousness.
            (c) Tachypneic.
        (2) Treatment.
            (a) Seal the wound(s), covering the larger wound first.
NOTE: All penetrating chest wounds should be treated as if they were sucking chest wounds.
            (b) Cut the dressing wrapper on one long and two short sides and remove the dressing.
NOTE: In an emergency, any airtight material can be used. It must be large enough so it is not sucked into the chest cavity.
            (c) Apply the inner surface of the wrapper to the wound when the casualty exhales.
            (d) Ensure that the covering extends at least two inches beyond the edges of the wound.

Performance Steps

        (e) If you do not have the ability to perform a needle chest decompression, seal by applying overlapping strips of tape to three sides of the plastic covering to provide a flutter-type valve.

        (f) If you have the ability to perform a needle chest decompression, ensure all four sides of the occlusive dressing are secured.

        (g) Cover the exit wound in the same way, if applicable.

NOTE: Assess the effectiveness of the flutter valve when the casualty breathes. When the casualty inhales, the plastic should be sucked against the wound, preventing the entry of air. When the casualty exhales, trapped air should be able to escape from the wound and out the untaped side of the dressing.

        (h) Supplement with oxygen if available.

WARNING: Complication - if tension pneumothorax is suspected, perform a needle chest decompression (NCD). (See task 081-833-3007.)

    b. Rib fracture - generally caused by a direct blow to the chest or compression of the chest. Severe coughing can also cause rib fracture.

      (1) Signs and symptoms.

        (a) Pain is aggravated by respirations and coughing.

        (b) Crepitus is present.

        (c) The casualty will guard to protect the injury.

      (2) Complications.

        (a) Internal bleeding (hemothorax).

        (b) Shock.

      (3) Treatment.

        (a) Use a sling and swathe to immobilize the affected side.

        (b) Administer oxygen as necessary.

NOTE: The broken rib may puncture the lung or the skin.

WARNING: Do not tape, strap, or bind the chest, these interventions increase the development of pneumonia.

    c. Flail chest - two or more ribs fractured in two or more places or a fractured sternum.

      (1) Signs and symptoms.

        (a) Severe pain at the site.

        (b) Rapid shallow breathing.

        (c) Paradoxical respirations.

      (2) Complications.

        (a) Respiratory insufficiency.

        (b) Pneumothorax with hemothorax.

      (3) Treatment.

        (a) Establish and maintain an airway.

        (b) Administer oxygen, if available.

        (c) Assist the casualty's respirations, if necessary.

        (d) Monitor the casualty for signs of hemothorax or tension pneumothorax, as necessary.

    d. Hemothorax - bleeding from lacerated blood vessels in the chest cavity and/or lungs. It results in the accumulation of blood in the chest cavity not outside the lungs.

      (1) Signs and symptoms.

        (a) Hypotension due to blood loss.

        (b) Shock.

        (c) Cyanosis.

        (d) Tightness in the chest.

Performance Steps

        (e) Mediastinal shift may produce deviated trachea away from the affected side.

        (f) Coughing up frothy red blood.

    (2) Complications.

        (a) Possibility of hypovolemic shock.

        (b) Frequently accompanies a pneumothorax.

    (3) Treatment.

        (a) Establish and maintain an airway.

        (b) Administer oxygen.

        (c) Assist the casualty's breathing, as necessary

  e. Tension pneumothorax.

NOTE: Condition in which air enters the chest cavity (pleural space) through a hole in the lung, expanding the space with every breath the casualty takes. The air becomes trapped and cannot escape.

    (1) Signs and symptoms.

        (a) Chest pain.

        (b) Increased pressure in the chest causes the lung(s) to collapse.

        (c) May result from the laceration of the lung by a broken rib or by spontaneous rupture of a bleb or lesion on the lung.

        (d) Position the casualty for evacuation. Conscious - in a comfortable position, preferably sitting. Unconscious - on the injured side.

    (2) Treatment.

        (a) Establish and maintain an airway.

        (b) Perform NCD if indicated. (See task 081-833-3007.)

        (c) Administer oxygen.

        (d) Assist the casualty's respirations, as necessary.

        (e) Monitor the casualty for progression of symptoms.

  5. Treat the injury.

  6. Treat the casualty for shock.

  7. Record the care provided on the FMC.

  8. Evacuate the casualty.

NOTE: Continue to assess the casualty. The casualty should be evacuated by the most expedient means.

Evaluation Preparation:

Setup: For training and evaluation, have another Soldier or mannequin act as the casualty. Use a moulage kit or similar materials to simulate entry and exit wounds.

Brief Soldier: Tell the Soldier to treat a casualty with a chest wound. Tell the Soldier whether the wound involves a simple rib fracture, a flail chest, a compression injury, an injury to the back of the chest, a pneumothorax, a hemothorax, or an open chest wound.

NOTE: Do not tell the Soldier whether an exit wound exists.

	GO	NO-GO

Performance Measures

1. Performed an initial assessment of the casualty.   ____ ____

2. Checked the casualty for signs and symptoms of chest injuries.   ____ ____

3. Checked for an exit wound or injury.   ____ ____

4. Determined the type of injury.   ____ ____

5. Treated the injury.   ____ ____

6. Treated the casualty for shock.   ____ ____

7. Recorded the care provided on the FMC.   ____ ____

8. Evacuated the casualty.   ____ ____

Evaluation Guidance: Score each Soldier according to the performance measures. Unless otherwise stated in the task summary, the Soldier must pass all performance measures to be scored GO. If the Soldier fails any steps, show what was done wrong and how to do it correctly.

References

Required	Related
DD FORM 1380	PHTLS

TREAT A CASUALTY WITH AN OPEN OR CLOSED HEAD INJURY
081-833-0052

Conditions: You need to treat a casualty with an open or closed head injury. All other more serious injuries have been assessed and treated. You will need field dressings, cravats, stethoscope, sphygmomanometer, cervical collar, oxygen tank set up, oropharyngeal airway, non-rebreather, intravenous (IV) setup, and DD Form 1380 (U.S. Field Medical Card). You are not in a CBRNE environment.

Standards: Treat the head injury and stabilize the casualty without causing additional injury.

Performance Steps
WARNING: Treat casualties with any type of traumatic head injury or loss of consciousness as if they have a spinal injury. (See task 081-833-0176.)

1. Take appropriate body substance isolation (BSI) precautions.

2. Check for the signs and symptoms of head injuries.
   a. Closed head injury--caused by a direct blow to the head.
WARNING: Brain injury, leading to a loss of function or death, often occurs without evidence of a skull fracture or scalp injury. Because the skull cannot expand, swelling of the brain or a collection of fluid pressing on the brain can cause pressure. This can compress and destroy the brain tissue.
   (1) Deformity of the head.
   (2) Clear fluid or blood escaping from the nose and/or ear(s).
   (3) Periorbital discoloration (raccoon eyes).
   (4) Bruising behind the ears, over the mastoid process (battle sign).
   (5) Lowered pulse rate if the casualty has not lost a significant amount of blood.
   (6) Signs of increased intracranial pressure.
      (a) Headache, nausea, and/or vomiting.
      (b) Possible unconsciousness.
      (c) Change in pupil size or symmetry.
      (d) Lateral loss of motor nerve function--one side of the body becomes paralyzed.
NOTE: Lateral loss may not happen immediately but may occur later.
      (e) Change in the casualty's respiratory rate or pattern.
      (f) A steady rise in the systolic blood pressure if the casualty hasn't lost significant amounts of blood.
      (g) A rise in the pulse pressure (systolic pressure minus diastolic pressure).
      (h) Elevated body temperature.
      (i) Restlessness--indicates insufficient oxygenation of the brain.
   b. Concussion--caused by a violent jar or shock.
NOTE: A direct blow to the skull may bruise the brain.
   (1) Temporary unconsciousness followed by confusion.
   (2) Temporary, usually short term, loss of some or all brain functions.
   (3) The casualty has a headache or is seeing double.
   (4) The casualty may or may not have a skull fracture.
   c. Contusion--an internal bruise or injury. It is more serious than a concussion. The injured tissue may bleed or swell. Swelling may cause increased intracranial pressure that may result in a decreased level of consciousness and even death.
   d. Open head injury.

Performance Steps

        (1) Penetrating wound--an entry wound with no exit wound.
        (2) Perforating wound--the wound has both entry and exit wounds.
        (3) Visibly deformed skull.
        (4) Exposed brain tissue.
        (5) Possible unconsciousness.
        (6) Paralysis or disability on one side of the body.
        (7) Change in pupil size.
        (8) Lacerated scalp tissue- may have extensive bleeding.

3. Direct manual stabilization of the casualty's head.

4. Check the casualty's vital signs.

5. Assess the casualty's level of consciousness using the AVPU scale.
    a. A--alert. The casualty responds spontaneously to stimuli and is able to answer questions in a clear manner.
    b. V--verbal. The casualty does not respond spontaneously but is responsive to verbal stimuli.
    c. P--pain. The casualty does not respond spontaneously or to verbal stimuli but is responsive to painful stimuli.
    d. U--unresponsive. The casualty is unresponsive to any stimuli.

6. Assess the casualty's pupil size.
    a. Observe the size of each pupil.
NOTE: A variation of pupil size may indicate a brain injury. In a very small percentage of people, unequal pupil size is normal.
    b. Shine a light into each eye to observe the pupillary reaction to light.
NOTE: The pupils should constrict promptly when exposed to bright light. Failure of the pupils to constrict may indicate brain injury.

7. Assess the casualty's motor function.
    a. Evaluate the casualty's strength, mobility, coordination, and sensation.
    b. Document any complaints, weakness, or numbness.
NOTE: Progressive loss of strength or sensation is an important indicator of brain injury.

8. Treat the head injury.
    a. Treat a superficial head injury.
        (1) Apply a dressing.
        (2) Observe for abnormal behavior or evidence of complications.
    b. Treat a head injury involving trauma.
        (1) Maintain a patent airway using the jaw thrust maneuver. (See task 081-831-0018.)
        (2) If the casualty is unconscious, insert an oropharyngeal airway without hyperextending the neck. (See task 081-833-0016.)
        (3) Administer high concentration oxygen by non-rebreather mask and evaluate the need for artificial ventilations with supplemental oxygen.
        (4) Apply a cervical collar. (See task 081-833-0092.)
        (5) Dress the head wound(s).
        (6) Control bleeding.
WARNING: Do not apply pressure to or replace exposed brain tissue.
        (7) Treat for shock.
        (8) Monitor the casualty for convulsions or seizures. (See task 081-831-0035.)

Performance Steps
        (9) Position the casualty with the head elevated 6 inches to assist with the drainage
            of blood from the brain.
CAUTION: Do not give the casualty anything by mouth.

  9. Continue to monitor the casualty and check and record the following at 5 minute intervals.
     a. Level of consciousness.
     b. Pupillary responsiveness and equality.
     c. Vital signs.
     d. Motor functions.

 10. Record the treatment on the FMC.

 11. Evacuate the casualty.

Evaluation Preparation:
Setup: For training and evaluation, have another Soldier act as the casualty. Use a moulage kit
or similar materials to simulate a head wound. To test steps 2 and 7, coach the simulated
casualty on how to answer the Soldier's questions regarding such symptoms as headache. Tell
the Soldier what signs, such as changes in pupil size, the casualty is exhibiting.

Brief Soldier: Tell the Soldier to identify the type of head injury and treat the casualty for a head
injury.

Performance Measures	<u>GO</u>	<u>NO-</u> <u>GO</u>
1. Took appropriate BSI procedures.	——	——
2. Checked for the signs and symptoms of head injuries.	——	——
3. Directed manual stabilization of the casualty's head.	——	——
4. Checked the casualty's vital signs.	——	——
5. Assessed the casualty's level of consciousness using the AVPU scale.	——	——
6. Assessed the casualty's pupil size.	——	——
7. Assessed the casualty's motor function.	——	——
8. Treated the head injury.	——	——
9. Continued to monitor the casualty at 5 minute intervals.	——	——
10. Recorded the treatment on the FMC.	——	——
11. Evacuated the casualty.	——	——
12. Did not cause further injury to the casualty.	——	——

Evaluation Guidance: Score each Soldier according to the performance measures. Unless
otherwise stated in the task summary, the Soldier must pass all performance measures to be
scored GO. If the Soldier fails any steps, show what was done wrong and how to do it correctly.

References
    Required
      DD FORM 1380

    Related
      ISBN 13:978-0-7637-4406-9

TREAT FOREIGN BODIES OF THE EYE
081-833-0056

Conditions: You have a casualty with a foreign body in his eye. All other more serious injuries have been assessed and treated. You have performed a patient care hand-wash. You will need cotton-tipped swabs, clean cloth, sterile irrigation solution (normal saline, water, or other prescribed solution), bandages, dry sterile dressings, eye patch, a paper cup or cardboard cone, and DD Form 1380 (U.S. Field Medical Card). You are not in a CBRNE environment.

Standards: Treat foreign bodies of the eye, minimizing the effects of the injury, without causing additional injury to the eye.

Performance Steps
WARNING: Wear gloves for self-protection against transmission of contaminants whenever handling body fluids.

1. Perform visual acuity testing. (See task 081-833-0193.)

2. Assess eyes: pupils, equal and round, regular in size, and react to light (PEARRL).

3. Locate the foreign body.
   a. Method one.
      (1) Pull the lower lid down.
      (2) Tell the casualty to look up and to both sides and check for foreign bodies.
      (3) Pull the upper lid up.
      (4) Tell the casualty to look down and to both sides and check for foreign bodies.
   b. Method two.
      (1) Tell the casualty to look down.
      (2) Grasp the casualty's upper eyelashes and gently pull the eyelid away from the eyeball.
      (3) Place a cotton-tipped swab horizontally along the outer surface of the upper lid and fold the lid back over the swab.
      (4) Look for the foreign bodies or damage on the globe.
CAUTION: If the foreign bodies cannot be located, bandage both eyes and seek further medical aid immediately.

4. Remove the foreign body.
CAUTION: Do not put pressure on the globe.
   a. Small foreign body on an anterior surface.
      (1) Hold the casualty's eye open.
      (2) Irrigate the eye. (See task 081-833-0054.)
   b. Foreign body stuck to the cornea or lying under the upper or lower eyelid.
      (1) For a foreign body under the lower eyelid, pull the lower lid down.
      (2) For a foreign body under the upper eyelid, pull the upper lid up.
      (3) Remove the foreign body with a moistened, sterile cotton-tipped swab.
CAUTION: Bandage both eyes if foreign bodies are not easily removed by these methods or if there is pain or loss of vision in the eye. Seek further medical aid immediately.

NOTE: In hazardous conditions, leave the good eye uncovered long enough to ensure the casualty's safety.

Performance Steps
  c. Foreign body stuck or impaled in the eye.
CAUTION: Do not attempt to remove a foreign body stuck to or sticking into the eyeball. A medical officer must remove such objects.
    (1) Apply dry sterile dressings to build around and support the object.
NOTE: This will help prevent further contamination and minimize movement of the object.
    (2) Cover the injured eye with a paper cup or cardboard cone.
    (3) Cover the uninjured eye with a dry dressing or eye patch.
    (4) Reassure the casualty by explaining why both eyes are being covered.
NOTE: The eyes move together. If the casualty uses (moves) the uninjured eye, the injured eye will move as well. Covering both eyes will keep them still and will prevent undue movement on the injured side.
    (5) Seek further medical aid immediately.

 5. Obtain details about the injury.
   a. Source and type of the foreign bodies.
   b. Whether the foreign bodies were wind-blown or high velocity.
   c. Time of onset and length of discomfort.
NOTE: In hazardous conditions, leave the good eye uncovered long enough to ensure the casualty's safety.
   d. Any previous injuries to the eye.

 6. Record the procedure on the FMC.

 7. Evacuate the casualty, as required.
   c. Do not remove an impaled object.

 8. Do not cause additional injury to the eye.
   a. Do not probe for foreign bodies.
   b. Do not put pressure on the globe.

Performance Measures	GO	NO-GO
1. Performed visual acuity testing.	____	____
2. Assessed PEARRL.	____	____
3. Located the foreign body.	____	____
4. Removed the foreign body.	____	____
5. Obtained details about the injury.	____	____
6. Recorded the procedure on the FMC.	____	____
7. Evacuated the casualty, as required.	____	____
8. Did not cause additional injury to the eye.	____	____

Evaluation Guidance: Score each Soldier according to the performance measures. Unless otherwise stated in the task summary, the Soldier must pass all performance measures to be scored GO. If the Soldier fails any steps, show what was done wrong and how to do it correctly.

References

Required	**Related**
DD FORM 1380	ISBN 0-7637-3901-4
	ISBN 0-07-065351-9

TREAT LACERATIONS, CONTUSIONS, AND EXTRUSIONS OF THE EYE
081-833-0057

Conditions: You have a casualty with a laceration, contusion, and/or extrusion of the eye. All other more serious injuries have been assessed and treated. You have performed a patient care hand-wash. You will need eye pads, field dressings, padding materials, roller gauze, scissors, paper cup or cone-shaped piece of cardboard, sterile water or sterile normal saline, and DD Form 1380 (U.S. Field Medical Card). You are not in a CBRNE environment.

Standards: Treat an eye injury, minimizing the effects of the injury, without causing additional injury to the eye.

Performance Steps

1. Position the casualty and remove his headgear, if necessary.
    a. Conscious--seated.
    b. Unconscious--lying on his back with the head slightly elevated.

2. If conscious, perform visual acuity testing. (See task 081-833-0193.)

3. Assess eyes: pupils, equal and round, regular in size, and react to light (PEARRL).

4. Examine the eyes for the following:
    a. Objects protruding from the globe.
    b. Swelling or lacerations on the globe.
    c. Bloodshot appearance of the sclera.
    d. Bleeding.
        (1) Surrounding the eye.
        (2) Inside the globe.
        (3) Coming from the globe.
    e. Contact lenses. Ask the casualty if he is wearing contact lenses but do not force the eyelids open. Record that they are being worn, if appropriate.
    f. Extrusion (the eye is protruding from the socket).

5. Categorize the injury.
    a. Injury to the tissue surrounding the eye (lacerations and contusions).
    b. Injury to the globe.
    c. Extrusion.
    d. Protruding (impaled) objects.

6. Treat the injury.
NOTE: Torn eyelids should be handled carefully. Wrap any detached fragments in a separate moist dressing and evacuate with the casualty.
    a. Lacerations and contusions of tissue surrounding the eye.
        (1) Close the lid of the affected eye. Do not exert pressure or manipulate the globe in any way.
        (2) Cover the injury with moist, sterile dressing to prevent drying.
        (3) Cover torn eyelids with a loose dressing.
        (4) Place a field dressing over the eye pad or dressing of the affected eye.
    b. Injury to the eyeball.
        (1) Cover the injured eye with a sterile dressing soaked in saline to keep the wound from drying.

Performance Steps
    (2) Place a field dressing over the eye pad.
    (3) Cover the uninjured eye to prevent sympathetic eye movement.
NOTE: In hazardous conditions, leave the good eye uncovered long enough to ensure the casualty's safety.
    (4) Tell the casualty not to squeeze the eyelids together.
  c. Extrusion.
CAUTION: Do not attempt to reposition the globe or replace it in the socket.
    (1) Position the casualty face up.
    (2) Cut a hole in several layers of dressing material, and then moisten it. Use sterile liquid, if available.
    (3) Place the dressing so the injured globe protrudes through the hole, but does not touch the dressing. The dressing should be built up higher than the globe.
NOTE: If available, place a paper cup or cone-shaped piece of cardboard over the eye. Do not apply pressure to the injury site. Apply roller gauze to hold the cup in place.
    (4) Cover the uninjured eye to prevent sympathetic eye movement.
NOTE: In hazardous conditions, leave the good eye uncovered long enough to ensure the casualty's safety.
  d. Protruding object. (See task 081-833-0056.)
CAUTION: Do not attempt to remove the protruding object.
    (1) Immobilize the object.
    (2) Dress the injured eye.
    (3) Cover the uninjured eye to prevent sympathetic movement.
NOTE: In hazardous conditions, leave the good eye uncovered long enough to ensure the casualty's safety.

  7. Record the procedure on the FMC.

  8. Evacuate the casualty.
    a. Transport the casualty on his back, with the head elevated and immobilized.
    b. Evacuate eyeglasses with the casualty, even if they are broken.

Evaluation Preparation:
Setup: For training and evaluation, have another Soldier act as the casualty. Use a moulage kit or similar material to simulate the injury, or describe the type of injury to the Soldier.

Brief Soldier: Tell the Soldier to treat the eye injury.

Performance Measures	<u>GO</u>	<u>NO-GO</u>
1. Positioned the casualty.	——	——
2. Performed visual acuity testing.	——	——
3. Assessed PEARRL.	——	——
4. Examined the eyes.	——	——
5. Categorized the injury.	——	——
6. Treated the injury.	——	——
7. Recorded the procedure on the FMC.	——	——

Performance Measures	GO	NO-GO
8. Evacuated the casualty.	____	____
9. Did not cause further injury to the casualty.	____	____

Evaluation Guidance: Score each Soldier according to the performance measures. Unless otherwise stated in the task summary, the Soldier must pass all performance measures to be scored GO. If the Soldier fails any steps, show what was done wrong and how to do it correctly.

References

Required	**Related**
DD FORM 1380	EMERG CARE AND TRANS 9
	ISBN 0-07-065351-9

TREAT BURNS OF THE EYE
081-833-0058

Conditions: You have a casualty with burns of the eye. All other more serious injuries have been assessed and treated. You have performed a patient care hand-wash. You will need irrigation equipment, irrigation solution (sterile water, sterile normal saline, or potable water), sterile dressings, gloves, field dressings, and DD Form 1380 (U.S. Field Medical Card). You are not in a CBRNE environment.

Standards: Treat burns of the eyes and stabilize the casualty without causing further injury to the casualty.

Performance Steps

1. Reassure the casualty and check for signs and symptoms to determine the type of burns.
   a. Chemical--such as acid, alkali, or petroleum.
CAUTION: The chemical may stick to the eye.
      (1) Pain and redness.
      (2) Watering or tearing.
      (3) Possible erosion of the corneal surface.
   b. Radiant burns.
      (1) Electric burns--electric welding processor.
         (a) Gritty feeling.
         (b) Severe pain.
         (c) Inability to tolerate light.
         (d) Redness, swelling.
         (e) Watering or tearing.
         (f) Immediate decrease in vision.
NOTE: Electrical burns often do not appear until several hours after exposure.
      (2) Laser burns--bright, visible light and invisible light such as ultraviolet or infrared.
         (a) Immediate decrease in vision.
         (b) No pain.
   c. Thermal burns.
      (1) Charred or swollen eyelids.
      (2) Singed eyelashes.
      (3) Pain or irritation.

2. Treat the burn.
   a. Chemical burn.
      (1) Brush dried chemicals off of the skin and clothing prior to irrigation.
      (2) Gently hold the casualty's eye(s) open.
      (3) Tilt the casualty's head toward the affected side if only one eye is involved.
      (4) Irrigate the eye(s) for a minimum of 20 minutes with copious amounts of water. Irrigation should continue for a minimum of 15 to 20 minutes after the casualty states that the burning has stopped.
NOTE: Irrigate the eye(s) with sterile water or sterile normal saline, if available. If not available, use any potable water.

CAUTION: Do not attempt to neutralize the chemical.
      (5) Cover the injured eye with a clean, sterile dressing.

Performance Steps

    b. Radiant energy burn (electric/laser).

      (1) No specific treatment is recommended.

      (2) Bandage both eyes with sterile, moist pads.

NOTE: In a combat environment, the eyes may have to remain uncovered so the casualty can see to get away from danger.

      (3) Avoid further light exposure.

      (4) Evacuate the casualty for further examination.

    c. Thermal burn.

      (1) Do not bandage the eyes.

NOTE: Burned eyelids swell to protect the underlying eyes. If the casualty can be evacuated immediately, the eyes may be loosely covered with sterile dressings moistened with sterile saline.

      (2) Protect the casualty from exposure to light.

WARNING: Casualties with severe burns to the eyes may have additional respiratory burns due to spontaneous inhalation.

  3. Record the treatment given on the FMC.

  4. Evacuate the casualty.

Performance Measures	GO	NO-GO
1. Reassured the casualty and checked for signs and symptoms to determine the type of burns.	____	____
2. Treated the burn.	____	____
3. Recorded the treatment given on the FMC.	____	____
4. Evacuated the casualty.	____	____

Evaluation Guidance: Score each Soldier according to the performance measures. Unless otherwise stated in the task summary, the Soldier must pass all performance measures to be scored GO. If the Soldier fails any steps, show what was done wrong and how to do it correctly.

References

    Required   **Related**

    DD FORM 1380   EMERG CARE AND TRANS 9

ADMINISTER INITIAL TREATMENT FOR BURNS
081-833-0070

Conditions: You have a casualty with burns. All other more serious injuries have been assessed and treated. You will need trauma dressings, sterile dressings, Kerlix, Ringer's lactate or normal saline, water, nonpetroleum liquid, oxygen, sterile sheet, or clean linen, large gauge (#16 or #18) needle, an intravenous (IV) setup, IV fluids, and DD Form 1380 (U.S. Field Medical Card). You are not in a CBRNE environment.

Standards: Administer initial treatment IAW the type and extent of the casualty's burns. Stabilize the casualty without causing further injury to the casualty or injuring self.

Performance Steps

1. Determine the cause of the burns.
   a. Assess the scene.
   b. Question the casualty and/or bystanders.
   c. Determine if the casualty has been exposed to smoke, steam, or combustible products.
   d. Determine if the cause was open flame, hot liquid, chemicals, or electricity.
   e. Determine whether the casualty was struck by lightning.
NOTE: If the burn was caused by an explosion or lightning, the casualty may also have been thrown some distance from the original spot of the incident. He may, therefore, have associated internal injuries, fractures, or spinal injuries.

2. Stop the burning process.
   a. Thermal burns.
     (1) Have the casualty STOP, DROP, and ROLL.
        (a) Do not permit the casualty to run, as this will fan the flames.
        (b) Do not permit the casualty to stand, as the flames may be inhaled or the hair ignited.
        (c) Place the casualty on the ground or floor and roll the casualty in a blanket or in dirt, and/or splash with water.
     (2) Remove all smoldering clothing and articles that retain heat, if possible.
     (3) Cut away clothing to expose the burned area.
CAUTION: Do not remove clothing that is stuck to the burned area. If the clothing and skin are still hot, irrigate with copious amounts of room-temperature water or cover with a wet dressing, if available.
   b. Electrical burns.
     (1) Turn off the current, if possible.
WARNING: Do not directly touch a casualty receiving a shock. To do so will conduct the current to you.
     (2) If necessary and/or possible, remove the electrical source from the casualty.
WARNING: Electrical shock may cause the casualty to go into cardiac arrhythmia or arrest. Initiate cardiopulmonary resuscitation (CPR) as appropriate. Casualties of lightning strikes may require prolonged CPR and extended respiratory support.
   c. Chemical burns.
WARNING: A chemical will burn as long as it is in contact with the skin.
     (1) Flush the area of contact immediately with water. Do not delay flushing by removing the casualty's clothing first.

Performance Steps

NOTE: If a solid chemical, such as lime, has been spilled on the casualty, brush it off before flushing. A dry chemical is activated by contact with water and will cause more damage to the skin.

      (2) Flush with cool water for 10 to15 minutes while removing contaminated clothing or other articles.

NOTES: 1. Flush longer for alkali burns because they penetrate deeper and cause more severe injury. 2. Many chemicals have a delayed reaction. They will continue to cause injury even though the casualty no longer feels pain.

WARNING: Do not use a hard blast of water. Extreme water pressure can add mechanical injury to the skin.

    d. White phosphorus burns.

NOTE: White phosphorus (WP) will stick to the skin and continue to burn until it is deprived of air. WP burns are usually multiple and deep, usually producing second and third degree burns.

      (1) Deprive the WP of oxygen.

        (a) Splash with a nonpetroleum liquid (such as water, mud, or urine).

        (b) Submerge the entire area.

        (c) Cover the affected area with a moistened cloth, if available, or mud.

      (2) Remove the WP particles from the skin by brushing with a wet cloth or using forceps, stick, or knife.

WARNING: Do not use any type of petroleum product to smother the WP. This will cause it to be more rapidly absorbed into the body.

   3. Maintain an open airway, if necessary. (See task 081-831-0018.)

NOTE: As long as 30 to 40 minutes may elapse before edema obstructs the airway and respiratory distress is noted. Always suspect an inhalation injury with a closed-space fire.

    a. Check for signs and symptoms of inhalation injury.

      (1) Facial burns.

      (2) Singed eyebrows, eyelashes, and/or nasal hairs.

      (3) Carbon deposits and/or redness in the mouth and/or oropharynx.

      (4) Sooty carbon deposits in the sputum.

      (5) Hoarseness, noisy inhalation, cough, or dyspnea.

    b. Check for signs and symptoms of carbon monoxide poisoning.

      (1) Dizziness, nausea, and/or headache.

      (2) Cherry-red colored skin and mucous membranes.

      (3) Tachycardia or tachypnea.

      (4) Respiratory distress or arrest.

    c. Administer humidified oxygen at a high flow rate. (See tasks 081-833-0018 and 081-833-0019.)

   4. Determine the percent of body surface area (BSA) burned.

    a. Cut the casualty's clothing away from the burned areas.

    b. Determine the percentage of BSA burned using the Rule of Nines. Refer to Figure 3-7.

Performance Steps

Rule of Nines		
1. Head and neck	=	9%
2. Anterior trunk	=	18%
3. Posterior trunk	=	18%
4. Upper extremities	=	18% (each 9%)
5. Lower extremities	=	36% (each 18%)
6. Perineum	=	1%

Figure 3-7. Rule of nines

5.Determine the degree of the burn.
  a. First degree.
    (1) Superficial skin only.
    (2) Red and painful, like a sunburn.
  b. Second degree.
    (1) Partial thickness of the skin.
    (2) Penetrates the skin deeper than first degree.
    (3) Blisters and pain.
    (4) Some subcutaneous edema.
  c. Third degree.
    (1) Damage to or the destruction of a full thickness of skin.
    (2) Involves underlying muscles, bones, or other structures.
    (3) The skin may look leathery, dry, and discolored (charred, brown, or white).
    (4) Nerve ending destruction causes a lack of pain.
    (5) Massive fluid loss.
    (6) Clotted blood vessels may be visible under the burned skin.
    (7) Subcutaneous fat may be visible.
CAUTIONS: 1. Check for entry and exit burns when treating electrical burns and lightning strikes. 2. The amount of injured tissue in an electrical burn is usually far more extensive than the appearance of the wound would indicate. Although the burn wounds may be small, severe damage may occur to deeper tissues. (High voltage can destroy skin and muscles to such an extent that amputation may eventually be necessary.)

  6. Treat for shock those casualties who have second or third degree burns of 20% BSA or more.
    a. Initiate treatment for hypovolemic shock. (See task 081-833-0047.)
    b. Keep the casualty flat.
    c. Initiate two IV infusions. (See task 081-833-0033.)
      (1) Use Ringer's lactate, if available. Normal saline is the second fluid of choice.
      (2) Use large gauge (#16 or #18) needles.
      (3) Initiate the IVs in an unburned area, if possible.
      (4) Use large peripheral veins.
NOTE: The presence of overlying burned skin should not deter the use of an accessible vein. The upper extremities are preferable to lower extremities.

Performance Steps
      d. Infuse fluids for a casualty based on fluid replacement calculations.
NOTE: The amount of fluids given in the first 24 hours after a burn should total 4mL/kg/%surface area burned. Half of this fluid is given in the first 8 hours and the second half is given over the remaining 16 hours.
        (1) Calculate the casualty's body weight in kilograms (kg).
          (a) Determine or estimate the casualty's body weight in pounds.
          (b) Divide the casualty's body weight by 2.2. For example, the casualty weighs about 165 pounds. 165/2.2 = 75 kg.
        (2) Calculate the amount of fluid to infuse for the next 24 hour period.
          (a) Determine the percentage of BSA burned (see step 4b). For example, the casualty's BSA burned is 36%.
          (b) Multiply 4 by the percentage of BSA burned. For example, 4.00 cc X 36 = 144 cc.
          (c) Multiply the above figure by the casualty's weight, found in step 6d(1). For example, 144 cc X 75 kg = 10,800 cc. The casualty will require this much fluid over the next 24 hour period.
          (d) Divide the above figure by 2 to determine the amount of fluid to give in the first 8 hours. For example, 10,800/2= 5400cc. 5400cc/8= 675cc/hour of fluid given in the first 8 hours.
          (e) The remainder of fluid will be given over the next 16 hours. 5400cc/16= 337cc/hour over the next 16 hour timeframe.
      e. Assess the circulatory blood volume.
NOTE: Urine output is a reliable guide to assess circulating blood volume.
        (1) Measure the casualty's urine output in cc per hour.
        (2) Adjust the IV fluid flow to maintain 30 to 50 cc of urine output per hour.

    7. Perform a secondary assessment. (See task 081-833-0155.)
NOTE: The secondary assessment of a burn casualty is no different than in any other trauma casualty.
      a. Measure and record the casualty's vital signs.
      b. Assess the casualty for associated injuries. (See task 081-833-0151.)
      c. Check the distal circulation by checking pulses in all extremities.

    8. Remove potentially constricting items such as rings and bracelets.
CAUTION: The swelling of burns on extremities can cause a tourniquet-like effect, and the swelling of a burned throat can impair breathing.

    9. Dress the burns.
      a. Apply a dry sterile dressing to the burns.
CAUTION: Do not put ointment on the burns and do not break blisters.
      b. Cover extensive burns with a sterile sheet, if available, or clean linen.

    10. Record the treatment given on the FMC.

    11. Evacuate the casualty.

Evaluation Preparation:
Setup: For training and evaluation, have another Soldier act as the casualty. You may use a moulage kit or similar material to simulate burns on the casualty, or you may describe to the Soldier the area(s) of the body burned. Create a scenario which describes the cause and depth of the burns. For step 2, have the Soldier describe what actions should be taken to prevent further injury. To test step 5, describe the depth of the burns and have the Soldier tell you if

they are first, second, or third degree. When testing step 6, have the Soldier describe what actions should be taken when administering IV therapy, if necessary. When testing step 7, have the Soldier describe what action is taken.

Brief Soldier: Tell the Soldier to determine the extent of the casualty's burns and the treatment required.

Performance Measures	GO	NO-GO
1. Determined the cause of the burns.	——	——
2. Stopped the burning process.	——	——
3. Maintained an open airway, if necessary.	——	——
4. Determined the percent of BSA burned.	——	——
5. Determined the degree of the burn.	——	——
6. Treated the casualty for shock, if necessary.	——	——
7. Performed a secondary assessment.	——	——
8. Removed potentially constricting items.	——	——
9. Dressed the burns.	——	——
10. Recorded the treatment given on the FMC.	——	——
11. Evacuated the casualty.	——	——
12. Did not cause further injury to the casualty.	——	——

Evaluation Guidance: Score each Soldier according to the performance measures. Unless otherwise stated in the task summary, the Soldier must pass all performance measures to be scored GO. If the Soldier fails any steps, show what was done wrong and how to do it correctly.

References

Required
DD FORM 1380

Related
EMERG CARE AND TRANS 9
PHTLS

PERFORM A TRAUMA CASUALTY ASSESSMENT
081-833-0155

Conditions: You encounter a casualty with multiple injuries. You will need sphygmomanometer, clean stethoscope, thermometer, airway adjuncts, oxygen, non-rebreathing mask, cervical collar, long spine board, pneumatic anti-shock garments (PASG), scoop stretcher, and DD Form 1380 (U.S. Field Medical Card). You are not in a CBRNE environment.

Standards: Assess the casualty, identify all life-threatening injuries, and manage him appropriately without causing further injury to the casualty. Perform the assessment in the correct sequence.

Performance Steps
CAUTION: All body fluids should be considered potentially infectious. Always observe body substance isolation (BSI) precautions by wearing gloves and eye protection as a minimal standard of protection.

1. Take BSI precautions.

2. Perform a Scene Assessment.
NOTE: It may be necessary to move the casualty out of the line of effective enemy fire and behind cover before properly assessing the trauma casualty.
    a. Determine the safest route to access the casualty.
    b. Determine the mechanism of injury (MOI).
    c. Determine the number of casualties.
    d. Request additional help, if necessary.
    e. Consider the need for spinal stabilization.
NOTE: If the MOI is significant, direct another Soldier to provide manual, in-line stabilization of the cervical spine.

3. Perform an Initial Assessment.
NOTE: Life-threatening injuries should be managed as they are identified.
    a. Form a general impression (global overview) of the casualty's condition and environment.
    b. Determine the chief complaint.
    c. Assess the airway.
        (1) Perform appropriate maneuver to open and maintain the airway. (See task 081-831-0018.)
        (2) Insert an appropriate airway adjunct, if necessary. (See tasks 081-833-0016, 081-833-0142, and 081-833-0169. Also if skill level 30, See task 081-830-3016.)
    d. Assess breathing.
        (1) Determine the rate, rhythm, and quality of respirations (breathing).
        (2) Administer supplemental oxygen by non-rebreathing mask, if available. (See tasks 081-833-0158 and 081-831-0048.)
    e. Assess circulation.
        (1) Skin color, condition, and temperature (CCT).
        (2) Assess the pulse for rate, rhythm, and strength.
            (a) Check the radial pulse in adults.
            (b) Check the radial pulse and capillary refill in children.
            (c) Check the brachial pulse and capillary refill in infants.
        (3) Check for significant hemorrhage (bleeding).

Performance Steps

(4) Control bleeding. (See tasks 081-833-0161 and 081-833-0212.)

(5) Treat for shock. (See task 081-833-0047.)

f. Assess the casualty's mental status using the Alert, Verbal, Pain, Unresponsive (AVPU) scale.

(1) A - Alert and oriented (eyes open spontaneously as you approach; casualty appears aware and responsive to the environment).

(2) V - Responsive to verbal stimuli (sound).

(3) P - Responsive to painful stimuli (touch, such as tapping the casualty on the shoulder or pinching the casualty's ear).

(4) U - Unresponsive (does not respond to any stimuli).

g. Determine casualty priority and make a transport decision.

NOTE: High priority conditions that require immediate transport include a poor general impression, unresponsive, responsive but not following commands, difficulty breathing, shock, complicated childbirth, chest pain with systolic blood pressure less than 100 mm Hg, uncontrolled bleeding, and severe pain.

4. If the MOI is significant, perform a Rapid Trauma Assessment.

NOTE: A significant MOI includes ejection from a vehicle, death in the same passenger compartment, falls three times the casualty's height, rollover of vehicle, high-speed vehicle collision, vehicle-pedestrian collision, motorcycle crash, unresponsive or altered mental status, and penetrations of the head, chest, or abdomen (e.g., stab and gunshot wounds). Additional significant MOI for a child include falls from more than 10 feet, bicycle collision, and vehicles in medium speed collision.

a. Head.

(1) Inspect for deformities, contusions, abrasions, punctures or penetration, burns, tenderness, lacerations, and swelling (DCAP-BTLS).

(2) Palpate for tenderness, instability, or crepitus (TIC).

b. Neck.

(1) Inspect for DCAP-BTLS.

(2) Palpate spinal step-offs.

(3) Inspect for jugular vein distention (JVD).

(4) Inspect to ensure the trachea is midline (without deviation).

(5) Apply a cervical collar, if necessary.

c. Chest.

(1) Inspect for DCAP-BTLS.

(2) Palpate for TIC.

(3) Inspect for the presence of paradoxical motion.

(4) Auscultate (listen) for breath sounds (present, diminished, absent, equal).

d. Abdomen.

(1) Inspect for DCAP-BTLS.

(2) Palpate for tenderness, rigidity, and distension (TRD).

e. Pelvis.

NOTE: If a conscious casualty complains of pain or if an unconscious casualty responds as if in pain at any time during the assessment, do not continue the exam. Treat for pelvic fracture.

CAUTION: Do not "log roll" casualties suspected of having a pelvic fracture.

(1) Inspect for DCAP-BTLS.

(2) Gently compress (downward or inward) to detect TIC.

(3) Inspect for priapism.

f. Extremities.

Performance Steps

        (1) Inspect for DCAP-BTLS.

        (2) Palpate for TIC.

        (3) Check the distal pulses, motor function, and sensation (PMS).

    g. Posterior.

NOTE: The casualty must be "log rolled" to do this portion of the assessment. If necessary, the casualty should be placed on a long spine board after assessment. If the PASG is deemed necessary, it should be positioned on the long spine board before casualty placement. If the casualty has a suspected pelvic fracture or bilateral femoral fractures, lift the casualty using a scoop stretcher, assess the posterior and place the casualty on the long spine board.

        (1) Inspect for DCAP-BTLS.

NOTE: If penetrating wounds were noted during the anterior assessment, check for posterior exit wounds while the casualty is log-rolled/lifted with the scoop stretcher.

        (2) Inspect for rectal bleeding.

  5. If there is no significant MOI, perform a Focused History and Physical Exam.

    a. Based on chief complaint.

    b. Focus on the areas the casualty tells you are painful or that you suspect may be painful due to the MOI.

  6. Obtain a baseline set of vital signs. (See tasks 081-831-0010, 081-831-0011, 081-831-0012, and 081-831-0013.)

  7. Obtain a SAMPLE history.

NOTE: A SAMPLE history is obtained by questioning the casualty. If the casualty is unable to answer, search the scene or ask bystanders or unit members that witnessed the event for information.

    a. Signs/symptoms.

        (1) Ask the casualty what is wrong.

        (2) Observe the casualty.

    b. Allergies.

        (1) Ask the casualty if there are any allergies to medications, foods, or environment.

        (2) Look for a medical (red) identification tag.

    c. Medications.

        (1) Ask the casualty if he is taking any medications (prescription, over the counter, or illegal).

        (2) Ask a female casualty if she is taking birth control pills.

        (3) Search for an identification tag with medications on it or medications in the area.

    d. Pertinent past history.

        (1) Ask the casualty if there are any medical problems (past and present).

        (2) Ask the casualty if he has been feeling ill.

        (3) Ask the casualty about recent surgery or injuries.

        (4) Ask the casualty if he is currently under the care of a medical officer and, if so, what's their name and what type of care is being provided.

    e. Last oral intake.

        (1) Ask the casualty when his last meal or drink was.

        (2) Ask the casualty what he ate or drank.

    f. Events leading to the injury or illness.

        (1) Ask about the sequence of events that led up to the current event.

        (2) If the casualty is unable to answer, search the scene for anything that may indicate what occurred.

Performance Steps

8. Perform a Detailed Physical Examination.
NOTE: This is done only if time permits during casualty evacuation (CASEVAC) or while waiting for CASEVAC, if not performing life-saving interventions. Do not delay evacuation to perform this exam.
    a. Assess the scalp and cranium.
        (1) Inspect for DCAP-BTLS.
        (2) Palpate for TIC.
    b. Assess the ears.
        (1) Inspect for DCAP-BTLS.
        (2) Inspect for drainage.
            (a) Blood or serous fluids.
            (b) Clear fluids.
    c. Assess the face for DCAP-BTLS.
    d. Assess the eyes.
        (1) Inspect for DCAP-BTLS.
        (2) Inspect for discoloration.
        (3) Inspect for unequal pupils.
        (4) Inspect for foreign bodies.
        (5) Inspect for blood in anterior chamber.
    e. Assess the nose.
        (1) Inspect for DCAP-BTLS.
        (2) Inspect for drainage of blood and/or clear fluid.
    f. Assess the mouth.
        (1) Inspect for DCAP-BTLS.
        (2) Inspect for loose or broken teeth.
        (3) Inspect for objects that could cause obstruction.
        (4) Inspect for swelling or laceration of the tongue.
        (5) Inspect for unusual breath odor (alcohol, acetone, etc.)
    g. Assess the neck.
        (1) Inspect for DCAP-BTLS.
        (2) Inspect for JVD.
        (3) Inspect to ensure the trachea is still midline (without deviation).
        (4) Palpate for TIC.
    h. Reassess the chest.
        (1) Inspect for DCAP-BTLS.
        (2) Palpate for TIC.
        (3) Auscultate breath sounds.
        (4) Assess for flail chest.
    i. Reassess the abdomen.
        (1) Inspect for DCAP-BTLS.
        (2) Palpate for TRD.
    j. Reassess the pelvis.
NOTE: If pain, instability or crepitus was noticed in the rapid trauma assessment, ensure the pelvis is properly stabilized and do not reassess.
        (1) Inspect for DCAP-BTLS.
        (2) Inspect for TIC.
    k. Reassess the extremities.
        (1) Inspect for DCAP-BTLS.
        (2) Palpate for TIC.

Performance Steps

    (3) Check the PMS.

    l. Reassess the posterior.

NOTE: If the casualty is secured to a long spine board, do not remove from the board. Reassess the flanks and as much of the spine as you can without moving the casualty unnecessarily.

    (1) Inspect for DCAP-BTLS.

    (2) Inspect for rectal bleeding.

    m. Manage secondary injuries and wounds appropriately. (See tasks 081-833-0045, 081-833-0046, 081-833-0049, 081-833-0050, 081-833-0052, 081-833-0056, 081-833-0057, 081-833-0058, 081-833-0060, 081-833-0062, 081-833-0064, 081-833-0154, and 081-833-3011).

    n. Reassess the casualty's vital signs every 5 minutes (if unstable), every 15 minutes (if stable).

  9. Document all assessment findings and care provided on the FMC. (See task 081-831-0033.)

Evaluation Preparation:

Setup: For training and evaluation, have another Soldier act as the casualty or use a trauma mannequin. Describe a general scenario to the Soldier. The casualty must have more than one injury or condition. Wounds may be simulated using moulage or other available materials. A "conscious" casualty can be coached to show signs of such conditions as shock, and to respond to the Soldier's questions about the location of pain and other symptoms of injury. The evaluator will cue the Soldier during the assessment of an "unconscious" casualty as to whether the casualty is breathing, and describe such conditions as shock to the Soldier as they are making the checks. Tell the casualty not to assist the Soldier in any other way.

Brief Soldier: Tell the Soldier to tell you what action he would take for each wound or condition identified.

Performance Measures	**GO**	**NO-GO**
1. Took BSI precautions.	——	——
2. Performed a Scene Assessment.	——	——
3. Performed an Initial Assessment.	——	——
4. If the MOI was significant, performed a Rapid Trauma Assessment.	——	——
5. If there was no significant MOI, performed a Focused History and Physical Exam.	——	——
6. Obtained baseline vital signs.	——	——
7. Obtained the SAMPLE history.	——	——
8. Performed a Detailed Physical Exam, if appropriate.	——	——
9. Documented medical findings and treatment provided on the FMC.	——	——

Evaluation Guidance: Score each Soldier according to the performance measures in the evaluation guide. Unless otherwise stated in the task summary, the Soldier must pass all performance measures to be scored GO. If the Soldier fails any step, show what was done wrong and how to do it correctly.

References

    Required
    DD FORM 1380

                         Related
                         EMERG CARE AND TRANS 9
                         PHTLS

PROVIDE BASIC EMERGENCY MEDICAL CARE FOR AN AMPUTATION
081-833-0157

Conditions: A casualty has experienced a partial or complete amputation. All other more serious injuries have been assessed and treated. You will need a tourniquet, marker, sterile gauze, intravenous (IV) equipment, saline, plastic bag, container, DD Form 1380 (U.S. Field Medical Card), ice, and cravats. You are not in a CBRNE environment.

Standards: Provide correct treatment based on severity and location of amputation without causing further injury to the casualty.

Performance Steps
NOTE: If an obvious amputation with significant hemorrhage, immediately apply a tourniquet. (See task 081-833-0210.)

1. Assess the casualty. (See task 081-833-0155.)

2. Treat immediate life threats as they are discovered.

3. Perform emergency management for traumatic amputations.
    a. Combat environment (under fire).
        (1) Move the casualty to a covered and concealed location.
        (2) Immediately apply a tourniquet.
    b. Noncombat environment (not under fire).
        (1) Implement airway, breathing, and circulation.
        (2) Initially apply a tourniquet to control bleeding.
        (3) The tourniquet may be loosened and bleeding should be controlled by attempting other means.
            (a) Direct pressure.
            (b) Pressure point.
            (c) Pressure dressing.
        (4) Using a marker or the casualty's blood, make a "T" on the casualty's forehead.
    NOTE: With either of the above methods, consider the time to definitive care and the severity of blood loss. A tourniquet that prevents excessive blood loss and that will be removed within a few hours at a definitive care facility is preferable to excessive blood loss while trying other methods.

4. Treat for shock. (See task 081-833-0047.)

5. Care for the amputated part.
    a. Rinse free of debris with saline if available.
    b. Wrap the part loosely in saline-moistened sterile gauze.
    c. Seal the amputated part inside a plastic bag or wrap it in a cravat. The amputated part should then be placed in another container containing ice. Keep it cool, but do not allow it to freeze.
    d. Avoid further injury to the amputated part.
        (1) Never warm an amputated part.
        (2) Never place an amputated part directly in water.
        (3) Never place an amputated part directly on ice.
        (4) Never use dry ice to cool an amputated part.
    e. Transport the part with the casualty to the hospital for possible reimplantation or skin graft.

Performance Steps

  6. Record all treatment on the casualty's FMC.

  7. Evacuate the casualty to a definitive care facility as soon as possible.

Performance Measures	**GO**	**NO-GO**
1. Assessed the casualty.	——	——
2. Treated immediate life threats as discovered.	——	——
3. Performed emergency management for traumatic amputations.	——	——
4. Treated for shock.	——	——
5. Cared for the amputated part.	——	——
6. Recorded all treatment on the casualty's FMC.	——	——
7. Evacuated the casualty to a definitive care facility as soon as possible.	——	——

Evaluation Guidance: Score each Soldier according to the performance measures in the evaluation guide. Unless otherwise stated in the task summary, the Soldier must pass all performance measures to be scored GO. If the Soldier fails any step, show what was done wrong and how to do it correctly.

References
  Required   **Related**
  DD FORM 1380   EMERG CARE AND TRANS 9
                        PHTLS

CONTROL BLEEDING
081-833-0161

Conditions: You have encountered a casualty who is bleeding externally and may also be bleeding internally. All other more serious injuries have been assessed and treated. Body substance isolation precautions have been taken. You will need field dressings, emergency bandages, cravats, gauze pads, gauze roller bandage, sterile dressing, Kerlix, ace wraps, air splints, hemostatic bandage, windlass device, materials for a tourniquet, and DD Form 1380 (U.S. Field Medical Card). You are not in a CBRNE environment.

Standards: Control bleeding without further harming the casualty.

Performance Steps

1. Determine if the bleeding is external or internal.
    a. External bleeding (go to step 2).
    b. Internal bleeding--suspicion should be based on mechanism of injury (MOI). (See tasks 081-833-0047, 081-833-0062, 081-833-0064, and 081-833-0154.)
        (1) Large bruises on the trunk or abdomen indicating injury to underlying organs.
        (2) Painful, swollen or deformed extremities indicating underlying fractures.
        (3) Rigid and/or tender abdomen may indicate bleeding into the abdomen.
        (4) Bleeding from the mouth, rectum, or other body orifice.
        (5) Vomiting bright red or dark (like coffee grounds) blood (hematemesis).
        (6) Bloody stool that is dark and tarry (melena) or bright red (hematochezia).

2. Apply direct pressure to the wound with a gauze pad or field dressing.
NOTE: If bleeding is profuse, apply direct pressure to the wound with your gloved hand. Do not waste time looking for a dressing.

NOTE: If in a tactical environment. (See task 081-833-0213.)

3. Elevate the affected extremity above the level of the heart.
CAUTION: Do not elevate if there are suspected musculoskeletal injuries, impaled objects in the extremity, or any suspected spinal injury.

4. Apply a pressure dressing or emergency bandage if the wound continues to bleed. (See task 081-833-0212.)
CAUTION: Once bleeding has been controlled it is important to check a distal pulse to make sure that the dressing has not been applied too tightly. If a pulse is not palpable, adjust the dressing to reestablish circulation.

5. Locate and apply pressure to the appropriate arterial pressure point, if the wound continues to bleed.
NOTE: Pressure points may not be effective if the wound is at the distal end of the limb. Blood is being sent to these areas from many smaller arteries.
    a. Brachial artery--used to control bleeding from the distal end of an upper extremity.
        (1) Hold the casualty's arm out at a right angle to his body with the palm facing up.
NOTE: Do not use force to raise the arm if the movement causes pain.
        (2) Locate the groove between the humerus and the biceps muscle.

Performance Steps

        (3) Hold the upper arm in the palm of your hand with your fingers positioned in the medial groove.

        (4) Press your fingers into the groove to compress the artery against the underlying bone.

NOTE: If pressure is applied properly, the radial pulse will not be palpable.

    b. Femoral artery--used to control bleeding of a lower extremity.

        (1) Locate the femoral artery on the medial side of the anterior thigh, just below the groin.

        (2) Place the heel of your hand over the site and apply pressure toward the bone.

NOTE: More pressure is needed to compress the femoral artery than the brachial artery due to the amount of tissue and muscle in the thigh. Greater force is needed for obese and muscular individuals. If pressure is applied properly, a distal pulse will not be palpable.

  6. Consider other conjunctive therapies to control bleeding if necessary.

    a. Splinting. (See task 081-831-0044.)

    b. Hemostatic bandage. (See task 081-833-0211.)

  7. Apply a tourniquet if the wound continues to bleed. (See task 081-833-0210.)

CAUTION: In combat while under enemy fire, a tourniquet is the primary means to control bleeding. It allows the individual, his battle buddy, or the combat medic to quickly control life threatening hemorrhage until the casualty can be moved away from the firefight. Always treat life-threatening hemorrhage while you and the casualty are behind cover.

NOTE: If the source of bleeding was due to a traumatic amputation--(See task 081-833-0157.)

  8. Initiate treatment for shock as needed. (See task 081-833-0047.)

  9. Record treatment given on the FMC.

10. Evacuate the casualty.

Performance Measures	GO	NO-GO
1. Determined the type of bleeding (external or internal).	——	——
2. Applied direct pressure to the wound with a gauze pad or field dressing.	——	——
3. Elevated the extremity above the level of the heart.	——	——
5. Located and applied pressure to the appropriate arterial pressure point, if the wound continued to bleed.	——	——
6. Considered other conjunctive therapies to control bleeding if necessary.	——	——

Performance Measures	GO	NO-GO
7. Applied a tourniquet if the wound continued to bleed.	____	____
8. Initiated treatment for shock as needed.	____	____
9. Recorded treatment given on the FMC.		
10. Evacuated the casualty.	____	____
	____	____

Evaluation Guidance: Score each Soldier according to the performance measures in the evaluation guide. Unless otherwise stated in the task summary, the Soldier must pass all performance measures to be scored GO. If the Soldier fails any step, show what was done wrong and how to do it correctly.

References

Required
DD FORM 1380

Related
EMERG CARE AND TRANS 9
PHTLS

PREPARE AN AID BAG
081-833-0194

Conditions: You need to prepare an aid bag for an upcoming mission. The type and length of the mission are given to you. You will need airway, breathing, fracture, and circulation supplies, antibiotics, pain medications, Nuclear Biological Chemical (NBC) medications, other miscellaneous supplies and a DD Form 1380 (U.S. Field Medical Card). You are not in a CBRNE environment.

Standards: Pack an aid bag with appropriate supplies for the mission.

Performance Steps
   1. Determine the type of aid bag for mission required.
NOTE: The contents of the aid bag will be based on the type and length of the mission, and the skill level of the combat medic. There is no standard packing list for an aid bag.
     a. M-5 aid bag.
     b. Blackhawk aid bag.
     c. London Bridge aid bag.
     d. Skedco aid bag.
     e. Others.
   2. Pack appropriate aid bag (based on the skill level of individual medic).
     a. Airway supplies.
       (1) Nasopharyngeal airways.
       (2) Oropharyngeal airways.
       (3) Combitube kit.
       (4) Surgical cricothyroidotomy kit.
     b. Breathing supplies.
       (1) Vaseline gauze pads (occlusive dressing).
       (2) Asherman chest seal.
       (3) 10-14 gauge 3.25 inch needle catheter unit (for needle chest decompression).
     c. Circulation supplies.
       (1) Kerlix.
       (2) Emergency bandages (Israeli bandage).
       (3) Cravats.
       (4) Tourniquets.
       (5) Improvised windlass devices (7to 10 tongue depressors wrapped in duct tape).
       (6) IV Infusion sets.
       (7) IV fluids.
       (8) FAST1 sternal intraosseous device.
       (9) Constricting band.
      (10) Alcohol pads.
      (11) Iodine pads.
      (12) Tegaderm dressings.
      (13) 18 ga intravenous catheters.
      (14) Saline locks.
      (15) Hemostatic bandages.
     d. Fracture supplies.
       (1) Sam splints.
       (2) Miscellaneous splints.
       (3) Ace wraps 2-4-6 inch.

Performance Steps
 e. Antibiotics.
  (1) Gatifloxacin tablets 400mg.
  (2) IV antibiotics.
 f. Pain medications.
  (1) Morphine tubex injectors 10 mg/ml.
  (2) Acetaminophen tablets 500mg.
 g. NBC medications.
  (1) NAAK injectors.
  (2) CANA injectors.
 h. Miscellaneous supplies.
  (1) Large abdominal pad.
  (2) Tape nylon 1-2-3 inch size.
  (3) Gauze pads 4x4 inch.
  (4) Gauze pads 2x2 inch.
  (5) Eye pads.
  (6) Cotton tipped applicators.
  (7) Band-aids.
  (8) ENT kit.
  (9) Stethoscope.
  (10) Burn packs.
  (11) Surgilube.
  (12) Tincture of Benzoin.
  (13) Exam gloves.
  (14) Adjustable C-collar.
  (15) Field Medical Card.
  (16) Bandage scissors.
  (17) Needles (various sizes).
  (18) Syringes (various sizes).
  (19) Chemlights.
  (20) Space blanket.
  (21) Oral hydration solution packs.
  (22) Tongue depressors.
  (23) Miscellaneous medications based on the medical officer's determination.
 3. Verify aid bag is packed appropriately.

Performance Measures	GO	NO-GO
1. Determined the type of aid bag for mission required.	____	____
2. Packed aid bag appropriate for the mission.	____	____
3. Verified aid bag was packed appropriately.	____	____

Evaluation Guidance: Score each Soldier according to the performance measures. Unless otherwise stated in the task summary, the Soldier must pass all performance measures to be scored GO. If the Soldier fails any steps, show what was done wrong and how to do it correctly.

References
    Required   **Related**
    DD FORM 1380   None

TREAT A CASUALTY FOR CONTUSIONS OR ABRASIONS
081-833-0209

Conditions: You have a casualty with a contusion or an abrasion. All other more serious injuries have been assessed and treated. You will need the casualty's medical record, normal saline, sterile water, gauze, exam gloves, marker, a needle, number 11 blade or tissue forceps, dressing materials, wrap, antibiotic ointment, splinting material, ice, and tape. You are not in a CBRNE environment.

Standards: Treat the casualty for contusions or abrasions without causing further injury.

Performance Steps

1. Solicit the casualty's history.
2. Assess injury for underlying complications.
   a. Abrasions.
      (1) Depth of wound (relates to method of anesthesia and cleaning).
      (2) Amount of body surface (fluid loss can be significant in children).
      (3) Amount of contamination (precursor to infection).
   b. Contusions.
      (1) Underlying fracture. Forceful impact of objects creating injury can result in fractures.
      (2) Vascular involvement (extensive bleeding into tissue).
      (3) Check distal circulation.
      (4) Measure or mark the outline of the contusion.
      (5) Measure circumference of injured extremity, and compare measurement to uninjured extremity.
      (6) Neurological involvement. Test the sensation and movement of the injured part. Any signs of neurologic deficit may indicate a serious complication.
3. Treat the abrasion.
   a. Principles of management are as follows:
      (1) Prevention of infection.
      (2) Promotion of rapid healing.
      (3) Prevention of "tattooing" from retained foreign bodies.
   b. Wound must be gently but thoroughly scrubbed with normal saline.
   c. Remove all foreign matter that cannot be scrubbed out by using a needle, number 11 blade, or tissue forceps.
   d. Apply antibiotic ointment.
   e. Administer antibiotic therapy, if needed. Antibiotic therapy may be indicated for prophylaxis (consult medical officer).
   f. Give casualty instructions on wound care and signs and symptoms of infection.
      (1) Topical antibiotic ointment applied three times or four times a day.
      (2) Dressing changed every 2 to 3 days with gentle cleaning.
      (3) Monitor abrasion for signs and symptoms of infection.
4. Treat the contusion.
   a. Ensure that there is no underlying fracture or evidence of any neurological or vascular involvement.
   b. Pad and splint injury, if needed.

Performance Steps

    c. Manage complications appropriately (consult medical officer if question of underlying injury).

        (1) Apply splint or cast to fractures (following medical officer's recommendation).

        (2) Refer vascular or neurologic injury to a medical officer.

    d. Prescribe rest, ice, compression, and elevation (RICE).

        (1) Wrap injured area with a roller bandage to compress the wound and slow bleeding into the tissue.

        (2) Apply ice to area over the wound

        (3) If wound is significant, have casualty keep area elevated

  5. Record all treatment in the casualty's medical record.

Evaluation Preparation: This task is best evaluated by performance of the steps. Give the Soldier a simulated casualty and a scenario in which he must manage contusions or abrasions.

Performance Measures	**GO**	**NO-GO**
1. Solicited the casualty's history.	——	——
2. Assessed the injury for underlying complications.	——	——
3. Treated the abrasion.	——	——
4. Treated the contusion.	——	——
5. Recorded all treatment in the casualty's medical record.	——	——

Evaluation Guidance: Score each Soldier according to the performance measures. Unless otherwise stated in the task summary, the Soldier must pass all performance measures to be scored GO. If the Soldier fails any steps, show what was done wrong and how to do it correctly.

References

  Required

  None

            Related

            ISBN 0-7637-4406-9

APPLY A TOURNIQUET TO CONTROL BLEEDING
081-833-0210

Conditions: You have encountered a casualty who is bleeding profusely from an extremity and needs a tourniquet to control the bleeding. All other more serious injuries have been assessed and treated. You will need an emergency bandage, clean cloth or sterile dressing, plastic, container, cravats, marker, DD Form 1380 (U.S. Field Medical Card), and materials to improvise a tourniquet or a combat application tourniquet (C.A.T.). You are not in a CBRNE environment.

Standards: Control the bleeding from the extremity without causing further harm to the casualty.

Performance Steps
CAUTION: All body fluids should be considered potentially infectious. Always observe body substance isolation (BSI) precautions by wearing gloves and eye protection as a minimal standard of protection.

1. Determine if the bleeding is life-threatening.

2. Apply a tourniquet if direct pressure and the emergency bandage fail to control the bleeding. (See task 081-833-0161.)
CAUTION: Under combat conditions, while under effective enemy fire, a temporary tourniquet may often be the primary means to control bleeding. A properly applied tourniquet will quickly control life-threatening hemorrhage until the casualty can be moved away from the effective fire.
    a. Improvised tourniquet.
        (1) Apply pressure to pressure point above the wound.
        (2) Prepare equipment.
        (3) Take BSI.
        (4) Expose the wound.
        (5) Place the prepared cravat and windlass 2-3 inches above the wound (not over a joint) and secure the cravat tightly against the extremity with a full non-slip knot.
        (6) Twist the windlass until the bleeding stops.
        (7) While holding tension on the windlass, place the windlass inside the half knot of the second cravat proximal to the tourniquet (if possible).
        (8) Tighten the second cravat around windlass and secure the second cravat to the extremity with a full non-slip knot.
        (9) Assess for the absence of a distal pulse (not indicated for amputations).
       (10) Place a "T" and the time of application on the casualty.
       (11) Secure the tourniquet in place with tape.
    b. C-A-T. (NSN 6515-01-521-7976.)
        (1) Apply pressure to pressure point above the wound.
        (2) Take BSI.
        (3) Expose the wound enough to ensure the tourniquet is placed above the injury.
        (4) Place C.A.T. between the heart and the wound on the injured extremity, 2-3 inches above the wound.
        (5) Pull the free end of the self adhering band through the buckle and route through the friction adapter buckle (it is not necessary to route through friction adapter on an arm wound).
        (6) Pull the self adhering band tight around the extremity and fasten it back on itself.
        (7) Twist the windlass until the bleeding stops.
        (8) Lock the windlass in place within the windlass clip.

Performance Steps

      (9) Secure the windlass with the windlass strap.

      (10) Assess for the absence of a distal pulse (not indicated for amputations).

      (11) Place a "T" and the time of application on the casualty.

      (12) Secure the C.A.T. in place with tape.

  3. Record the treatment on the FMC.

  4. Reassess the injury to ensure bleeding has been controlled.

  5. If the source of bleeding was due to a traumatic amputation--

     a. Wrap the amputated part in a clean cloth or sterile dressing (if available).

     b. Wrap or bag the amputated part in plastic.

     c. Label the plastic bag with the casualty's information.

     d. Transport the amputated part in a cool container (if available) with the casualty.

CAUTION: Do not place the amputated part directly in contact with ice. Do not submerge the part directly in water. Do not allow the part to freeze.

  6. Evacuate the casualty.

Evaluation Preparation:

Setup: For training and evaluation, have another Soldier act as the casualty. Have an emergency bandage, cravats, materials to improvise a tourniquet or a combat application tourniquet present. After the emergency bandage is applied, tell the Soldier that bleeding is not controlled. Once the tourniquet (improvised or C.A.T.) has been applied and the windlass device has been tightened, tell the Soldier that the bleeding has stopped. Tell the casualty not to assist the Soldier in any way.

CAUTION: Do not allow the Soldier to fully tighten the windlass on the tourniquet.

Brief the Soldier: Tell the Soldier to control the bleeding from a casualty's extremity.

Performance Measures	**GO**	**NO-GO**
1. Determined if the bleeding was life threatening.	——	——
2. Applied direct pressure and a pressure dressing to the wound with an emergency bandage.	——	——
3. Applied an improvised tourniquet or C.A.T.	——	——
4. Recorded the treatment on the FMC.	——	——
5. Reassessed the injury to ensure bleeding was controlled.	——	——
6. Evacuated the casualty.	——	——
7. Caused no further injury to the casualty.	——	——

Evaluation Guidance: Score each Soldier according to the performance measures in the evaluation guide. Unless otherwise stated in the task summary, the Soldier must pass all performance measures to be scored GO. If the Soldier fails any step, show what was done wrong and how to do it correctly.

References
    Required
        DD FORM 1380

Related
EMERG CARE AND TRANS 9
PHTLS

APPLY A HEMOSTATIC DRESSING
081-833-0211

Conditions: You have encountered a casualty who is bleeding externally. All other more serious injuries have been assessed and treated. The wound is either not amendable to a tourniquet or a tourniquet is in place and alternate means of hemorrhage control is necessary. Body substance isolation precautions have been taken. You will need a roll of Combat Gauze, cotton gauze wad and rolled, an emergency bandage or elastic bandage, and a DD Form 1380 (U.S. Field Medical Card). You are not in a CBRNE environment.

Standards: Apply a hemostatic dressing to control bleeding without causing further harm the casualty.

Performance Steps

1. Remove all clothing or equipment to obtain access to the wound.

2. Apply pressure to a pressure point proximal to the bleeding site.

3. Identify the point of bleeding within the wound.
   a. Remove any pooled blood from the wound cavity with your hand or a wad of cotton gauze.
   b. Locate the bleeding vessel(s).

4. Pack combat gauze directly over the source of bleeding.

5. Pack the wound with the entire dressing

6. Apply direct pressure for 3 minutes.
   a. Periodically check the dressing to ensure proper placement and bleeding control.
   b. If the bandage becomes completely soaked through and there is still active bleeding.
      (1) Remove all of the gauze from the wound.
      (2) Replace with a new package of combat gauze.
      (3) Repeat steps starting at step 2.

7. Bandage wound to secure the dressing in place.
   a. If the wound cavity is deep, apply cotton gauze (either wad or rolled) over the dressing.
   b. Secure dressing in place with either an emergency bandage or an elastic bandage.

8. Secure the bandage in place with tape.

9. Document treatment on the FMC.

Evaluation Preparation:
Setup: For training and evaluation, have another Soldier act as the casualty. Have a roll of combat gauze, cotton gauze wad and rolled, and an emergency bandage or elastic bandage. Tell the casualty not to assist the Soldier in any way. The bleeding may be from the extremity (upper or lower), axillary area, inguinal area or neck. Use a training package of Combat Gauze.

Brief Soldier: Tell the Soldier the bleeding is bright red and spurting. The wound is not amendable to a tourniquet or a tourniquet is in place and alternate means of hemorrhage control is necessary; a hemostatic dressing is available. Tell the Soldier that they are not in a CBRNE environment and to control the bleeding from the casualty's wound using the hemostatic dressing.

Performance Measures <u>GO</u>   <u>NO-GO</u>

1. Removed all clothing or equipment to obtain access to the wound. ____ ____

2. Applied pressure to a pressure point proximal to the bleeding site. ____ ____

3. Identified the point of bleeding within the wound. ____ ____
    a. Removed any pooled blood from the wound cavity with your hand or a wad of cotton gauze.
    b. Located the bleeding vessel(s).

4. Packed Combat Gauze directly over the source of bleeding. ____ ____

5. Packed the wound with the entire dressing. ____ ____

6. Applied direct pressure for 3 minutes. ____ ____
    a. Periodically checked the dressing to ensure proper location and bleeding control.
    b. If the bandage became completely soaked through and there is still active bleeding
        (1) Removed all of the gauze from the wound.
        (2) Replaced with a new package of Combat Gauze.
        (3) Repeated steps starting at step 2.

7. Bandaged wound to secure the dressing in place. ____ ____
    a. If the wound cavity is deep, applied cotton gauze (either wad or rolled) over the dressing.
    b. Secure dressing in place with either an emergency bandage or an elastic bandage.

8. Secured the bandage in place with tape. ____ ____

9. Documented treatment on the FMC. ____ ____

Evaluation Guidance: Score each Soldier according to the performance measures in the evaluation guide. Unless otherwise stated in the task summary, the Soldier must pass all performance measures to be scored GO. If the Soldier fails any step, show what was done wrong and how to do it correctly.

References
    Required
    DD FORM 1380

    Related
    STP 21-1-SMCT

APPLY A PRESSURE DRESSING TO AN OPEN WOUND
081-833-0212

Conditions: You have encountered a casualty who is bleeding externally from an open wound. All other more serious injuries have been assessed and treated. You will need the casualty's emergency bandage, bandage scissors, and a DD Form 1380 (U.S. Field Medical Card). You are not in a CBRNE environment.

Standards: Apply the emergency bandage to control the bleeding without causing further harm to the casualty.

Performance Steps
CAUTION: All body fluids should be considered potentially infectious. Always observe body substance isolation (BSI) precautions by wearing gloves and eye protection as a minimal standard of protection.

1. Fully expose the injury unless clothing is adhered to the wound.

2. Maintain direct (manual) pressure to limit the blood loss.

3. Apply the casualty's emergency bandage.
NOTE: If the casualty is operating in a CBRNE environment, mask the casualty and apply the emergency bandage as you would in a conventional environment. Take care not to contaminate the wound while applying the emergency bandage and do not kneel or come into unnecessary contact with the chemically-contaminated ground. Skin decontamination (See STP 21-1-SMCT, task 031-503-1012) and placement of the casualty in MOPP 4 should be performed as soon as possible. Record the circumstances in which the casualty was initially found and all treatment provided in the FMC.
    a. Open the plastic dressing package.
    b. Apply the dressing, white (sterile, non-adherent pad) side down, directly over the wound.
    c. Wrap the elastic tail (bandage) around the extremity and run the tail through the plastic pressure bar.
    d. Reverse the tail while applying pressure and continue to wrap the remainder of the tail around the extremity, continuing to apply pressure directly over the wound.
    e. Secure the plastic closure bar to the last turn of the wrap.
    f. Check the emergency bandage to make sure that it is applied firmly enough to prevent slipping without causing a tourniquet-like effect.
WARNING: The emergency bandage must be loosened if the skin distal to the injury becomes cool, blue, numb, or pulseless.

4. Reassess the wound to ensure bleeding has been controlled and treat for shock, if necessary. (See task 081-833-0047.)

5. Apply manual pressure and elevate to reduce bleeding, if necessary.
    a. Apply firm manual pressure over the emergency bandage for 5 to 10 minutes.
    b. Elevate the extremity above the level of the heart unless a fracture is suspected and has not been splinted.

6. Record the treatment on the FMC.

7. Evacuate the casualty.

Evaluation Preparation:
Setup: For training and evaluation, have another Soldier act as the casualty. Use moulage on the casualty's arm or leg to simulate the wound. Have an emergency bandage present. After the emergency bandage has been applied, tell the Soldier the bleeding has not been controlled. Once manual pressure and elevation have been applied, tell the Soldier the bleeding is now controlled. Tell the casualty not to assist the Soldier in any way.

Brief Soldier: Tell the Soldier that they are not in a CBRNE environment and to control the bleeding from a casualty's extremity.

Performance Measures	GO	NO-GO
1. Exposed the wound.	___	___
2. Maintained direct (manual) pressure to limit the blood loss.	___	___
3. Applied an emergency bandage.	___	___
4. Reassessed the wound to ensure bleeding is controlled and treated for shock, if necessary.	___	___
5. Applied manual pressure and elevated the extremity, if necessary.	___	___
6. Recorded the treatment on the FMC.	___	___
7. Evacuated the casualty.	___	___
8. Caused no further injury to the casualty.	___	___

Evaluation Guidance: Score each Soldier according to the performance measures in the evaluation guide. Unless otherwise stated in the task summary, the Soldier must pass all performance measures to be scored GO. If the Soldier fails any step, show what was done wrong and how to do it correctly.

References

Required	Related
DD FORM 1380	STP 21-1-SMCT

PERFORM A TACTICAL CASUALTY ASSESSMENT
081-833-0213

Conditions: You have a combat casualty under tactical conditions. You will need a combat medic aid bag, tourniquet, nasopharyngeal airway, Combitube or King LT, four-sided occlusive dressing, 14 gauge 3-1/4 inch catheter-over-needle, tape, 18 gauge needle, intravenous (IV) emergency trauma bandage, combat pill pack, morphine, phenergan, and a DD Form 1380 (U.S. Field Medical Card). You are not in a CBRNE environment.

Standards: Tactically manage a casualty to prevent additional injuries to the casualty or additional casualties without endangering the mission.

Performance Steps

1. Perform care under fire.
NOTE: Care under fire is care rendered at the scene of the injury while the combat medic and the casualty are still under effective hostile fire.
    a. Return fire as directed or required before providing medical treatment. This may include wounded Soldiers still able to fight.
    b. Determine if casualty is alive or dead.
        (1) Provide care to live casualty tactically.
            (a) Suppress enemy fire.
            (b) Use cover or concealment (smoke).
            (c) Direct casualty to return fire, move to cover, and administer self-aid, (stop bleeding) if possible. If unable to move casualty to cover and still under direct enemy fire, have casualty "Play Dead".
            (d) If casualty is unresponsive, move casualty and his equipment to cover as the tactical situation permits.
            (e) Keep the casualty from sustaining additional wounds.
            (f) Reassure the casualty.
    c. Administer only life-saving care while still under enemy fire.
        (1) Identify and control life-threatening hemorrhage with a tourniquet.
        (2) Cervical spine control is not necessary if the casualty has sustained penetrating head or neck wounds. If suffering from blunt head or neck trauma, the medic must determine if the tactical situation will allow attempts at cervical spine control while under enemy fire.
NOTE: The combat medic rendering care decides treatment on the basis of the relative risk of further injury versus that of exsanguination.
    d. Communicate medical situation to team leader.
    e. Tactically transport casualty, his weapon, and mission-essential equipment to cover.
    f. Recheck bleeding control measures as the tactical situation permits.

2. Perform tactical field care.
NOTE: Tactical field care is care rendered by the medic when no longer under effective hostile fire. Tactical field care also applies to situations in which an injury has occurred on a mission but there has been no hostile fire. Available medical equipment is limited to that carried into the field by the combat medic and mission personnel.
    a. In the following situations, communicate medical situation to patrol leader.
        (1) Upon determining that casualty will not be able to continue mission.
        (2) Before initiating any medical procedures (ensure tactical situation allows for time required).

Performance Steps

        (3) Upon any significant change in casualty's status.

    b. Note general impression of the casualty by determining responsiveness or level of consciousness (AVPU).

        (1) A - Alert.

        (2) V - Responds to verbal commands.

        (3) P - Responds to painful stimuli.

        (4) U - Unresponsive.

NOTE: If the casualty has suffered from a blast or penetrating trauma and has no signs of life (no pulse, no blood pressure, no respirations), do not perform CPR. These casualties will not survive and you may expose yourself to enemy fire and delay care to other casualties.

    c. Assess and secure the airway.

        (1) If the casualty is conscious and not in respiratory distress, do not administer airway intervention.

        (2) If the casualty is unconscious, use a chin-lift or jaw-thrust to open the airway. Use a nasopharyngeal airway (NPA) to maintain the airway.

        (3) Roll the casualty into the recovery position. This allows for accumulated blood and mucus to drain and not choke the casualty

        (4) If more advanced airway support is needed, insert a Combitube or King LT. (See task 081-833-0230.)

        (5) For an unconscious casualty with an obstructed airway or severe maxillofacial trauma, perform a surgical cricothyroidotomy. (See task 081-833-3005.)

    d. Assess the chest and perform medical care to correct problems in breathing or respiration.

        (1) Immediately seal any penetrating injuries to the chest with a four-sided occlusive dressing, or apply a commercial device such as the Asherman®, Hyphen®, or Bolin® Chest Seals.

        (2) Monitor casualty for progressive severe respiratory distress (breathing becomes more labored and faster).

        (3) If respiration becomes progressively worse, consider this a tension pneumothorax and decompress affected chest side with a 14 gauge 3-1/4 inch catheter-over-needle inserted at second intercostal space (ICS) at midclavicular line (MCL). Secure the catheter in place with tape.

    e. Identify and control major bleeding not previously controlled.

        (1) Apply direct pressure and/or an emergency trauma bandage, as appropriate.

        (2) If a tourniquet was previously applied, consider changing the tourniquet to a pressure dressing and/or using a hemostatic bandage to control bleeding.

        (3) Leave the tourniquet in place while doing this. Loosen it, but do not remove it.

        (4) If after using other conventional methods (direct pressure, pressure dressing, or hemostatic bandages) you are not able to control hemorrhage, retighten the tourniquet until bleeding stops.

    f. Determine if the casualty requires fluid resuscitation.

        (1) A palpable radial pulse and normal mental status should be used to determine who needs fluid resuscitation. These can be determined in the typical noisy and chaotic battlefield environment. Blood pressure measurement is not necessary. If a casualty has a radial pulse, his equivalent blood pressure is at least 80 mmHg.

        (2) If the casualty has a superficial wound, IV resuscitation is not necessary but oral fluid hydration should be encouraged.

        (3) If the casualty has a significant wound, either extremity or truncal (neck, chest, abdomen, or pelvis), and the casualty is coherent and has a palpable radial pulse:

Performance Steps

        (a) Start an 18 gauge IV catheter and place a saline lock. Hold fluids but reevaluate as frequently as the tactical situation allows. If unable to start a peripheral IV, consider starting a F.A.S.T. 1 sternal intraosseous line. (See task 081-833-0185.)

        (b) Upper extremity is first choice. Do not start an IV on an extremity distal to a significant wound.

        (c) If the casualty does not have a radial pulse, ensure that the bleeding has stopped using whatever means available--direct pressure, pressure dressings, hemostatic bandage, or tourniquets as needed.

        (d) Once the bleeding has stopped, give 500 ml of Hextend as rapidly as possible. Recheck in 30 minutes. If the radial pulse has returned, do not give any additional fluids but monitor as frequently as possible.

        (e) If the radial pulse does not return, give an additional 500 ml of Hextend.

        (f) Recheck in 30 minutes. If the radial pulse returns, hold additional fluids and evacuate ASAP. If the radial pulse does not return, then triage your supplies and equipment to other casualties.

  g. Expose any wounds.

     (1) Remove the minimum amount of clothing required to expose and treat injuries. Dress the wounds to prevent contamination and help with hemostasis. An emergency bandage is ideal for this. Search for exit wounds.

     (2) Protect the casualty against the environment.

NOTE: High velocity projectiles from modern military weapons may travel great distances from where they entered. You may have to search for the exit wound.

  h. Splint obvious long-bone fractures. (See STP 21-1-SMCT, task 081-831-1034.)

  i. Administer pain medications as needed to any Soldier wounded in combat.

     (1) If the casualty is still able to fight, Mobic 15 mg po qd with two 650 mg caplets of acetaminophen every 8 hours, will control mild to moderate pain and not cause drowsiness. These medications and an antibiotic make up the "Combat Pill Pack", and should be issued to each Soldier prior to deployment.

     (2) If the casualty is unable to fight--

        (a) Morphine 5 mg given IV (through the saline lock) and repeated every 10 minutes as necessary is very effective in controlling severe pain. If a saline lock is used, it should be flushed with 5 ml of saline after the morphine administration.

        (b) Phenergan 25 mg IV or IM may be necessary to combat the nausea and vomiting associated with morphine.

NOTE: Medics who carry morphine must be familiar with its side effects and trained in the use of Naloxone to counter these side effects.

        (c) Pain relief can also be attained with the use of fentanyl transmucosal lozenges. These lozenges are placed between the cheek and gum and will be absorbed through the oral mucosa. This method allows for narcotic pain control without IV access.

        (d) Ensure there is visible evidence of the amount and time of pain medication given.

     (3) Soldiers should avoid aspirin and some of the older anti-inflammatory medications because of their detrimental effects on blood clotting.

     (4) Antibiotics should be considered in all Soldiers wounded in combat who have a 3 hour delay in evacuation time since these wounds are prone to infection.

        (a) In Soldiers who are awake and alert, give an oral antibiotic.

Performance Steps

        (b) In unconscious Soldiers or those who may not be able to take an oral antibiotic, IV antibiotics may be given through the saline lock every 12 hours.

NOTE: Soldiers who may have allergies to these medications must be identified in the pre-deployment planning phase and alternate medications provided.

      j. Initiate medical evacuation request lines 1 through 5 (lines 6 through 9, as appropriate).

      k. Complete FMC.

      l. Transport the casualty to the site where evacuation is anticipated. (See task 081-833-4680.)

3. Perform combat casualty evacuation care (CASEVAC). Care in the CASEVAC phase does not differ significantly from the tactical field care phase. However, there are two significant differences.

      a. Additional medical personnel may accompany the evacuation asset to assist the medic.

      b. Additional medical supplies and equipment may also accompany the evacuation asset. This equip may consist of--

        (1) Oxygen.

        (2) Electronic monitoring devices.

        (3) Additional IV fluids.

        (4) Blood (may be available).

NOTE: Combat casualty evacuation care is care rendered once the casualty is awaiting pickup or has been picked up by an aircraft, vehicle, or boat. If evacuating on a nonmedical vehicle, a combat medic may need to accompany an unconscious casualty to monitor the airway, breathing, bleeding, and IV Infusion.

Evaluation Preparation:

Setup: Give the Soldier a simulated casualty and a scenario in which he must perform tactical combat casualty care.

Brief Soldier: Tell the Soldier to perform tactical combat casualty care.

Performance Measures	<u>GO</u>	<u>NO-GO</u>
1. Performed care under fire.	——	——
2. Performed tactical field care.	——	——
3. Performed combat casualty evacuation care.	——	——

Evaluation Guidance: Score each Soldier according to the performance measures. Unless otherwise stated in the task summary, the Soldier must pass all performance measures to be scored GO. If the Soldier fails any steps, show what was done wrong and how to do it correctly.

References

Required	**Related**
DD FORM 1380	PHTLS

APPLY KERLIX TO AN OPEN WOUND
081-833-0229

Conditions: You have a casualty with a penetrating wound; it is bleeding severely. All other more serious injuries have been assessed and treated. You are unable to apply a tourniquet above the wound, you must apply Kerlix to the wound. You will need packages of roller gauze, an emergency bandage, ace wrap or pressure bandage, and tape. The tactical situation is controlled, and you are not in a CBRNE environment.

Standards: Stop life-threatening hemorrhage by packing the wound with gauze and applying a pressure bandage.

Performance Steps

1. Have an assistant apply pressure to the artery site above the wound and control the bleeding.

2. Pack the wound.
   a. Open a package of roller gauze (Kerlix).
   b. Unroll the gauze and begin packing the gauze into the wound.
   c. Continue to pack the wound with gauze until the wound is completely filled. If necessary get another roll and continue packing.
   d. When complete, form the remainder of the roll directly over the wound and hold it in place with pressure.

3. While keeping the roller gauze focused directly over the wound, begin wrapping an emergency bandage, ace wrap, or similar pressure bandage over the wound.
NOTE: This bandage must be wrapped extremely tight. You must stretch all the elastic out of the bandage while wrapping. Do not allow the roller gauze to leak out from under the pressure bandage.

4. Apply three inch tape over the finished product to secure the pressure bandage in place, even if a closure bar is already engaged.

5. Ensure that bleeding is controlled.

Evaluation Preparation:
Setup: For training and evaluation, have another Soldier act as the casualty. You will need one Soldier to act as the assistant. The Soldier being tested is to act as the team leader and direct the actions of the assistant. The casualty may be placed in a vehicle or other scenario, depending on available resources and the technique you are testing.

Brief the casualty not to assist the Soldier being evaluated.

Performance Measures	GO	NO-GO
1. Had assistant apply pressure to the pressure point above the wound to control bleeding.	——	——
2. Packed the wound with gauze (Kerlix) filling the entire wound.	——	——
3. Applied a pressure bandage over the gauze.	——	——

Performance Measures	GO	NO-GO
4. Applied tape to secure the pressure bandage.	——	——
5. Ensured that bleeding was controlled.	——	——

Evaluation Guidance: Score each Soldier according to the performance measures in the evaluation guide. Unless otherwise stated in the task summary, the Soldier must pass all performance measures to be scored GO. If the Soldier fails any step, show what was done wrong and how to do it correctly.

References

Required	Related
None	ISBN13:978-0-7637-4406-9

Subject Area 4: Airway Management

OPEN THE AIRWAY
081-831-0018

Conditions: You are evaluating a casualty who is not breathing. You need to open his airway. You are not in a CBRNE environment.

Standards: Complete all of the steps required to open the casualty's airway without causing unnecessary injury.

Performance Steps
CAUTION: All body fluids should be considered potentially infectious. Always observe body substance isolation (BSI) precautions by wearing gloves and eye protection as a minimal standard of protection.

1. Roll the casualty onto his back if necessary.
   a. Kneel beside the casualty.
   b. Raise the near arm and straighten it out above the head.
   c. Adjust the legs so that they are together and straight or nearly straight.
   d. Place one hand on the back of the casualty's head and neck.
   e. Grasp the casualty under the arm with your free hand.
   f. Pull steadily and evenly toward yourself, keeping the casualty's head and neck in line with his torso.
   g. Roll the casualty as a single unit.
   h. Place the casualty's arms at his side.

2. Establish the airway using the head-tilt/chin-lift or jaw thrust method.
   a. Head-tilt/chin-lift maneuver.
CAUTION: Do not use this method if a spinal injury is suspected.

NOTE: Remove any foreign material or vomitus seen in the mouth as quickly as possible.
   (1) With the casualty in a supine position, position yourself beside the casualty's head.
   (2) Place one hand on the casualty's forehead and apply firm, backward pressure with your palm to tilt the head back.
   (3) Place the tips of your fingers of your other hand under the lower jaw near the boney part of the casualty's chin.
   (4) Lift the chin upward, bringing the entire lower jaw with it, helping to tilt the head back.
CAUTIONS: 1. Do not use the thumb to lift the lower jaw. 2. Do not press deeply into the soft tissue under the chin with the fingers. 3. Do not completely close the casualty's mouth.
   b. Jaw-thrust maneuver.
CAUTION: Use this method if a spinal injury is suspected.
   (1) Kneel above the supine casualty's head. Rest your elbows on the surface on which the casualty is lying.
   (2) Carefully reach forward and gently place one hand on each side of the casualty's lower jaw, at the angles of the jaw below the ears.
   (3) Stabilize the casualty's head with your forearms.
   (4) Using your index fingers, push the angles of the casualty's lower jaw forward.

Performance Steps

        (5) Use your thumbs to help position the lower jaw to allow breathing through the mouth as well as the nose.

        (6) The completed maneuver should open the airway with the mouth slightly open and the jaw jutting forward.

CAUTION: Do not tilt or rotate the casualty's head.

  3. Check for breathing within 3 to 5 seconds. While maintaining the open airway position, place an ear over the casualty's mouth and assess the breathing using the "look, listen and feel" technique.

    a. Look for the chest to rise and fall.

    b. Listen for air escaping during exhalation.

    c. Feel for the flow of air on the side of your face.

  4. Take appropriate action.

    a. If the casualty resumes breathing on his own, maintain the airway and (if no spinal injury is assessed or suspected) place the casualty in the recovery position.

      (1) Roll the casualty as a single unit onto his side.

      (2) Place the hand of his upper arm under his chin.

      (3) Flex his upper leg.

NOTE: Continue the initial assessment to check the casualty for other injuries. (See task 081-833-0156.)

    b. If the casualty does not resume breathing, perform rescue breathing. (See task 081-831-0048.)

Evaluation Preparation:

Setup: Place a cardiopulmonary resuscitation (CPR) mannequin or another Soldier acting as the casualty face down on the ground. For training and evaluation, you may specify to the Soldier whether the casualty has a spinal injury to test step 2, or you may create a scenario in which the casualty's condition will dictate to the Soldier how to treat the casualty. After step 3 tell the Soldier whether the casualty is breathing or not and ask what should be done.

Brief Soldier: Tell the Soldier to open the casualty's airway.

Performance Measures

	GO	NO-GO
1. Rolled the casualty onto his back, if necessary.	____	____
2. Established the airway using the head-tilt/chin-lift or jaw-thrust maneuver.	____	____
3. Checked for breathing within 3 to 5 seconds.	____	____
4. Took appropriate action.	____	____
5. Did not cause further injury to the casualty.	____	____

Evaluation Guidance: Score each Soldier according to the performance measures. Unless otherwise stated in the task summary, the Soldier must pass all performance measures to be scored GO. If the Soldier fails any steps, show what was done wrong and how to do it correctly.

References
    Required   **Related**
    None   EMERG CARE AND TRANS 9

CLEAR AN UPPER AIRWAY OBSTRUCTION
081-831-0019

Conditions: You are evaluating a casualty who is not breathing or is having difficulty breathing, and you suspect the presence of an upper airway obstruction. You are not in a CBRNE environment.

Standards: Complete, in order, all the steps necessary to clear an object from a casualty's upper airway. Continue the procedure until the casualty is able to speak and breathe normally or until relieved by a qualified person.

Performance Steps
CAUTION: All body fluids should be considered potentially infectious. Always observe body substance isolation (BSI) precautions by wearing gloves and eye protection as a minimal standard of protection.

1. Determine whether the casualty needs your assistance.
   a. Conscious casualty.
      (1) Ask the casualty if he is choking and if he can speak.
         (a) If the casualty has adequate air exchange (is able to speak, coughs forcefully or wheezes between coughs), do not interfere except to encourage the casualty to continue coughing.
NOTE: Abdominal thrusts are usually not effective for dislodging a partial obstruction; attempts to remove the object manually could cause a complete obstruction.
         (b) If the casualty has poor air exchange (weak, ineffective cough; high-pitched noise while inhaling; increased respiratory difficulty; and possible cyanosis), continue with step 1a(2).
         (c) If the casualty has a complete airway obstruction (is unable to speak, breathe or cough and may clutch the neck between the thumb and fingers), continue with step 1a(2).
2. If the casualty is lying down, bring him to a sitting or standing position.
3. Apply abdominal or chest thrusts.
NOTE: Use abdominal thrusts unless the casualty is in the advanced stages of pregnancy, is very obese, or has a significant abdominal wound.
         (a) Abdominal thrusts.
           1) Stand behind the casualty and wrap your arms around his waist.
           2) Make a fist with one hand and place the thumb side of your fist against the casualty's abdomen in the midline, slightly above the navel and well below the tip of the xiphoid process.
           3) Grasp your fist with your other hand and press your fist into the casualty's abdomen with quick inward and upward thrusts.
           4) Continue abdominal thrusts until the blockage is expelled or the casualty becomes unconscious.
NOTE: Make each thrust a separate, distinct movement given with the intent of relieving the obstruction.

Performance Steps

(b) Chest thrusts.

1) Stand behind the casualty and encircle his chest with your arms just under the armpits.

2) Make a fist with one hand and place the thumb side of the fist against the middle of the casualty's sternum.

3) Grasp your fist with your other hand and give backward thrusts, pressing your fist into the casualty's chest.

4) Continue giving thrusts until the blockage is expelled or the casualty becomes unconscious.

CAUTION: Do not position your hand on the xiphoid process or the lower margins on the rib cage.

NOTES: 1. Administer each thrust with the intent of relieving the obstruction. 2. If the casualty becomes unconscious, position the casualty on his back, open the airway (see task 081-831-0018), observe the airway to look for the obstruction, and begin the steps of cardiopulmonary resuscitation (CPR).

4. Take appropriate action if the casualty becomes unconscious. Begin the steps of CPR.

a. Determine unresponsiveness.

b. Open the airway.

NOTE: Perform the head-tilt/chin-lift maneuver to clear an obstruction that is caused by the tongue and throat muscles relaxing back into the airway in any person who is found unconscious.

c. Observe the airway to look for the obstruction.

NOTE: If the obstruction is visible, you may attempt to remove the object.

d. Attempt ventilation.

e. If the first ventilation does not produce visible chest rise, reposition the head and reattempt to ventilate.

f. If both breaths fail to produce visible chest rise, perform 30 chest compressions and then open the airway and look into the mouth for the obstruction.

g. Reattempt ventilation.

5. When the object is dislodged, check for breathing. Perform rescue breathing, if necessary (See task 081-831-0048.) or continue to evaluate the casualty for other injuries.

Evaluation Preparation:

NOTE: Only the procedure for clearing an airway obstruction in a conscious casualty will be evaluated. The procedure for an unconscious casualty can be evaluated as a part of task 081-831-0048.

Setup: You will need another Soldier to play the part of the casualty. Instruct the casualty how to appear as though he has an obstructed airway; with consciousness progressing to unconsciousness. Instruct the casualty not to assist the Soldier in any way.

Brief Soldier: Describe the signs and symptoms of a casualty with good air exchange, poor air exchange, or a complete airway obstruction. Ask the Soldier what should be done and score step 1 based on the answer. Then tell the Soldier to clear an upper airway obstruction. Tell the Soldier to demonstrate how to position the casualty, where to stand and how to position his hands for the thrusts. The Soldier must tell you how they should be done and how many thrusts should be performed. Ensure that the Soldier understands that he must not actually perform the thrusts. After completion of step 5, ask the Soldier what must be done if the casualty becomes unconscious.

Performance Measures	GO	NO-GO
1. Determined whether the casualty needed your assistance.	____	____
2. Moved the casualty to a sitting or standing position, if necessary.	____	____
3. Applied abdominal or chest thrusts.	____	____
4. Took appropriate action if the casualty became unconscious. Began the steps of CPR.	____	____

    a. Repositioned the casualty.
    b. Opened the airway.
    c. Observed the airway looking for the obstruction, and removed it if seen.
    d. Attempted ventilation.
    e. If the first ventilation did not produce visible chest rise, repositioned the head and reattempted to ventilate.
    f. If both breaths failed to produce visible chest rise, performed 30 chest compressions, and then opened the airway and looked into the mouth for the obstruction.
    g. Attempted ventilation.
    h. Performed rescue breathing procedures, if necessary. (See task 081-831-0048.)

5. When the object was dislodged, checked for breathing. Performed rescue breathing, if necessary.	____	____

Evaluation Guidance: Score each Soldier according to the performance measures. Unless otherwise stated in the task summary, the Soldier must pass all performance measures to be scored GO. If the Soldier fails any steps, show what was done wrong and how to do it correctly.

References
    Required
    None

    Related
    EMERG CARE AND TRANS 9

PERFORM RESCUE BREATHING
081-831-0048

Conditions: You are treating a casualty who is unconscious and is not breathing. You must perform rescue breathing. You will need a bag-valve-mask (BVM) system or flow-restricted oxygen-powered ventilation device (FROPVD), and airway adjuncts. You are not in a CBRNE environment.

Standards: Complete in order, all the steps necessary to restore breathing. Continue the procedure until the casualty starts to breathe, you are relieved by another qualified person or medical officer, are required to perform cardiopulmonary resuscitation (CPR), are too exhausted to continue, or enemy fire prevents you from continuing until you can move the casualty behind effective cover.

Performance Steps
CAUTION: All body fluids should be considered potentially infectious. Always observe body substance isolation (BSI) precautions by wearing gloves and eye protection as a minimal standard of protection.

1. Position yourself at the casualty's head.

2. Open the airway. (See task 081-831-0018.)
    a. Use the head-tilt/chin-lift maneuver where no trauma is suspected.
    b. Use the jaw thrust maneuver when trauma is observed or suspected.

3. Ventilate the casualty using the mouth-to-mouth, mouth-to-nose, mouth-to-mask, bag-valve-mask (BVM) system, or flow-restricted oxygen-powered ventilation device (FROPVD or demand-valve), as appropriate.
    a. Mouth-to-mouth method.
       (1) Maintain the chin-lift while pinching the nostrils closed using the thumb and index fingers of your hand on the casualty's forehead.
       (2) Take a deep breath and make an airtight seal around the casualty's mouth with your mouth.
       (3) Give one slow breath (lasting 1 second) into the casualty's mouth, watching for the chest to rise and fall and listening and feeling for air to escape during exhalation.
       (4) If the chest rises and air escapes--
          (a) Give a second slow breath.
          (b) Go to step 6.
       (5) If the chest does not rise or air does not escape, go to step 4.
    b. Mouth-to-nose method.
NOTE: The mouth-to-nose method is recommended when you cannot open the casualty's mouth, there are jaw or mouth injuries, or you cannot maintain a tight seal around the casualty's mouth.
       (1) Maintain the head-tilt with the hand on the forehead while using the other hand to lift the casualty's jaw and close the mouth.
       (2) Take a deep breath and make an airtight seal around the casualty's nose with your mouth.
       (3) Blow one full breath (lasting 1 second) into the casualty's nose while watching for the chest to rise and fall and listening and feeling for air to escape during exhalation.

Performance Steps

NOTE: It may be necessary to open the casualty's mouth or separate the lips to allow air to escape.

        (4) If the chest rises--

           (a) Give a second full breath.

           (b) Go to step 6.

        (5) If the chest does not rise, go to step 4.

      c. Mouth-to-mask.

NOTE: The face mask is an important part of infection control to the rescuer. Rescuer breaths are delivered to the casualty through the one-way valve of the mask. There is no direct contact with the casualty's mouth.

NOTE: Kneel above the casualty's head to perform this.

        (1) Insert an airway adjunct as necessary.

        (2) Place the mask on the casualty.

           (a) Position the apex of the mask on the bridge of the nose.

           (b) Place the base of the mask at the chin between the lower lip and the chin prominence.

        (3) Create a seal while maintaining the airway.

           (a) Place your thumbs over the top half of mask.

           (b) Place your index and middle fingers over the bottom half of the mask.

           (c) Use your fourth and fifth fingers to bring the jaw toward the mask.

        (4) Take a deep breath and exhale into the mask.

NOTES: 1. Remove your mouth from the valve to allow for exhalation. 2. Some masks have oxygen inlets. Providing supplemental oxygen significantly increases the concentration of oxygen delivered to the casualty. Oxygen concentrations can reach 55% when the flow is set to 15 liters per minute (LPM).

           (a) If the breath goes in, give a second breath and go to step 6.

           (b) If the breath fails to go in, go to step 4.

      d. Bag-valve-mask (BVM) system.

NOTE: Supplemental oxygen can be given while using the BVM to increase oxygen concentration levels to 55%. When BVM systems have a reservoir supply, oxygen concentrations approach 100%.

NOTE: Kneel above the casualty's head to perform this.

        (1) Insert an airway adjunct as needed.

        (2) Select the proper size of mask.

        (3) Position the mask on the casualty's face.

        (4) Form a "C" around the ventilation port. Hold your index finger over the lower part of the mask and your thumb over the upper part of the mask. Use the third, fourth and fifth fingers under the casualty's jaw to hold the mask in place.

NOTE: The most difficult part of performing rescue breathing using the BVM system is maintaining an adequate seal. The American Heart Association recommends two rescuer BVM ventilation; in this method, one rescuer maintains a two-hand seal while the other rescuer squeezes the bag.

        (5) Squeeze the bag every 5 to 6 seconds. Deliver each breath over 1 second, just enough to produce visible chest rise.

        (6) Release pressure from the bag and allow the casualty to exhale passively.

           (a) If the chest rises and air goes in, squeeze the bag again to give a second breath and then go to step 6.

           (b) If the chest fails to rise, go to step 4.

Performance Steps
  e. Flow-restricted oxygen-powered ventilation device (FROPVD).
CAUTION: When using the FROPVD on casualties with chest injuries, be careful not to force excess air into the stomach instead of the lungs. This may cause gastric distention and vomiting. Do not use this device on infants or children.
    (1) Follow the same steps to position and seal the mask as with the BVM system.
    (2) Push the trigger/button on the device once.
      (a) If the chest rises, push the trigger/button again and proceed to step 6.
      (b) If the chest fails to rise, go to step 4.

 4. Reposition the head to ensure an open airway and attempt the breath again.
NOTE: When using a BVM system or FROPVD, it is also important to check the mask seal.
  a. If the chest rises, give another breath and go to step 6.
  b. If the chest does not rise, continue with step 5.

 5. Clear an airway obstruction, if necessary. (See task 081-831-0019.) When the obstruction has been cleared, continue with step 6.

 6. Check the carotid pulse for at least 5 seconds but no longer than 10 seconds.
  a. While maintaining the airway, place the index and middle fingers of your hand on the casualty's throat.
  b. Slide the fingers into the groove beside the casualty's Adam's apple and feel for a pulse for no longer than 10 seconds.
  c. If a pulse is present, go to step 7.
  d. If a pulse is not found, initiate CPR. (See task 081-831-0046.)

 7. Continue rescue breathing.
  a. Ventilate the casualty at the appropriate rate.
    (1) Adult: 12-20 breaths per minute.
    (2) Children (one year of age to the onset of puberty): 25-30 breaths per minute (mouth-to-mouth or mouth-to-nose).
    (3) Infants (less than one year of age): 25-50 breaths per minute (mouth-to-nose).
  b. Watch for rising and falling of the chest.

Evaluation Preparation:
Setup: For training and evaluation, a CPR mannequin must be used. Position the mannequin on its back. To test step 1, create a trauma or non-trauma scenario that will dictate which maneuver should be used. To test step 2, create a scenario in which the casualty's condition dictates which method is to be used. You may determine how much of the task is tested by telling the Soldier whether the airway is clear or a pulse is found as the Soldier proceeds through the task. However, you should ensure that the Soldier is routed through the task far enough to continue rescue breathing after checking the carotid pulse.

Brief Soldier: Tell the Soldier to perform rescue breathing on the casualty.

Performance Measures	GO	NO-GO
1. Positioned yourself at the casualty's head.	——	——
2. Opened the airway.	——	——
a. Used head-tilt/chin-lift maneuver (non trauma scenario).		
b. Used jaw thrust maneuver (trauma scenario).		

Performance Measures	GO	NO-GO
3. Ventilated the casualty using the mouth-to-mouth, mouth-to-nose, mouth-to-mask, BVM, or FROPVD method, as appropriate.	——	——
4. Repositioned the head to ensure an open airway and repeated ventilation attempt, if necessary.	——	——
5. Cleared an airway obstruction, if necessary.	——	——
6. Checked the carotid pulse for at least 5 seconds but no longer than 10 seconds.	——	——
7. Continued rescue breathing until breathing was restored or stopped IAW the task standard.	——	——
8. Completed all necessary steps in order.	——	——

Evaluation Guidance: Score each Soldier according to the performance measures. Unless otherwise stated in the task summary, the Soldier must pass all performance measures to be scored GO. If the Soldier fails any steps, show what was done wrong and how to do it correctly.

References
>   **Required**
>   None

    Related
    EMERG CARE AND TRANS 9

INSERT AN OROPHARYNGEAL AIRWAY (J TUBE)
081-833-0016

Conditions: You are assessing an unconscious casualty who requires insertion of an oropharyngeal airway (OPA). You will need three sizes of OPAs and gauze pads or tongue blades. You are not in a CBRNE environment.

Standards: Insert an OPA without causing further injury to the casualty.

Performance Steps

CAUTION: All body fluids should be considered potentially infectious. Always observe body substance isolation (BSI) precautions by wearing gloves and eye protection as a minimal standard of protection.

WARNING: Use an OPA for an unconscious casualty only. Do not use an OPA on a conscious or semiconscious casualty because he may still have an active gag reflex. In such cases, a nasopharyngeal airway (NPA) would be more appropriate. An OPA should not be used in children who may have ingested a caustic or petroleum-based product, as it may induce vomiting.

    1. Select the appropriate size of OPA.
        a. Place the airway beside the outside of the casualty's jaw.
        b. Measure from the casualty's ear lobe to the corner of the mouth.
NOTE: The measurement from the ear lobe to the corner of the casualty's mouth is equivalent to the depth of insertion in the airway.

    2. Perform the head-tilt/chin-lift or jaw thrust maneuver to open the airway. (See task 081-831-0018.)
WARNING: If a neck or spinal injury is suspected, use the jaw thrust maneuver to open the airway.

    3. Open the casualty's mouth.
        a. Place the crossed thumb and index finger of one hand on the casualty's upper and lower teeth at the corner of the mouth.
        b. Use a scissors motion to pry the casualty's teeth apart.
NOTE: If the teeth are clenched, wedge the index finger behind the casualty's back molars to open the mouth.

    4. Insert the OPA.
        a. Insert the airway with the tip facing the roof of the mouth.
        b. Slide the OPA along the roof of the mouth. Follow the natural contour of the tongue past the soft palate.
        c. Rotate the airway 180° as the tip reaches the back of the tongue.
NOTE: The airway may be difficult to insert. If so, use a gauze pad to pull the tongue forward or a tongue blade to depress the tongue.
        d. Gently advance the airway and adjust it so the flange rests against the casualty's lips or teeth.

Performance Steps
NOTES: 1. The tip of the airway should rest just above the epiglottis. 2. If the flange of the airway did not seat correctly on the lips or if the casualty gags, the airway may be the wrong size. Repeat the procedure using a different size of airway.

WARNING: If the casualty starts to regain consciousness and gags or vomits, remove the airway immediately.

   5. Insert the OPA using a tongue blade.
      a. Use the tongue blade to depress the tongue, ensuring the tongue remains forward.
      b. Insert the OPA sideways from the corner of the mouth until the flange reaches the teeth.
      c. Rotate the OPA at a 90° angle, removing the tongue blade as you exert gentle backward pressure on the OPA until it rests securely in place against the lips or teeth.

   6. Evacuate the casualty.
NOTE: The airway may need to be taped or tied in place to avoid dislodgement during evacuation. If so, the casualty must be constantly monitored for the return of consciousness.

   7. Did not cause further injury to the casualty.

Evaluation Preparation:
Setup: For training and evaluation, use a cardiopulmonary resuscitation (CPR) mannequin capable of accepting an OPA.

Brief Soldier: Tell the Soldier the simulated casualty is unconscious and breathing. The casualty does not have an active gag reflex. Tell the Soldier to insert an oropharyngeal airway.

Performance Measures	GO	NO-GO
1. Selected the appropriate size of OPA.	——	——
2. Performed the head-tilt/chin-lift or jaw thrust maneuver.	——	——
3. Opened the casualty's mouth using scissors technique.	——	——
4. Inserted the OPA.	——	——
5. Inserted the OPA using a tongue blade.	——	——
6. Evacuated the casualty.	——	——
7. Did not cause further injury to the casualty.	——	——

Evaluation Guidance: Score each Soldier according to the performance measures. Unless otherwise stated in the task summary, the Soldier must pass all performance measures to be scored GO. If the Soldier fails any steps, show what was done wrong and how to do it correctly.

References
    Required
    None

    Related
    EMERG CARE AND TRANS 9
    PHTLS

VENTILATE A PATIENT WITH A BAG-VALVE-MASK SYSTEM
081-833-0017

Conditions: You encounter an unconscious patient. The patient has signs of difficulty breathing. You must ventilate the patient with a bag-valve-mask (BVM) system. You will need a DD Form 1380 (U.S. Field Medical Card), oropharyngeal airway (OPA), BVM system, and supplemental oxygen (if available). You are not in a CBRNE environment.

Standards: Ventilate the patient with a BVM system until spontaneous breathing returns, until a normal rate and depth of respiration is achieved, or until directed to stop by a medical officer. Perform the procedure without causing further injury to the casualty.

Performance Steps
CAUTION: All body fluids should be considered potentially infectious. Always observe body substance isolation (BSI) precautions by wearing gloves and eye protection as a minimal standard of protection.

1. Insert an OPA if the patient is unconscious. (See task 081-833-0016.)
WARNING: Do not attempt to use an OPA on a conscious or semiconscious patient.

2. Assemble the BVM system, selecting the correct size of mask for the patient.

3. Ensure the bag is operational.
NOTE: An operational BVM should have a self-refilling bag, a non-rebreathing outlet valve, oxygen reservoir, a one-way inlet valve, and a transparent face mask.

4. Kneel above the patient's head facing the patient's feet.

5. Place the patient's head in an extended position unless you suspect a spinal injury.

6. Fit the mask to the patient.
   a. Stretch the mask with the thumb and index finger on both sides of the mask.
   b. Place the mask over the patient's face with the apex of the mask over the bridge of the patient's nose and the base of the mask in the groove between the lower lip and the chin to form a tight seal.
NOTE: As the stretched mask resumes its original shape, pull the patient's skin taut to help form a leak proof seal.

7. Ventilate the patient using the one-rescuer method, if appropriate.
   a. Hold the mask in place with one hand.
      (1) Place your little, ring, and middle fingers along the mandible.
      (2) Place your thumb on the upper portion of the mask above the valve connection.
      (3) Place your index finger on the lower portion of the mask under the valve connection.
NOTE: This is known as the "C-clamp" method and will maintain the seal.
   b. Maintain a leak proof mask seal with one hand. Use firm pressure to hold the mask in position and to maintain a seal on the patient's face.
   c. Continue squeezing the bag once every 5 seconds for an adult, once every 3 seconds for infants and children.

8. Ventilate the patient using the two-rescuer method, if appropriate.
   a. Hold the mask in place with two hands.
      (1) Place your little, ring, and middle fingers along the mandible.

Performance Steps

        (2) Place your thumb on the upper portion of the mask above the valve connection.
        (3) Place your index finger on the lower portion of the mask under the valve connection.
        (4) With your other hand, duplicate the above steps (mirror image) to achieve a leak proof seal.
    b. Have your assistant continue squeezing the bag with two hands until the chest rises; squeeze once every 5 seconds for an adult, once every 3 seconds for infants and children.

  9. Observe for rise and fall of the patient's chest.
    a. If the chest does not rise, reposition the airway.
    b. If the chest rises and falls, continue with step 10.

  10. Continue ventilations.
    a. Observe for spontaneous respirations.
    b. Periodically check the pulse.
    c. Observe for vomiting or secretions in or around the mouth or mask.

NOTE: If an oxygen source is available, it should be attached to the oxygen reservoir (at 10 to 15 L/min to increase the percentage of oxygen from 55% to approaching 90-100%).

  11. Continue ventilations until spontaneous breathing returns, until a more normal rate and depth of respiration is achieved, or until directed to stop by a medical officer.

  12. Document all medical care and procedures on the FMC.

  13. Evacuate the patient.

Evaluation Preparation:

Setup: For training and evaluation, use a cardiopulmonary resuscitation (CPR) mannequin capable of accepting an OPA. Have another Soldier act as an assistant. If oxygen will be used, prepare the oxygen source. Tell the Soldier if oxygen is to be used and whether the patient is conscious or unconscious. Have the Soldier insert an OPA, and ventilate the patient with a BVM using the one-rescuer and two-rescuer methods. After 2 minutes of ventilation, tell the Soldier that the patient has resumed normal breathing. Tell the assisting Soldier to only perform those actions the Soldier being evaluated directs.

Brief Soldier: Tell the Soldier to ventilate the patient with a BVM system.

Performance Measures	GO	NO-GO
1. Inserted an OPA if the patient was unconscious.	——	——
2. Assembled the BVM, selecting the correct size of mask for the patient.	——	——
3. Ensured that the bag was operational.	——	——
4. Knelt above the patient's head facing the patient's feet.	——	——
5. Placed the patient's head in an extended position (if no spinal injury is suspected).	——	——
6. Fit the mask to the patient.	——	——

Performance Measures	<u>GO</u>	<u>NO- GO</u>
7. Ventilated the patient using the one-rescuer method, if appropriate.	——	——
8. Ventilated the patient using the two-rescuer method, if appropriate.	——	——

9. Observed for rise and fall of the patient's chest.
   a. If the chest did not rise, repositioned the airway.
   b. If the chest rose and fell, continued with step 10.
	——	——

10. Continued ventilations.	——	——
11. Continued ventilations until spontaneous breathing returned, until a more normal rate and depth of respiration was achieved or until directed to stop by a medical officer.	——	——
12. Recorded the procedure on a FMC.	——	——
13. Evacuated the patient.	——	——
14. Did not cause further injury to the patient.	——	——

Evaluation Guidance: Score each Soldier according to the performance measures. Unless otherwise stated in the task summary, the Soldier must pass all performance measures to be scored GO. If the Soldier fails any steps, show what was done wrong and how to do it correctly.

References
  Required
  DD FORM 1380

                 Related
                 EMERG CARE AND TRANS 9
                 PHTLS

SET UP A D-SIZED OXYGEN TANK
081-833-0018

Conditions: You need to set up a D-sized oxygen tank. You have already performed a patient care hand-wash. You will need a full oxygen cylinder with a regulator/flowmeter, non-sparking cylinder wrench, oxygen regulator/flowmeter for D cylinders, yoke attachment, humidifier, sterile water, oxygen cylinder transport carrier and/or stand oxygen, oxygen administration device, and warning signs. You are not in a CBRNE environment.

Standards: Set up the oxygen tank without violating safety precautions or endangering patients or yourself.

Performance Steps
CAUTION: All body fluids should be considered potentially infectious. Always observe body substance isolation (BSI) precautions by wearing gloves and eye protection as a minimal standard of protection.

   1. Obtain the necessary equipment.
      a. Oxygen cylinder. Refer to Figure 3-8.

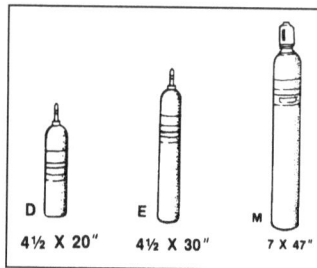

Figure 3-8. Oxygen cylinder

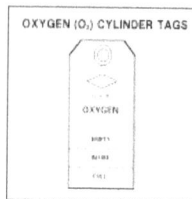

Figure 3-9. Oxygen cylinder tag

NOTE: Check the oxygen cylinder tag to determine whether the tank is "FULL", "IN USE" (partially full), or "EMPTY". Refer to Figure 3-9.

CAUTION: Always ensure that the cylinder selected contains oxygen and not some other compressed gas. United States oxygen cylinders are color coded green, silver or chrome with a green area around the valve stem on top. The international color code is white.

Performance Steps
    b. Cylinder with regulator/flowmeter. Refer to Figure 3-10.

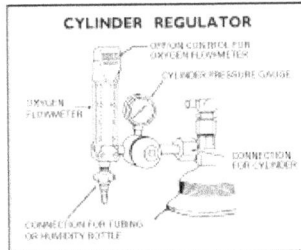

Figure 3-10. Cylinder with regulator

NOTES: 1. When the cylinder regulator pressure gauge reads 200 psi or lower, the oxygen tank is considered empty. 2. The pressure-compensated flowmeter is affected by gravity and must be maintained in an upright position. The Bourdon gauge flowmeter is not affected by gravity and can be used in any position.
    c. Humidifier.
    d. Sterile water.
    e. Non-sparking cylinder wrench.
    f. Oxygen cylinder transport carrier and/or stand.
    g. Oxygen administration device appropriate for the patient or as ordered by the medical officer (nasal cannula, non-rebreather mask, or BVM device with reservoir).
    h. Warning signs.
      (1) "NO SMOKING".
      (2) "OXYGEN IN USE".
CAUTION: Because of the extreme pressure in oxygen tanks, they should be handled with great care. Do not allow tanks to be banged together, dropped, or knocked over.

  2. Secure the oxygen cylinder.
    a. Upright position or IAW local SOP.
    b. Secured with straps or in a stand.
    c. Away from doors and areas of high traffic.

  3. Remove the cylinder valve cap.
NOTE: The cylinder valve cap may be noisy or difficult to remove; however, the threads of the cylinder cap should never be oiled.

  4. Use either the hand wheel or a non-sparking wrench to "crack" (slowly open and quickly close) the cylinder to flush out any debris.

  5. Attach the regulator/flowmeter to the cylinder.
    a. D cylinder.
      (1) Locate the three holes on the oxygen cylinder stem and ensure that an "O" ring is present. Refer to Figure 3-11.

Performance Steps

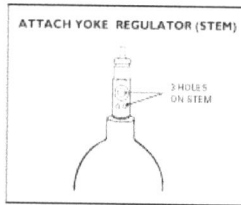

Figure 3-11. Three holes on the oxygen cylinder stem

NOTE: If the "O" ring is not present, an oxygen leak will occur.

    (2) Examine the yoke attachment and locate the three corresponding pins on the yoke attachment. Refer to Figure 3-12.

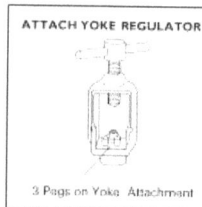

Figure 3-12. Three corresponding pins

NOTE: The compressed gas industry uses a "pin-indexing system" for portable gas cylinders. The locations of the pins on the yoke match only the regulator/flowmeter for an oxygen cylinder.

    (3) Slide the yoke attachment over the cylinder stem, ensuring that the pins are seated in the proper holes.

    (4) Turn the vise-like screw on the side of the yoke attachment to secure it.

    (5) Open the valve to test for leaks, and then close it.

NOTES: 1. If there is a leak, check the regulator connection and obtain a new regulator/flowmeter and/or cylinder, if necessary. 2. When in-wall oxygen is available, the flowmeter will be attached to the oxygen outlet as follows: a. Turn the flow adjusting valve of the flowmeter to the OFF position. b. Insert the flowmeter adapter into the opening outlet and press until a firm connection is made.

  6. Fill the humidifier bottle to the level indicated (about two-thirds full) with sterile water.

  7. Attach the humidifier to the flowmeter.

NOTE: If an oxygen tube connector adapter is present, remove it from the flowmeter by turning the wing nut.

    a. Attach the humidifier to the flowmeter with the wing nut on the humidifier.

NOTE: Not all humidifiers have "wing" style nuts. Some have regular "bolt" style nuts.

    b. Secure the nut by hand-tightening it.

NOTE: Humidifiers and tubing should be changed at least once every 24 hours (or more often IAW local SOP).

Performance Steps

8. Post warning signs.
CAUTION: "OXYGEN" and "NO SMOKING" signs should be posted in the areas where oxygen is in use or stored.

9. Report and/or record completion of the procedure.

Evaluation Preparation:
Setup: Place all necessary materials and equipment including a full oxygen cylinder, non-sparking cylinder wrench, appropriate oxygen regulator/flowmeter for designated oxygen cylinder, yoke attachment, humidifier, sterile water, oxygen administration devices, and warning signs. Create a trauma or non trauma scenario that will dictate which oxygen administration device should be applied.

Brief Soldier: Tell the Soldier to assemble the oxygen system to include appropriate oxygen administration device.

Performance Measures	GO	NO-GO
1. Obtained the necessary equipment.	——	——
2. Secured the oxygen cylinder.	——	——
3. Removed the cylinder valve cap.	——	——
4. Used either the hand wheel or a non-sparking wrench to "crack" (slowly open and quickly close) the cylinder to flush out any debris.	——	——
5. Attached the regulator/flowmeter to the cylinder.	——	——
6. Filled the humidifier bottle to the level indicated (about two-thirds full) with sterile water.	——	——
7. Attached the humidifier to the flowmeter.	——	——
8. Posted warning signs.	——	——
9. Reported and/or recorded completion of the procedure.	——	——

Evaluation Guidance: Score each Soldier according to the performance measures. Unless otherwise stated in the task summary, the Soldier must pass all performance measures to be scored GO. If the Soldier fails any steps, show what was done wrong and how to do it correctly.

References
    Required   **Related**
    None   EMERG CARE AND TRANS 9
                             PHTLS

PERFORM ORAL AND NASOPHARYNGEAL SUCTIONING OF A PATIENT
081-833-0021

Conditions: You are managing a patient that requires suctioning. You will need a portable suction apparatus, suction tubing, a rigid or flexible suction catheter, saline solution, a basin, and a collection bottle. You are not in a CBRNE environment.

Standards: Perform oral or nasopharyngeal suctioning to clear the airway without causing injury to the patient.

Performance Steps
CAUTION: All body fluids should be considered potentially infectious. Always observe body substance isolation (BSI) precautions by wearing gloves and eye protection as a minimal standard of protection.

1. Position the patient in a semi-Fowler's (semi-sitting) position or, in the case of severe trauma, roll the patient onto his side to allow gravity to assist in clearing the airway.
NOTE: In some cases, such as spinal injuries, the patient must remain in whatever position they are initially found in or must be managed while they are immobilized on a long spine board.

2. Check the suction unit for proper assembly of all its parts.

3. Turn on the assembled unit and check to see if it is operational.
NOTE: Inspect the suction unit regularly to ensure it is in working condition. Switch on the suction, clamp the tubing, and make certain the unit generates a vacuum of more than 300 mm Hg. Check that a battery-charged unit has charged batteries.

4. Select the appropriate catheter and attach it to the suction tubing.
   a. Tonsil-tip (Yankauer) catheters are best for suctioning in the field, as they have wide diameter tips and are somewhat rigid.
   b. Flexible (French, or whistle-tip) catheters are used in situations where rigid catheters cannot be used, such as a patient with clenched teeth or for use in nasopharyngeal suctioning.

5. Prepare equipment.
   a. Open the basin package.
   b. Pour the saline solution into the basin.
   c. Open the suction catheter package.

6. Explain to the patient the reason for suctioning.

7. Pre-oxygenate the patient with 100% oxygen.
   a. If the patient is receiving oxygen therapy, increase the oxygen to 100% for 1 minute.
   b. Monitor the patient's pulse oximeter reading during the entire procedure. (See task 081-833-0164.)
   c. If the patient is not receiving oxygen therapy, have him take a minimum of five deep breaths or administer the breaths with a bag-valve-mask (BVM) system.
NOTE: After each suctioning attempt or suctioning period, re-oxygenate the patient.

8. Remove the catheter from the package using your dominant hand.

9. Test the patency of the catheter.
   a. Turn the suction unit on with your nondominant hand.

Performance Steps
  b. Insert the catheter tip into the saline solution using your dominant hand.
  c. Occlude the suction control port with your nondominant thumb and observe the saline entering the drainage bottle.
NOTE: If no saline enters the bottle, check the suction unit and/or replace the catheter and retest for patency.

10. Suction the patient.
  a. Oral route.
    (1) Rigid catheter.
      (a) Instruct a conscious patient to cough to help bring secretions up to the back of his throat.
      (b) If the patient is unconscious, use the cross finger method of opening the airway. (See task 081-831-0019.)
      (c) Place the convex (outward curving) side of the rigid tip against the roof of the mouth and insert to the base of the tongue.
NOTE: A rigid tip does not need to be measured. Only insert the tip as far as you can see it. Be aware that advancing the catheter too far may stimulate the patient's gag reflex and cause him to vomit.
      (d) Apply suction by placing the thumb of your nondominant hand over the suction control port.
WARNING: Never suction for more than 15 seconds at one time for adults, 10 seconds for children, and 5 seconds for infants. Longer periods of continuous suctioning may cause oxygen deprivation.
      (e) Clear the secretions from the catheter between each suctioning interval by inserting the tip into the saline solution and suction the solution through the catheter until the catheter is clear of secretions.
      (f) Repeat steps 10a(1)(a) through 10a(1)(e) until all secretions have been removed or until the patient's breathing becomes easier. Noisy, rattling or gurgling sounds should no longer be heard.
    (2) Flexible catheter.
      (a) Measure the catheter from the patient's earlobe to the corner of the mouth or the center of the mouth to the angle of the jaw.
      (b) Insert the catheter into the patient's mouth to the correct depth, without the suction applied.
NOTE: If an oropharyngeal airway (OPA) is in place, insert the catheter alongside the airway and then back into the pharynx.
      (c) Place the thumb of your nondominant hand over the suction control port on the catheter, applying intermittent suction by moving your thumb up and down over the suction control port.
      (d) Apply suction in a circular motion as you withdraw the catheter.
      (e) Suction for no longer than 15 seconds removing secretions from the back of the throat, along outer gums, cheeks, and base of tongue.
WARNING: Advancing the catheter too far into the back of the patient's throat may stimulate the gag reflex. This could cause vomiting and the aspiration of stomach contents.
      (f) Clear the secretions from the catheter between suctioning by inserting the tip into the saline solution and suction the solution through the catheter until the catheter is clear of secretions.
      (g) Repeat steps 10a(2)(a) through 10a(2)(f) until all secretions have been removed or until the patient's breathing becomes easier. Noisy, rattling or gurgling sounds should no longer be heard.

Performance Steps

NOTE: If the patient is uncooperative or oral entry is not possible due to facial trauma, nasopharyngeal suctioning may be required.

      b. Nasopharyngeal route.

          (1) Measure the flexible catheter from the tip of the earlobe to the nose.

          (2) Lubricate the catheter by dipping the tip into the saline solution.

          (3) Insert the catheter into one nostril without suction applied. If an obstruction is met, try the other nostril.

          (4) Quickly and gently advance the catheter 3 to 5 inches.

          (5) Perform steps 10a(2)(c) through 10a(2)(e) to suction secretions.

11. Re-oxygenate the patient and/or ventilate for at least five assisted ventilations.

12. Observe the patient for hypoxemia.

    a. Color change.

    b. Increased or decreased pulse rate.

WARNING: Discontinue suctioning immediately if severe changes in color or pulse rate occur.

13. Place the patient in the recovery (lateral recumbent, coma) position.

14. Record the procedure.

Evaluation Preparation:

Setup: For training and evaluation, use a cardiopulmonary resuscitation (CPR) mannequin capable of accepting oral and nasopharyngeal suction catheters.

Brief Soldier: Tell the Soldier the simulated patient is conscious and there are gurgling noises whenever the patient attempts to breathe. The patient has an active gag reflex. Tell the Soldier to suction the airway.

Performance Measures	GO	NO-GO
1. Positioned the patient.	____	____
2. Checked the suction unit for proper assembly.	____	____
3. Turned on the assembled unit and checked to see if it is operational.	____	____
4. Selected the appropriate catheter and attached it to the suction tubing.	____	____
5. Prepared the equipment.	____	____
6. Explained to the patient the reason for suctioning.	____	____
7. Pre-oxygenated the patient with 100% oxygen.	____	____
8. Removed the catheter from the package.	____	____
9. Tested the patency of the catheter.	____	____
10. Suctioned the patient.	____	____
11. Re-oxygenated and/or ventilated the patient.	____	____

Performance Measures	<u>GO</u>	<u>NO-GO</u>
12. Observed the patient for hypoxemia.	——	——
13. Placed the patient in the recovery position.	——	——
14. Recorded the procedure.	——	——
15. Did not cause further injury to the patient.	——	——

Evaluation Guidance: Score each Soldier according to the performance measures. Unless otherwise stated in the task summary, the Soldier must pass all performance measures to be scored GO. If the Soldier fails any steps, show what was done wrong and how to do it correctly.

References
 Required
 None

        Related
        EMERG CARE AND TRANS 9
        PHTLS

INSERT A NASOPHARYNGEAL AIRWAY
081-833-0142

Conditions: You are assessing a patient with a reduced level of consciousness who is unable to maintain his airway. You must insert a nasopharyngeal airway. You will need a nasopharyngeal airway (NPA), water-soluble lubricant, and DD Form 1380 (U.S. Field Medical Card). You are not in a CBRNE environment.

Standards: Insert the appropriate size of NPA, without causing further injury to the patient.

Performance Steps
CAUTION: All body fluids should be considered potentially infectious. Always observe body substance isolation (BSI) precautions by wearing gloves and eye protection as a minimal standard of protection.

   1. Place the patient supine with the head in a neutral position.
CAUTION: Do not use a NPA if the patient has maxillofacial or head trauma.

   2. Select the appropriate size NPA by measuring the tip of the patient's nose to earlobe.

   3. Coat the distal tip (non-flanged end) of the NPA with a water-soluble lubricant.
CAUTION: Do not use a petroleum-based or non-water-based lubricant. These substances can cause damage to the tissues lining the nasal cavity and pharynx thus increasing the risk for infection.

   4. Insert the NPA.
      a. Push the tip of the nose upward gently.
      b. Position the tube so that the bevel of the airway faces toward the septum.
NOTE: Most NPAs are designed to be placed in the right nostril.
      c. Gently advance the lubricated NPA into the nostril with the curvature of the device following the curve of the floor of the nose. Advance it until the flange rests against the nostril.
CAUTION: Never force the NPA into the patient's nostril. If resistance is met, pull the tube out and attempt to insert it in the other nostril. If the patient becomes intolerant of the airway, gently withdraw it from the nasal passage.

NOTE: NPA insertion may cause nasal bleeding.

   5. Place the patient in the recovery (lateral recumbent, coma) position to prevent aspiration of blood, mucus, or vomitus.

   6. Record the procedure on a FMC.

Evaluation Preparation:
Setup: For training and evaluation, use a cardiopulmonary resuscitation (CPR) mannequin capable of accepting a NPA.

Brief Soldier: Tell the Soldier the simulated patient is unconscious and unable to maintain his airway. The patient has an active gag reflex. Tell the Soldier to insert a nasopharyngeal airway.

Performance Measures	<u>GO</u>	<u>NO- GO</u>
1. Positioned the patient.	——	——
2. Measured and selected the appropriate size of NPA.	——	——
3. Lubricated the NPA.	——	——
4. Inserted the NPA.	——	——
5. Placed the patient in the recovery position.	——	——
6. Recorded the procedure on the FMC.	——	——

Evaluation Guidance: Score each Soldier according to the performance measures. Unless otherwise stated in the task summary, the Soldier must pass all performance measures to be scored GO. If the Soldier fails any steps, show what was done wrong and how to do it correctly.

References
    Required
    None

                     Related
                     EMERG CARE AND TRANS 9
                     PHTLS

ADMINISTER OXYGEN
081-833-0158

Conditions: You have a patient requiring oxygen administration. You will need an oxygen tank, regulator/flowmeter, water, non-rebreather (NRB) mask with extension tubing, and a nasal cannula. You are not in a CBRNE environment.

Standards: Administer oxygen therapy using a NRB mask or nasal cannula to assist the patient's breathing without causing further harm to the patient. Calculate the duration of flow of the oxygen.

Performance Steps
CAUTION: All body fluids should be considered potentially infectious. Always observe body substance isolation (BSI) precautions by wearing gloves and eye protection as a minimal standard of protection.

1. Explain the procedure to the patient.

2. Assemble and prepare the equipment.
   a. Inspect the oxygen cylinder and its markings.
NOTE: Ensure the cylinder is labeled for medical oxygen; the bottles may be completely green, silver, or chrome with a green area around the valve stem on top.
   b. Attach the regulator/flowmeter.
   c. Open the oxygen cylinder.
   d. Check for leaks.
   e. Check oxygen cylinder pressure.
NOTE: The safe residual level of the oxygen at which the tank should be replaced has been established to be 200 pounds per square inch (psi).

3. Position the patient in the position of comfort to facilitate his breathing unless contraindicated by the mechanism of injury (MOI).

4. Determine the delivery device to use.
NOTE: Humidifiers can be connected to flowmeters to provide moisture to dry oxygen; oxygen can dry out mucous membranes with prolonged use. Humidified oxygen is usually more comfortable to the patient and is particularly helpful for children and for chronic obstructive pulmonary disease (COPD) patients.
   a. A bag-valve-mask (BVM) system is the delivery device of choice for patients with signs of inadequate breathing. (See task 081-833-0017.)
   b. A NRB mask is usually the delivery device of choice in the prehospital setting for patients with signs of inadequate breathing, or who are cyanotic, have chest pain, severe trauma, signs of shock, or an altered mental status.
   c. A nasal cannula is appropriate for patients unable to tolerate the NRB.

5. Apply the NRB mask.
   a. Select the correct size of mask.
NOTE: The apex of the mask should fit over the bridge of the patient's nose and extend to rest on the chin, covering the mouth and nose completely. NRB masks come in different sizes for adults, children, and infants.
   b. Attach the extension tubing to the regulator/flowmeter.
   c. Initiate the oxygen flow and adjust it to the prescribed rate of 10-15 liters/minute (LPM) to deliver up to 90% oxygen.

Performance Steps

       d. Pre-fill the reservoir bag using your fingers to cover the connection between the mask and the reservoir, if applicable.

       e. Place the mask on the patient and adjust the straps.

       f. Instruct the patient to breathe normally.

6. Apply the nasal cannula.

       a. Attach the cannula tubing to the regulator/flowmeter.

       b. Adjust the oxygen flow to the prescribed rate of 1-4 LPM to deliver 24-44% oxygen.

       c. Position the cannula so the two small, tube-like prongs fit in the patient's nostrils.

       d. Adjust the nasal cannula to hold in place.

7. Continue to monitor the patient for signs of confusion, restlessness, level of consciousness, skin color, or changes in vital signs.

8. Check the equipment for security of tubing connections and administration device, oxygen flow, and humidified water level as indicated.

NOTE: Change the delivery device and tubing every 24 hours, or more often IAW local protocols. Humidifier water should be changed every shift, or more often IAW local protocols.

9. Calculate the duration of flow of the oxygen cylinder.

       a. Determine the remaining pressure in the tank by reading the regulator gauge.

       b. Determine the safe residual level of the oxygen tank.

NOTE: The safe residual level of the oxygen at which the tank should be replaced has been established to be 200 psi.

       c. Determine the available cylinder pressure by subtracting the safe residual level from the remaining pressure. Example: 2000 psi remaining pressure minus 200 psi safe residual level = 1800 psi available pressure.

       d. Determine the conversion factor for the oxygen cylinder in use.

NOTE: Each type of oxygen cylinder, depending on its size, employs a specific conversion factor.

          (1) D size oxygen cylinder--0.16.

          (2) E size oxygen cylinder--0.28.

          (3) G size oxygen cylinder--2.41.

          (4) H size oxygen cylinder--3.14.

          (5) K size oxygen cylinder--3.14.

          (6) M size oxygen cylinder--1.56.

       e. Determine the available liters by multiplying the conversion factor by the amount of available pressure. Example: A "D" size cylinder is being used. A 0.16 conversion factor x 1800 psi available pressure = 288 liters of oxygen available for use.

       f. Determine the flow rate as prescribed by medical direction.

       g. Determine the duration of the oxygen by dividing the available liters by the flow rate. Example: 288 available liters divided by the prescribed flow rate of 10 LPM = 28.8 (29) minutes duration of oxygen flow.

10. Follow safety precautions.

       a. Ensure "OXYGEN" and "NO SMOKING" signs are posted wherever oxygen is used or stored.

       b. Inform the patient and visitors about the restrictions.

WARNING: The principle danger in using oxygen is fire. The presence of oxygen in increased concentrations makes all materials more combustible. Materials that burn slowly in ordinary air, burn violently and even explosively in the presence of oxygen.

       c. Use only non-sparking wrenches on oxygen cylinders.

Performance Steps
    d. Ensure all electrical equipment is properly grounded.
    e. Position oxygen cylinders away from doors and high traffic areas.
    f. Do not use oil or grease around oxygen fittings.
    g. Secure and store oxygen cylinders in an upright position.

Evaluation Preparation:
Setup: For training and evaluation, have another Soldier act as the patient; tell the patient not to assist the Soldier in any way. Place the unassembled oxygen cylinder and regulator/flowmeter, along with a non-sparking wrench, NRB mask, extension tubing, and nasal cannula next to the patient. Once the NRB mask has been placed on the patient, inform the Soldier that the patient is not able to tolerate the NRB mask. The Soldier should then substitute the nasal cannula for the NRB.

Brief the Soldier: Tell the Soldier to assemble and prepare the equipment and administer oxygen with the proper delivery device to a patient requiring oxygen therapy.

Performance Measures	GO	NO-GO
1. Explained the procedure to the patient.	____	____
2. Assembled and prepared the oxygen equipment.	____	____
3. Positioned the patient.	____	____
4. Selected the proper NRB and adjusted the oxygen flow appropriately.	____	____
5. Properly applied the NRB delivery device to the patient.	____	____
6. Applied the nasal cannula.	____	____
7. Monitored the patient.	____	____
8. Checked the equipment.	____	____
9. Calculated the duration of flow of the oxygen cylinder correctly.	____	____
10. Followed safety precautions.	____	____

Evaluation Guidance: Score each Soldier according to the performance measures. Unless otherwise stated in the task summary, the Soldier must pass all performance measures to be scored GO. If the Soldier fails any steps, show what was done wrong and how to do it correctly.

References
    Required   **Related**
    None   EMERG CARE AND TRANS 9

INSERT A COMBITUBE
081-833-0169

Conditions: An unconscious casualty requires the insertion of an esophageal tracheal Combitube. An assistant is performing resuscitative measures. No cervical spine injury is present. You will need a Combitube, 140 cc syringe, 10 cc syringe, gloves, eye protection, suction equipment, stethoscope, pulse oximeter, and bag-valve-mask (BVM). You are not in a CBRNE environment.

Standards: Insert the Combitube and successfully ventilate the casualty without causing further injury.

Performance Steps

1. Take body substance isolation (BSI) precautions.

2. Inspect upper airway for visible obstruction.

3. Inspect and test equipment.

4. Verbalize lubricating distal end of tube.

5. Perform a tongue-jaw lift.

6. Insert device until casualty's teeth sit between printed black rings, within 3 attempts.

7. Inflate #1 (blue) cuff with appropriate amount of air based on size of tube.

8. Inflate #2 (white) cuff with appropriate amount of air based on size of tube.

9. Direct assistant to ventilate casualty with a BVM through primary tube.

10. Perform steps 5-9 in less than 30 seconds.

11. Watch for rise and fall of the chest, auscultate for breath sounds and over the epigastrium to confirm tube placement.

12. Assess casualty for spontaneous respirations. (For 10 seconds.)

13. Attach pulse oximeter to casualty, if available.

14. Assist when respirations are <8 or >30 or a pulse oximeter reading <90%.

15. Secure device to the casualty around casualty's neck.

Performance Measures	GO	NO-GO
1. Took BSI precautions.	——	——
2. Inspected upper airway for visible obstruction.	——	——
3. Inspected and tested equipment.	——	——
4. Verbalized lubricating distal end of tube.	——	——
5. Performed a tongue-jaw lift.	——	——

Performance Measures	GO	NO-GO
6. Inserted device until casualty's teeth were between printed black rings, within 3 attempts.	——	——
7. Inflated #1 (blue) cuff with appropriate amount of air based on size of tube.	——	——
8. Inflated #2 (white) cuff with appropriate amount of air based on size of tube.	——	——
9. Directed assistant to ventilate casualty with a BVM through primary tube.	——	——
10. Performed steps 5-9 in less than 30 seconds.	——	——
11. Watched for rise and fall of the chest, auscultated for breath sounds and over the epigastrium to confirm tube placement.	——	——
12. Assessed casualty for spontaneous respirations. (For 10 seconds.)	——	——
13. Attached pulse oximeter to casualty, if available.	——	——
14. Assisted when respirations were <8 or >30 or a pulse oximeter reading <90%.	——	——
15. Secured device to the casualty around casualty's neck.	——	——
16. Did not cause further injury.	——	——

Evaluation Guidance: Score each Soldier according to the performance measures. Unless otherwise stated in the task summary, the Soldier must pass all performance measures to be scored GO. If the Soldier fails any steps, show what was done wrong and how to do it correctly.

References

Required	**Related**
None	PHTLS

PERFORM ENDOTRACHEAL SUCTIONING OF A PATIENT
081-833-0170

Conditions: You have done an assessment and determined your patient needs suctioning. You have already done a patient care hand-wash. You will need suction unit, suction catheter, sterile basin, sterile water, sterile gloves, or a disposable suction kit. You are not in a CBRNE environment.

Standards: Perform endotracheal suctioning without violating aseptic technique or causing injury to the patient.

Performance Steps

1. Explain the procedure to the patient.

2. Position the patient in the semi-Fowler's position.
NOTE: In some cases, such as spinal injuries, the patient will have to remain in whatever position he is in at the time.

3. Check the pressure on the suction apparatus.
   a. Turn the unit on, place a thumb over the end of the suction connecting tube, and observe the pressure gauge.
   b. Ensure that the pressure reading is within the limits specified by local SOP and the recommendations of the equipment manufacturer.
   c. Notify the supervisor if the pressure is not within the recommended limits.
   d. Turn the unit off after verifying the correct pressure.
WARNING: If the suction pressure is too low, the secretions cannot be removed. If the pressure is too high, the mucous membranes may be forcefully pulled into the catheter opening.

4. Prepare the sterile materials. (See task 081-833-0007.)
   a. Open the sterile solution basin package on the bedside stand or table to create a sterile field.
   b. Pour sterile saline solution into the basin.
   c. Open the suction catheter package to expose the suction port of the catheter.
   d. Open the sterile glove package.
NOTE: Disposable suctioning kits contain the same items.

5. Pre-oxygenate the patient.
   a. Pre-oxygenate the patient for 1 to 2 minutes.
   b. Monitor the patient's pulse oximeter reading during the entire procedure. (See task 081-833-0164.)

6. Put on sterile gloves. (See task 081-831-0008.)

7. Remove the catheter from the package using the dominant hand, keeping the catheter coiled to prevent contamination.
NOTE: This hand must remain sterile.

8. Measure the length of the suction catheter so that it will be approximately at the carina.
CAUTION: Do not touch the patient while measuring the length of the catheter. This will violate aseptic technique.
   a. Tip of catheter to the ear.
   b. From the ear to the nipple line.

Performance Steps

9. Attach the tubing to the catheter with the nondominant hand.
NOTE: This hand does not have to remain sterile. The glove is for your protection.

10. Test the patency of the catheter.
    a. Turn the suction unit on with the non-sterile hand.
    b. Insert the catheter tip into the sterile saline solution using the sterile hand.
    c. Place the non-sterile thumb over the suction port to create suction. Observe the saline entering the drainage bottle.
NOTE: If no saline enters the bottle, check the suction unit and/or replace the catheter and retest for patency.

11. Suction the patient.
    a. Remove the oxygen delivery device with the nondominant hand.
    b. Lubricate the catheter tip by dipping it into the saline solution.
    c. Gently insert the catheter into the airway to the measured length without suctioning.
    d. Apply intermittent suction by placing and releasing the nondominant hand over the vent of the catheter while withdrawing the catheter in a twisting motion.
CAUTION: Do not suction any longer than 15 seconds.
    e. Replace the oxygen delivery device and re-oxygenate the patient.
    f. Repeat steps 10a through 10e until secretions are removed.

12. Observe the patient for hypoxemia.
WARNING: Discontinue suctioning immediately if severe changes in color or pulse rate occur.

13. Disconnect the catheter and remove the gloves.
    a. Hold the catheter in one hand.
    b. Remove that glove by turning it inside out over the catheter to prevent the spread of contaminants.
    c. Remove the other glove.
    d. Discard them in contaminated trash.

14. Make the patient comfortable.

15. Discard, or clean and store, used items.

16. Record the procedure on the appropriate form.
    a. Respirations (rate and breath sounds before and after suctioning).
    b. Type and amount of secretions.
    c. Patient's toleration of the procedure.

Performance Measures	GO	NO-GO
1. Explained the procedure to the patient.	____	____
2. Positioned the patient.	____	____
3. Checked the pressure on the suctioning apparatus.	____	____
4. Prepared the sterile materials.	____	____
5. Pre-oxygenated the patient.	____	____
6. Put on sterile gloves.	____	____

Performance Measures	GO	NO-GO
7. Removed the catheter from the package using the dominant hand, keeping the catheter coiled to prevent contamination.	——	——
8. Measured the length of the suction catheter so that it will be approximately at the carina.	——	——
9. Attached the tubing to the catheter using the nondominant hand.	——	——
10. Tested the patency of the catheter.	——	——
11. Suctioned the patient.	——	——
12. Observed the patient for hypoxemia.	——	——
13. Disconnected the catheter and removed the gloves.	——	——
14. Made the patient comfortable.	——	——
15. Discarded, or cleaned and stored, used items.	——	——
16. Recorded the procedure on the appropriate form.	——	——
17. Did not violate aseptic technique, or cause further injury to the patient.	——	——

Evaluation Guidance: Score each Soldier according to the performance measures. Unless otherwise stated in the task summary, the Soldier must pass all performance measures to be scored GO. If the Soldier fails any steps, show what was done wrong and how to do it correctly.

References
Required
None

Related
PHTLS

INSERT A KING LT
081-833-0230

Conditions: An unconscious casualty requires the insertion of an esophageal King LT. An assistant is performing resuscitative measures. No cervical spine injury is present. You will need a King LT, syringe provided in kit based on size, gloves, stethoscope, and bag-valve-mask (BVM). You are not in a CBRNE environment.

Standards: Insert the King LT without causing further injury.

Performance Steps

1. Take body substance isolation (BSI) precautions.

2. Inspect the upper airway for visible obstruction.

3. Direct the assistant to pre-oxygenate the casualty for a minimum of 30 seconds.

4. Inspect and test equipment.

5. Lubricate the distal end of the tube.

6. Perform a tongue-jaw lift.

7. Insert the device until the base connector is aligned with the casualty's teeth.

8. Inflate the cuffs with the appropriate amount of air based on the size of the tube.
    a. Use size 3 if the casualty is less than 61 inches in height. Inflate with 60 ml of air.
    b. Use size 4 if the casualty is 61 inches to 71 inches in height. Inflate with 80 ml of air.
    c. Use size 5 if the casualty is taller than 71 inches in height. Inflate with 80 ml of air.

9. Direct the assistant to ventilate the casualty with a BVM.

10. Auscultate the lung fields and epigastrium, and watch for rise and fall of the chest to confirm tube placement.

11. Assess casualty for spontaneous respirations, if any, for 10 seconds.

12. Attach pulse oximeter to casualty, if available.

13. Ventilate casualty when respirations are <8 or > 30 or a pulse oximeter reading <90%.

14. Secure the device to the casualty.

Performance Measures	**GO**	**NO-GO**
1. Took BSI precautions.	___	___
2. Inspected the upper airway for visible obstruction.	___	___
3. Directed the assistant to pre-oxygenate the casualty for a minimum of 30 seconds.	___	___
4. Inspected and tested the equipment.	___	___
5. Lubricated the distal end of the tube.	___	___

Performance Measures	GO	NO-GO
6. Performed a tongue-jaw lift.	——	——
7. Inserted the device until the base connector was aligned with the casualty's teeth.	——	——
8. Inflated the cuffs with the appropriate amount of air based on the size of the tube.	——	——
9. Directed the assistant to ventilate the casualty with a BVM.	——	——
10. Auscultated the lung fields and epigastrium, and watched for rise and fall of the chest to confirm tube placement.	——	——
11. Assessed casualty for spontaneous respirations, if any, for 10 seconds.	——	——
12. Attached pulse oximeter to casualty, if available.	——	——
13. Ventilated casualty when respirations were <8 or > 30 or a pulse oximeter reading <90%.	——	——
14. Secured the device to the casualty.	——	——

Evaluation Guidance: Score each Soldier according to the performance measures. Unless otherwise stated in the task summary, the Soldier must pass all performance measures to be scored GO. If the Soldier fails any steps, show what was done wrong and how to do it correctly.

References
    Required
    None

    Related
    ISBN 0-07-065351-8

PERFORM A SURGICAL CRICOTHYROIDOTOMY
081-833-3005

Conditions: You have a casualty requiring a surgical cricothyroidotomy. You will need a cutting instrument (scalpel, knife blade), airway tube (endotracheal (ET) tube, tracheotomy tube, or any non collapsible tube, suctioning apparatus, alcohol swabs, knife handle, gloves, and tape. You are not in a CBRNE environment.

Standards: Perform a surgical cricothyroidotomy without causing unnecessary injury to the casualty.

Performance Steps
CAUTION: Casualties with a total upper airway obstruction, inhalation burns, or massive maxillofacial trauma who cannot be ventilated by other means are candidates for a surgical cricothyroidotomy.

　1. Gather cricothyroidotomy kit or minimum essential equipment.
NOTE: Because of the need for speed, every medic should have an easily accessible cricothyroidotomy kit that contains all required items.
　　a. Cutting instrument: number 10 or 15 scalpel or knife blade.
　　b. Airway tube: ET tube, tracheotomy tube, or any noncollapsible tube that will allow enough airflow to maintain oxygen saturation.
NOTE: In a field setting, an ET tube is preferred because it is easy to secure. Use a size 6.0 to 7.0 ET tube, and ensure the cuff will hold air.

　2. Hyperextend the casualty's neck.
WARNING: Do not hyperextend the casualty's neck if a cervical injury is suspected.
　　a. Place the casualty in the supine position.
　　b. Place a blanket or poncho rolled up under the casualty's neck or between the shoulder blades to hyperextend the neck.

　3. Put on gloves.

　4. Locate the cricothyroid membrane.
　　a. Place a finger of the nondominant hand on the thyroid cartilage (Adam's apple), and slide the finger down to the cricoid cartilage.
　　b. Palpate for the soft cricothyroid membrane below the thyroid cartilage and just above the cricoid cartilage.
　　c. Slide the index finger down into the depression between the thyroid and cricoid cartilage.
　　d. Prepare the skin over the membrane with an alcohol swab.

　5. Stabilize the larynx with the nondominant hand.

　6. With the cutting instrument in the dominant hand, make a 1 1/2 inch vertical incision through the skin over the cricothyroid membrane.
NOTE: A vertical incision will allow visualization of the cricothyroid membrane, but keep the scalpel blade away from the lateral aspect of the neck. This is important because of the large blood vessels located in the lateral areas of the neck.

CAUTION: Do not cut the cricothyroid membrane with this incision.

Performance Steps

7. Maintain the opening of the skin incision by pulling the skin taut with the fingers of the nondominant hand.

8. Stabilize the larynx with one hand and cut horizontally through the cricothyroid membrane.

9. Insert a commercially designed cricothyroidotomy hook or improvise with the tip of an 18-gauge needle formed into a hook through the opening; hook the cricoid cartilage, and lift to stabilize the opening.

10. Insert the end of the ET tube or tracheotomy tube through the opening and towards the lungs. The tube should be in the trachea and directed toward the lungs. Inflate the cuff 10 cubic centimeters (cc) of air.

11. Assess the casualty for spontaneous respirations (10 seconds).

12. Attach a pulse oximeter to the casualty, if available.

13. Assist with ventilations when respirations are <8 or >30 or a pulse oximeter reading <90% Direct an assistant to ventilate the casualty with a BVM, if necessary.

14. Auscultate lung fields and watch for rise and fall of the chest to confirm tube placement.

15. Secure the tube, using tape, cloth ties, or other measures, and apply a dressing to further protect the tube and incision.

16. Monitor the casualty's respirations on a regular basis.
    a. Reassess air exchange and placement every time the casualty is moved.
    b. Assist with respirations if the respiratory rate falls below 8 or rises above 30 per minute.

Evaluation Preparation:
Setup: For training and evaluation, use a mannequin or have another Soldier act as the casualty. Under no circumstances will the skin be incised. Have the Soldier demonstrate and explain what he would do.

Brief Soldier: Tell the Soldier to perform a surgical cricothyroidotomy.

Performance Measures	GO	NO-GO
1. Gathered cricothyroidotomy kit or minimum essential equipment.	——	——
2. Hyperextended the casualty's neck.	——	——
3. Put on gloves.	——	——
4. Located the cricothyroid membrane and decontaminated with an alcohol swab.	——	——
5. Stabilized the larynx with the nondominant hand.	——	——
6. Made a 1 1/2 inch vertical incision over the cricothyroid membrane, with the cutting instrument in the dominant hand.	——	——

Performance Measures	GO	NO-GO

7. Maintained the opening of the skin incision by pulling the skin taut with the fingers of the nondominant hand. —— ——

8. Stabilized the larynx with one hand and cut through the cricothyroid membrane. —— ——

9. Inserted a commercial cricothyroidotomy hook or the tip of an 18 gauge needle bent into a hook through the opening of the cricoid cartilage lifting to stabilize the opening. —— ——

10. Inserted the end of the ET tube or tracheotomy tube into the opening. The tube was entered into the trachea and directed toward the lungs and entered approximately 1-2 cm distal to the proximal tip of the cuff. —— ——

11. Assessed the casualty for spontaneous respirations (10 seconds). —— ——

12. Attached a pulse oximeter to the casualty, if available. —— ——

13. Assisted with ventilations when respirations were <8 or >30 or a pulse oximeter reading <90% Directed an assistant to ventilate the casualty with a BVM, if necessary. —— ——

14. Auscultated lung fields and watched for rise and fall of the chest to confirm tube placement. —— ——

15. Secured the tube and applied a dressing to further protect the tube and incision. —— ——

16. Monitored the casualty's respirations on a regular basis. —— ——

Evaluation Guidance: Score each Soldier according to the performance measures in the evaluation guide. Unless otherwise stated in the task summary, the Soldier must pass all performance measures to be scored GO. If the Soldier fails any step, show what was done wrong and how to do it correctly.

References

Required	**Related**
None	PHTLS

PERFORM A NEEDLE CHEST DECOMPRESSION
081-833-3007

Conditions: You have a breathing casualty with chest trauma who requires needle chest decompression. You will need a large bore needle (3.25 inch, 14 gauge) and tape. You are not in a CBRNE environment.

Standards: Complete all the steps necessary to perform a needle chest decompression, without causing unnecessary injury to the casualty.

Performance Steps

NOTE: Pneumothorax is defined as the presence of air within the chest cavity. Air may enter the chest cavity either from the lungs through a rupture, laceration, or from the outside through a sucking chest wound. Trapped air in the chest cavity under pressure called a (tension pneumothorax) compresses the lung beneath it. Unrelieved pressure will push and compress the contents of the chest in the opposite direction, away from the side of the tension pneumothorax. This, in turn, will prevent the heart from filling with blood and beating correctly and the good lung from providing adequate respirations.

CAUTION: This procedure should ONLY be performed if the casualty has a chest trauma and progressive respiratory distress.

1. Locate the insertion site. Locate the second intercostal space approximately two finger widths below the clavicle (between the second and third ribs) at the midclavicular line (approximately in line with the nipple) on the same side of the casualty's chest as the injury.

2. Insert a large bore (3.25 gauge) needle and catheter unit.
   a. Firmly insert the needle into the skin over the top of the third rib into the second intercostal space, until the chest cavity has been penetrated, as evidenced by feeling a "pop" as the needle enters the chest cavity.
   b. A hiss of escaping air may be heard.

WARNING: Proper positioning of the needle is essential to avoid puncturing blood vessels and/or nerves. Blood vessels and nerves run along the bottom of each rib.

3. Withdraw the needle while holding the catheter still.

NOTE: The casualty's respiration should improve.

4. Secure the catheter to the chest wall using tape.

5. Monitor the casualty until medical care arrives.

Evaluation Preparation:

Setup: For training and evaluation, use a mannequin and have the Soldier practice needle insertion. Have the Soldier demonstrate and explain what he would do.

Brief Soldier: Tell the Soldier to perform needle chest decompression.

	GO	NO-GO
Performance Measures		
1. Located the insertion site.	——	——
2. Inserted a large bore needle.	——	——
3. Withdrew the needle while holding the catheter still.	——	——
4. Secured the catheter to the chest.	——	——
5. Monitored the casualty.	——	——

Evaluation Guidance: Score each Soldier according to the performance measures. Unless otherwise stated in the task summary, the Soldier must pass all performance measures to be scored GO. If the Soldier fails any steps, show what was done wrong and how to do it correctly.

References

Required	**Related**
None	PHTLS

Subject Area 5: Venipuncture and IV Therapy

OBTAIN A BLOOD SPECIMEN USING A VACUTAINER
081-833-0032

Conditions: You have a patient that needs to have a blood specimen drawn. You must obtain a blood specimen using a vacutainer. You will need blood specimen tubes, constricting band, vacutainer adapter, vacutainer needles, disinfectant pads, sterile 2 x 2 gauze sponges, betadine or alcohol, adhesive bandage strips, protective pad, labels, and gloves. You are not in a CBRNE environment.

Standards: Obtain a blood specimen without causing injury to the patient or violating aseptic technique.

Performance Steps

1. Verify the request to obtain a blood specimen. Select the proper blood specimen tube for the test to be performed.

2. Label the blood specimen tube with the information necessary to identify the patient.

3. Perform a patient care hand-wash.
WARNING: Gloves should be worn for self-protection against transmission of contaminants whenever handling body fluids.

4. Assemble the vacutainer adapter, the needle, and the blood specimen tube.
   a. Inspect the needle for nicks or barbs. Replace the needle if it is flawed or dull.
   b. Insert the rubber stoppered end of the specimen tube into the vacutainer holder and advance the tube until it is even with the guideline.
NOTE: The needle is now partially imbedded into the stopper. If the tube is pushed beyond the guideline, the vacuum of the tube may be broken.

5. Identify the patient.
   a. Ask the patient his name and compare the name to the bed card and identification band or tags.
   b. If the specimen is being obtained from an outpatient, identify the patient by asking his name and comparing the name with the medical records or the laboratory request.
NOTE: Ask the patient about allergies to such things as iodine or alcohol.

6. Explain the procedure and purpose for collecting the blood specimen to the patient.

7. Position the patient.
   a. Assist the patient into a comfortable sitting or lying position.
WARNING: Never attempt to draw blood from a standing patient.
   b. The patient should be positioned so the arm is well supported and stabilized by using a pillow, table, or other flat surface.
   c. Place a protective pad under the elbow and forearm.

8. Expose the area for venipuncture.

9. Select and palpate one of the prominent veins in the bend of the arm (antecubital space).
   a. The first choice is the median cubital vein. It is well supported and least apt to roll.
   b. The second choice is the cephalic vein.

Performance Steps

      c. The third choice is the basilic vein. Although it is often the most prominent, it tends to roll easily and makes venipuncture difficult.

WARNINGS: 1. Avoid veins that are infected, irritated, injured, or have an IV running distal to the proposed venipuncture site. 2. Do not use the vacutainer to draw blood from small or fragile veins, because this can cause the vein walls to collapse. Use a needle and syringe instead.

10. Prepare the sponges for use.
      a. Open the betadine or alcohol and 2 X 2 gauze sponge packages.
      b. Place them within easy reach (still in the packages).

11. Apply the constricting band with enough pressure to stop venous return without stopping the arterial flow (a radial pulse will be present).
      a. Wrap latex tubing around the limb approximately 2 inches above the proposed venipuncture site.
      b. Stretch the tubing slightly and pull one end so that it is longer than the other.
      c. Form a loop with the longer end and draw the loop under the shorter end so that the tails of the tubing are turned away from the proposed site.

NOTE: If a commercial band is used, wrap it around the limb as in step 11a and then secure the band by overlapping the Velcro ends.

      d. Instruct the patient to form a fist, clench and unclench several times, and then hold the fist in a clenched position.

12. Palpate the selected vein lightly with the index finger, moving an inch or two in either direction so that the size and direction of the vein can be determined. The vein should feel like a spongy tube.

13. With a disinfectant soaked pad, cleanse the area around the puncture site using an outward circular motion.

CAUTION: After cleansing the skin, do not re-palpate the area.

WARNING: Do not leave the constricting band on for more than 2 minutes.

14. Prepare to puncture the vein.
      a. Grasp the vacutainer unit and remove the protective needle cover.
      b. Position the needle directly in line with the vein. Using the free hand, grasp the patient's arm below the expected point of entry.
      c. Place the thumb of the free hand approximately 1 inch below the expected point of entry and pull the skin taut toward the hand.

15. Puncture the vein.
      a. Place the needle, bevel up, in line with the vein and pierce the skin at a 15 to 30 degree angle.
      b. Decrease the angle until the needle is almost parallel to the skin surface. Direct it toward the vein and pierce the vein wall.

NOTE: A faint "give" will be felt when the vein is entered and blood will appear in the hub of the needle.

      (1) If the venipuncture is unsuccessful, pull the needle back slightly (not above the skin surface) and attempt to pierce the vein again.

CAUTION: If the needle is withdrawn above the skin surface, quickly release the constricting band and stop the procedure. Begin again with a new needle.

Performance Steps

        (2) If the venipuncture is still unsuccessful, release the constricting band, place a gauze sponge lightly over the site, quickly withdraw the needle, and immediately apply pressure to the site.

        (3) Notify the supervisor before attempting to enter another vein.

     c. Instruct the patient to unclench the fist.

16. Collect the specimen.

     a. Single specimen sample.

        (1) With the dominant hand, hold the vacutainer unit and the needle steady.

        (2) Place the index and middle fingers of the free hand behind the flange of the vacutainer and ease the tube as far forward as possible. Blood will enter the tube.

WARNING: If the unit and needle are not held steady while pushing in the tube, the needle may either slip out of the vein or puncture the opposing vein wall.

        (3) After the tube is approximately two-thirds full of blood or the flow of blood stops, prepare to withdraw the needle.

     b. Multiple specimen samples (multiple tubes).

        (1) Follow steps 16a(1) and 16a(2) for collecting a single specimen.

        (2) Remove the first tube and insert another tube into the vacutainer.

        (3) Repeat this procedure until the desired number of tubes are filled or blood stops flowing.

        (4) Release the constricting band using the nondominant hand.

        (5) After the last tube is approximately two-thirds full of blood or the flow stops, prepare to withdraw the needle.

NOTE: If the blood flow starts to slow down between samples, remove the constricting band.

17. Withdraw the needle.

     a. Release the constricting band by pulling on the long, looped end of the tubing or pulling the Velcro fasteners open.

WARNING: Never withdraw the needle prior to removing the constricting band because this will cause blood to be forced out of the venipuncture site with resulting blood loss and/or hematoma formation.

     b. Place a gauze sponge lightly over the venipuncture site.

     c. Keeping the patient's arm fully extended, withdraw the needle smoothly and quickly. Immediately apply firm manual pressure over the venipuncture site with the sponge.

     d. Instruct the patient to elevate the arm slightly and keep the arm fully extended. Continue to apply firm manual pressure to the site for 2 to 3 minutes.

18. Remove the specimen tube from the vacutainer.

     a. Replace the protective cover over the needle.

NOTE: Dispose of the uncapped needle IAW local SOP.

WARNING: If accidentally punctured by a used needle, force the puncture site to bleed, wash it thoroughly, and report the incident to your supervisor immediately.

     b. Pull the tube from the vacutainer.

     c. If the tube contains an anticoagulant, gently invert the tube several times to mix it with the blood.

19. Apply an adhesive bandage strip to the venipuncture site after the bleeding has stopped. Adhesive bandage strips do not take the place of pressure and therefore, are not applied until the bleeding has stopped.

Performance Steps

20. Provide for the patient's safety and comfort.
    a. Remove the protective pad.
    b. Assist the patient to assume a comfortable position.

21. Dispose of and/or store the equipment.
    a. Collect all the equipment and remove it from the area.
    b. Place the used gauze sponge, alcohol or betadine sponge, and the protective pad in the trash receptacle.
    c. Store the constricting band and vacutainer adapter IAW local SOP and dispose of the needle and syringe IAW local SOP.

22. Remove the gloves.

23. Perform a patient care hand-wash.

24. Complete the laboratory request.
    a. Patient identification.
    b. Requesting medical officer's name.
    c. Ward number or clinic.
    d. Date and time of specimen collection.
    e. Test(s) requested.
    f. Specimen source--blood.
    g. Remarks. Write in the admission diagnosis or the type of surgery in this section.
    h. Complete the "urgency" box. (Routine, today, preop, STAT, or ASAP.)
NOTE: There are many lab request slips which are used for requesting specific blood tests. All slips must be checked for the minimum information, as given.

25. Forward the specimen to the laboratory.
    a. Attach the lab request to the specimen tube(s) with a rubber band or paper clip.
NOTE: Ensure that the lab requests and blood tubes are appropriately labeled with infectious warning labels IAW local SOP.
    b. Arrange for the specimen to be sent to the lab or transport the specimen to the lab IAW local SOP.

26. Perform a patient care hand-wash.

27. Record the procedure on the appropriate form.

Performance Measures	GO	NO-GO
1. Verified the request and selected the proper blood specimen tube.	——	——
2. Labeled the blood specimen tube.	——	——
3. Performed a patient care hand-wash.	——	——
4. Assembled the vacutainer unit, needle, and blood specimen tube.	——	——
5. Identified the patient.	——	——
6. Explained the procedure and purpose for collecting the blood.	——	——
7. Positioned the patient.	——	——

Performance Measures	GO	NO-GO
8. Exposed the venipuncture site.	——	——
9. Selected and palpated the vein.	——	——
10. Prepared sponges for use.	——	——
11. Applied the constricting band.	——	——
12. Palpated the selected vein.	——	——
13. Cleaned the venipuncture site.	——	——
14. Prepared to puncture the vein.	——	——
15. Punctured the vein.	——	——
16. Collected the specimen.	——	——
17. Withdrew the needle.	——	——
18. Removed the specimen tube from the vacutainer.	——	——
19. Applied an adhesive bandage strip to the site.	——	——
20. Provided for the patient's safety and comfort.	——	——
21. Disposed of and/or stored equipment.	——	——
22. Removed the gloves.	——	——
23. Performed a patient care hand-wash.	——	——
24. Completed the laboratory request.	——	——
25. Forwarded the specimen to the laboratory.	——	——
26. Performed a patient care hand-wash.	——	——
27. Recorded the procedure on the appropriate form.	——	——
28. Did not violate aseptic technique, or cause further injury to the patient.	——	——

Evaluation Guidance: Score each Soldier according to the performance measures. Unless otherwise stated in the task summary, the Soldier must pass all performance measures to be scored GO. If the Soldier fails any steps, show what was done wrong and how to do it correctly.

References

Required
None

Related
BASIC NURSING 7

INITIATE AN INTRAVENOUS INFUSION
081-833-0033

Conditions: You need to initiate an intravenous (IV) infusion on a patient. You have performed a patient care hand-wash. You will need an IV injection set, IV solution, catheter-over-needle, constricting band, antiseptic sponges, 2 x 2 gauze sponges, tape, IV stand or substitute, armboard, and gloves. You are not in a CBRNE environment.

Standards: Initiate an IV without causing further injury or unnecessary discomfort to the patient. Do not violate aseptic technique.

Performance Steps
1. Identify the patient and explain the procedure.
    a. Ask the patient's name.
    b. Check the identification band against the patient's chart, as appropriate.
    c. Explain the reason for IV therapy.
    d. Explain the procedure and caution the patient against manipulating the equipment.
    e. Ask about any known allergies to such things as betadine or medication.
2. Select and inspect the equipment for defects, expiration date, and contamination.
    a. IV fluid of choice (check medical officer's order). Discard containers that have cracks, scratches, leaks, sedimentation, condensation, or fluid which is not crystal clear and colorless.
    b. IV injection set.
       (1) Spike, drip chamber, tubing, and needle adapter. Discard them if there are cracks or holes or if any discoloration is present.
       (2) Flow regulator. Inspect the flow regulator and ensure that it tightens.
       (3) Catheter-over-needle. Discard them if they are flawed with barbs or nicks.
NOTE: Place the stand to the side of the patient and close to the IV site.

3. Prepare the equipment.
    a. Move the flow regulator 6 to 8 inches below the drip chamber and tighten/close it.
    b. Remove the protective covers from the spike and from the outlet of the IV container.
CAUTION: Do not touch the spike or the outlet of the IV container.
    c. Push the spike firmly into the container's outlet tube.
    d. Hang the container at least 2 feet above the level of the patient's heart, if possible.
NOTE: An IV bag container may be placed under the patient's body if there is no way to hang it. You must completely fill the drip chamber if you place it under the patient's body to prevent air from entering the tubing.
    e. Squeeze the drip chamber until it is half full of the IV fluid.
    f. Prime the tubing.
NOTE: Ensure that all air is expelled from the tubing.
       (1) Hold the tubing above the level of the bottom of the container.
       (2) Loosen the protective cover from the needle adapter to allow the air to escape.
       (3) Release the clamp on the tubing.
       (4) Gradually lower the tubing until the solution reaches the end of the needle adapter.
       (5) Tighten the flow regulator to stop the flow of IV fluids.
       (6) Retighten the needle adapter's protective cover.
       (7) Loop the tubing over the IV stand or holder.
    g. Cut several pieces of tape and hang them in a readily accessible place.

Performance Steps

  4. Select the infusion site.
     a. Put on gloves for body substance isolation.
     b. Choose the most distal and accessible vein of an uninjured arm or hand.
     c. Avoid sites over joints.
     d. Avoid veins in infected, injured, or irritated areas.
     e. Use the nondominant hand or arm, whenever possible.
CAUTION: Do not use an arm that may require an operative procedure.
     f. Select a vein large enough to accommodate the size of needle/catheter to be used.

  5. Prepare the infusion site.
     a. Apply the constricting band.
NOTE: When applying the constricting band, use soft-walled latex tubing about 18 inches in length.
       (1) Place the tubing around the limb, about 2 inches above the site of venipuncture. Hold one end so that it is longer than the other, and form a loop with the longer end.
       (2) Pass the looped end under the shorter end of the constricting band.
NOTE: When placing the constricting band, ensure that the tails of the tubing are turned away from the proposed site of venipuncture.
       (3) Apply the constricting band tight enough to stop venous flow but not so tightly that the radial pulse cannot be felt.
       (4) Tell the patient to open and close his fist several times to increase circulation.
CAUTION: Do not leave the constricting band in place for more than 2 minutes.
     b. Select a prominent vein.
     c. Tell the patient to close his fist and keep it closed until instructed to open the fist.
     d. Clean the skin over the selected area with 70% alcohol or betadine, using a firm circular motion from the center outward.
     e. Allow the skin to dry and discard the gauze.
     f. Put on gloves for self-protection against transmission of contaminants.

  6. Prepare to puncture the vein.
     a. Pick up the assembled needle and remove the protective cover with the other hand.
       (1) Ensure the needle is bevel up.
       (2) Place the forefinger on the needle hub to guide it during insertion through the skin and into the vein.
     b. Position yourself so as to have a direct line of vision along the axis of the vein to be entered.

  7. Puncture the vein.
CAUTION: Keep the needle at the same angle to prevent through-and-through penetration of the vein walls.
NOTE: You may position the needle directly above the vein or slightly to one side of the vein.
     a. Draw the skin below the cleaned area downward to hold the skin taut over the site of venipuncture.
     b. Position the needle point, bevel up, parallel to the vein and about 1/2 inch below the site of venipuncture.
     c. Hold the needle at a 20 to 30 degree angle and insert it through the skin.
     d. Decrease the angle of the needle until it is almost parallel to the skin surface and direct it toward the vein.
     e. Move the needle forward about 1/2 inch into the vein.

Performance Steps

8. Confirm the puncture.

NOTE: A faint "give" will be felt as the needle enters the lumen of the vein.

    a. Check for blood in the flash chamber. If successful, proceed to step 9.

    b. If the venipuncture is unsuccessful, pull the needle back slightly (not above the skin surface) and attempt to pierce the vein again.

    c. If the venipuncture is still unsuccessful, release the constricting band and tell the patient to open and relax his clinched fist.

        (1) Place a sponge lightly over the site and quickly withdraw the needle.

        (2) Immediately apply pressure to the site.

    d. Notify your supervisor before attempting a venipuncture at another site.

9. Advance the catheter-over-needle.

    a. Grasp the hub and with a slight twisting motion fully advance the catheter.

    b. While continuing to hold the hub, press lightly on the skin over the catheter tip with the fingers of the other hand.

NOTE: This prevents the backflow of blood from the hub.

    c. Remove the needle from inside the catheter.

10. Remove the protective cover from the needle adapter on the tubing. Quickly and tightly connect the adapter to the catheter or needle hub.

WARNING: Do not allow air to enter the blood stream.

11. Tell the patient to unclench the fist, and then release the constricting band.

12. Loosen the flow regulator and adjust the flow rate to keep the vein open (KVO) or to keep open (TKO).

NOTE: A rate of about 30 cc per hour, or 7 to 10 drops per minute using standard drip tubing, is adequate to keep the vein open.

13. Check the site for infiltration. If it is painful, swollen, red, cool to the touch, or if fluid is leaking from the site, stop the infusion immediately.

14. Secure the site IAW local SOP.

    a. Apply a sterile dressing over the puncture site, leaving the hub and tubing connection visible.

    b. Loop the IV tubing onto the extremity and secure the loop with tape.

    c. Splint the arm loosely on a padded splint, if necessary, to reduce movement.

15. Readjust the flow rate.

    a. Determine the total time over which the patient is to receive the dosage.

Example: The patient is to receive the dosage over a 3 hour period.

    b. Determine the total IV dosage the patient is to receive by checking the doctor's orders.

Example: The patient is to receive 1000 cc of IV fluid.

    c. Check the IV tubing package to determine the number of drops of IV fluid per cc the set has been designed to deliver.

Example: The set is designed to give 10 drops of IV fluid per cc (10 gtts/cc).

    d. Multiply the total hours (step 15a) by 60 minutes to determine the total minutes over which the IV dosage is to be administered.

Example: 3 hours X 60 min = 180 min.

    e. Divide the total IV dosage (step 15b) by the total minutes over which the IV dosage is to be administered (step 15d) to determine the cc of fluid to be administered per minute.

Performance Steps
Example: 1000 cc / 180 min = 5.5 cc/min.
  f. Multiply the cc/min (step 15e) by the number of drops of IV fluid per cc delivered by the
   tubing (step 15c) to determine the number of drops per minute to be administered.
Example: 5.5 cc/min X 10 drops/cc = 55 drops/min.
NOTE: Always round drops per minute off to the nearest whole number. If drops per minute
equal .5, round up to the next whole number.

16. Prepare and place the appropriate label.
  a. Dressing.
   (1) Print the information on a piece of tape.
    (a) Date and time the IV was started.
    (b) Initials of the person initiating the IV.
   (2) Secure the tape to the dressing.
  b. IV solution container.
   (1) Print the information on a piece of tape.
    (a) Patient's identification.
    (b) Drip rate.
    (c) Date and time the IV infusion was initiated.
    (d) Initials of the person initiating the IV.
   (2) Secure the tape to the IV container.
  c. IV tubing.
   (1) Wrap a strip of tape around the tubing, leaving a tab.
   (2) Print the date and time the tubing was put in place and the initials of the person
    initiating the IV.
NOTE: Place disposable items in an appropriate receptacle and clean and store equipment
IAW local SOP.

17. Recheck the site for infiltration.

18. Perform a patient care hand-wash.

19. Record the procedure on the appropriate form.
  a. Date and time the IV infusion was initiated.
  b. Type and amount of IV solution initiated.
  c. Drip rate and total volume to be infused.
  d. Type and gauge of needle or cannula.
  e. Location of the infusion site.
  f. Patient's condition.
  g. Name of the person initiating the IV.

Performance Measures	GO	NO-GO
1. Identify the patient and explain the procedure.	——	——
2. Inspected the equipment.	——	——
3. Prepared the equipment.	——	——
4. Selected the infusion site.	——	——
5. Prepared the infusion site.	——	——

Performance Measures	<u>GO</u>	<u>NO- GO</u>
6. Prepared to puncture the vein.	——	——
7. Punctured the vein.	——	——
8. Confirmed the puncture.	——	——
9. Advanced the catheter.	——	——
10. Connected the adapter to the catheter hub.	——	——
11. Released the constricting band.	——	——
12. Loosened the flow regulator and adjusted the flow rate TKO.	——	——
13. Checked the site for infiltration.	——	——
14. Secured the site.	——	——
15. Readjusted the flow rate.	——	——
16. Prepared and placed the appropriate labels.	——	——
17. Rechecked the site for infiltration.	——	——
18. Performed a patient care hand-wash.	——	——
19. Recorded the procedure on the appropriate form.	——	——
20. Did not violate aseptic technique, or cause further injury to the patient.	——	——

Evaluation Guidance: Score each Soldier according to the performance measures. Unless otherwise stated in the task summary, the Soldier must pass all performance measures to be scored GO. If the Soldier fails any steps, show what was done wrong and how to do it correctly.

References

Required
None

Related
BASIC NURSING 7

MANAGE AN INTRAVENOUS INFUSION
081-833-0034

Conditions: You have a patient who has an intravenous (IV) infusion. You will need dressings, antiseptic swabs, sterile gauze, IV tubing, IV solution, tape, antimicrobial ointment, exam gloves, and DD Form 792 (Nursing Service – Twenty-Four Patient Intake and Output Worksheet). You are not in a CBRNE environment.

Standards: Properly manage a patient with an IV infusion, accurately document the IV therapy, properly assess for the complications of IV therapy, and initiate appropriate interventions when necessary. Do not violate aseptic technique and do not cause further injury to the patient.

Performance Steps

1. Assess for signs and symptoms of IV therapy complications.
   a. Infiltration is an accumulation of fluids in the tissue surrounding an IV needle site. It is caused by penetration of the vein wall by the needle/catheter or later dislodgement of the catheter.
      (1) Solution flows sluggishly or not at all.
      (2) Discoloration or cool feeling around the infusion site.
      (3) Swollen extremity.
      (4) Fluid leaking from the infusion site.
      (5) Patient complains of pain, tenderness, irritation, or burning at the infusion site.
   b. Phlebitis is an inflammation of the wall of the vein. It is caused by injury to the vein during puncture or from irritation to the vein caused by long term therapy, incompatible additives, or use of a vein that is too small to handle the amount or type of solution.
      (1) Swelling, redness, and/or tenderness around the venipuncture site.
      (2) Sluggish flow rate.
   c. Infection is a yellowish, foul-smelling discharge (pus) from the venipuncture site.
   d. Air embolism is the obstruction of a blood vessel by air carried via the bloodstream (usually occurring in the lungs or heart). It is caused by conditions such as air bubbles in the IV tubing, a solution container that has run dry, or disconnected IV tubing.
      (1) Abrupt drop in blood pressure.
      (2) Chest pain.
      (3) Weak, rapid pulse.
      (4) Cyanosis.
      (5) Loss of consciousness.
   e. Circulatory overload is an increased blood volume that is caused by excessive IV fluid infused too rapidly into the vein (over hydration).
      (1) Elevated blood pressure.
      (2) Distended neck veins.
      (3) Rapid breathing, shortness of breath, tachycardia.
      (4) Fluid intake is much greater than urine output.

2. Perform the nursing interventions for IV therapy complications.
   a. Infiltration.
      (1) Stop the infusion.
      (2) Notify your supervisor.
      (3) Record observations and action taken.
   b. Phlebitis.
      (1) Stop the infusion.

Performance Steps
    (2) Report observations to your supervisor.
    (3) Record observations and actions taken.
  c. Infection.
    (1) Report observations to your supervisor.
    (2) Record observations and actions taken.
  d. Air embolism.
    (1) Report observations to your supervisor.
    (2) Record observations and actions taken.
  e. Circulatory overload.
    (1) Slow the infusion rate to TKO.
    (2) Place the patient in the semi-Fowler's position.
    (3) Notify the medical officer or supervisor.
    (4) Record observations and actions taken.

3. Document the IV therapy.
  a. Frequency.
    (1) When the IV is initiated.
    (2) Each time any part of the IV equipment is changed.
  b. Label the dressing.
    (1) Cut adhesive tape and place it on a flat surface.
NOTE: Never write on the tape after it has been placed on the dressing.
    (2) Record the information on the piece of tape.
      (a) The gauge of the catheter.
      (b) The time and date the dressing was applied.
      (c) Your initials.
    (3) Place the labeled tape over the dressing.
  c. Label the solution container.
    (1) Cut adhesive tape and place it on a flat surface.
    (2) Record the information on the piece of tape.
      (a) The patient's name.
      (b) The patient's identification number and room/ward number, as appropriate.
      (c) The infusion rate.
      (d) The time and date the solution container was hung.
      (e) Your initials.
    (3) Place the label on the solution container.
    (4) Prepare the timing label.
      (a) Place a strip of adhesive tape vertically along the length of the solution container.
      (b) Determine how long the solution container will last. (See task 081-833-0033.)
      (c) Write on the tape the approximate times at which the solution level will reach the volume markings on the solution container.
      (d) At the bottom of the label write the approximate time the solution container will be empty.
  d. Label the tubing.
    (1) Place a strip of adhesive tape around the tubing, leaving a tab.
    (2) Write on the tab the date and time the tubing was changed.
  e. Record the information on the appropriate forms (Nursing Notes/Field Medical Card).
    (1) The date and time the IV was initiated.
    (2) The amount and type of solution.

Performance Steps

      (3) The infusion rate.

      (4) The type and gauge of the catheter.

      (5) The insertion site.

      (6) The patient's condition.

      (7) Your name.

    f. Record the amount of infusion on DD Form 792, if applicable.

  4. Replace the solution container (only).

NOTE: Change the solution container every 24 hours when running a slow infusion in which the container may not be depleted in 24 hours.

    a. Perform a patient care hand-wash.

    b. Select or prepare the new solution. (See task 081-833-0033.)

    c. Tighten the flow regulator and stop the infusion flow.

    d. Remove the used container from the IV hanger.

    e. Remove the spike from the used container.

    f. Insert the IV spike into a new IV container.

CAUTION: The old tubing is still connected to the catheter. Use care to maintain sterility. To prevent backflow of blood, keep the spike and tubing elevated.

    g. Hang the new container.

    h. Adjust the infusion rate.

    i. Label the solution container and prepare a timing label.

    j. Record the amount of solution received from the previous container, and the time, type, and amount of new solution.

  5. Change the dressing.

NOTE: Change the dressing every 24 hours or IAW local SOP.

    a. Perform a patient care hand-wash.

    b. Remove the tape and the old dressing without dislodging the catheter.

NOTE: Tubing should remain taped in place to reduce the chance of accidental dislodgement of the catheter or needle.

    c. Clean the area around the infusion site IAW local SOP.

    d. Examine the site for infiltration.

    e. Cover the infusion site with sterile gauze and secure with tape, or dress IAW local SOP.

    f. Secure the dressing to the site without encircling the wrist or arm.

    g. Label the dressing.

  6. Replace the solution container and tubing.

NOTE: Change the tubing every 48 hours or IAW local SOP. Time the tubing change to coincide with the time the solution container will be changed.

    a. Perform a patient care hand-wash.

    b. Tighten the flow regulator 6-8 inches below the drip chamber.

    c. Spike the new tubing into a new solution container and hang it from the IV pole.

    d. Fill the drip chamber ½ full and prime the tubing/bleed air from the IV line.

    e. Connect the new tubing to the needle hub.

WARNING: Wear gloves for self-protection against transmission of contaminants whenever handling body fluids.

      (1) Loosen the tape on the old tubing without dislodging the catheter and needle.

      (2) Place a sterile gauze pad under the catheter or needle hub to provide a small sterile field for the needle hub.

      (3) Grasp the new tubing between the fingers of one hand.

Performance Steps

        (4) Grasp the catheter or needle hub with a sterile gauze pad between the thumb and index finger and carefully disconnect the old adapter.

        (5) Press the fingers over the catheter to help prevent dislodgement and backflow of blood.

        (6) Remove the protective cap from the new tubing adapter and quickly connect it to the catheter hub.

CAUTION: Do not remove the protective cap with your teeth.

        (7) Remove the pressure over the catheter tip.

        (8) Remove the gauze pad from under the catheter hub and clean the site, if necessary.

        (9) Secure the tubing to the arm and reinforce the dressing, as necessary.

        (10) Adjust the infusion rate.

  7. Discontinue the infusion.

     a. Perform a patient care hand-wash.

     b. Put on exam gloves.

     c. Tighten the flow regulator on the IV tubing to stop the infusion.

     d. Remove the tape and dressing without dislodging the needle and catheter.

     e. Place a sterile gauze pad over the injection site.

     f. Smoothly pull out the catheter, following the course of the vein.

WARNING: Do not twist, raise, or lower the needle.

     g. Apply pressure to the site with the gauze.

     h. Examine the catheter to ensure that it was removed intact.

     i. Apply an adhesive bandage to the site, if necessary.

     j. Dispose of the used equipment IAW local SOP.

  8. Record the procedure on the appropriate form.

NOTE: Ensure that the fluids received have been recorded on the appropriate form(s).

Evaluation Preparation:

Setup: If the performance of this task must be simulated for training or evaluation, assemble the IV materials and equipment as indicated in task 081-833-0033. It is not necessary to have the catheter or needle inserted into a person. A simulated arm or other material may be used.

Brief Soldier: Tell the Soldier to manage a patient with an intravenous infusion.

Performance Measures	<u>GO</u>	<u>NO-</u> <u>GO</u>
1. Assessed for signs and symptoms of IV therapy complications.	____	____
2. Performed the nursing interventions for IV therapy complications.	____	____
3. Documented the IV therapy.	____	____
4. Replaced the solution container, as necessary.	____	____
5. Changed the dressing, as required.	____	____
6. Replaced the solution container and tubing, as necessary.	____	____
7. Discontinued the infusion, as required.	____	____

Performance Measures	GO	NO-GO
8. Recorded the procedure on the appropriate form.	——	——
9. Did not violate aseptic technique, or cause further injury to the patient.	——	——

Evaluation Guidance: Score each Soldier according to the performance measures. Unless otherwise stated in the task summary, the Soldier must pass all performance measures to be scored GO. If the Soldier fails any steps, show what was done wrong and how to do it correctly.

References

Required	**Related**
DD FORM 792	BASIC NURSING 7

INITIATE A FAST 1
081-833-0185

Conditions: You have a casualty who needs a First Access Shock and Trauma (FAST 1). You will need alcohol swabs, a FAST 1, intravenous (IV) administration set, IV solution, and DD Form 1380 (U.S. Field Medical Card). You are not in a CBRNE environment.

Standards: Observe universal precautions, clean the administration site, introduce the FAST 1 device, and administer fluids via the intraosseous route without breaking aseptic technique or causing subcutaneous infiltration of the IV solution.

Performance Steps

1. Prepare the site
   a. Undo or cut away outer clothing to expose the sternum.
   b. Identify the suprasternal notch.
   c. Use aseptic technique to prepare the site

2. Place the target patch
   a. Remove the top half of the backing (labeled remove 1) from the patch.
   b. Locate the sternal notch using your index finger.
   c. Holding your index finger perpendicular to the sternum, align the locating notch in the target patch with the sternal notch, keeping your index finger perpendicular.
   d. Verify that the target zone (circular hole) on the patch is directly over the casualty's midline.
   e. Secure the top half of the patch to the body by pressing firmly downward on the patch, engaging the adhesive.
   f. Remove the remaining backing (labeled remove 2) and secure patch to the casualty.
   g. Verify correct patch placement by checking the alignment of the locating notch with the casualty's sternal notch and making sure the target zone is over the midline of the casualty's body.

3. Insert the introducer.
   a. Remove sharps cap from the introducer.
   b. Place the bone probe cluster needles in the target zone of the target patch, and ensure that all the bone probe needles are within the target zone.
   c. Hold the introducer perpendicular to the sternum of the casualty to ensure proper functioning of the depth-control mechanism.
   d. Pressing straight along the introducer axis, with hand and elbow in line, push with firm constant force until a distinct release is heard and felt.

WARNING: Apply the force perpendicular to the skin and along the long axis of the introducer. Avoid extreme force, twisting and jabbing motions.

   e. After the release, expose the infusion tube by gently withdrawing the introducer along the same path used to insert it (perpendicular to the skin). The stylet supports will fall away.
   f. Locate the orange sharps plug, and place it on a flat surface with the foam facing up. Keep both hands behind the needles, and push the bone probe cluster straight into the foam. After the sharps plug has been engaged and the sharps are safely covered, reattach the clear sharps cap to the introducer.
   g. Dispose of the introducer using contaminated sharps protocols.

Performance Steps

4. Connect the infusion tube.
    a. Connect the infusion tube to the right-angle female connector on the target patch.
NOTE: This connection is a slip luer.
    b. Attach the straight female connector to the source of fluids or drugs. Fluid can now flow to the site.

5. Secure the protector dome.
    a. Place the protector dome directly over the target patch and press down firmly to engage the Velcro fastening. Ensure that the infusion tubing and the right angle female connector are contained under the dome.
    b. The dome can be removed by holding the patch against the skin and peeling back the dome Velcro.

6. Attach the remover package, if contained in system to the casualty for transport.
WARNING: The remover package must be transported with the casualty. It will be used later to remove the FAST 1 system.

CAUTION: Do not breach the packaging since the remover is sterile.

7. Record all treatment on the FMC.

Performance Measures	<u>GO</u>	<u>NO- GO</u>
1. Prepared the site.	——	——
2. Placed the target patch.	——	——
3. Inserted the introducer.	——	——
4. Connected the infusion tube.	——	——
5. Secured the protector dome.	——	——
6. Attached the remover package to the casualty for transport.	——	——
7. Recorded all treatment on the FMC.	——	——

Evaluation Guidance: Score each Soldier according to the performance measures. Unless otherwise stated in the task summary, the Soldier must pass all performance measures to be scored GO. If the Soldier fails any steps, show what was done wrong and how to do it correctly.

References
Required	**Related**
DD FORM 1380	PHTLS

INITIATE A SALINE LOCK
081-835-3025

Conditions: You need to establish a saline lock or convert a saline lock to an intravenous infusion (IV). You will need 18 gauge IV catheter/needles, 21 gauge 1 1/4" needle, saline lock adapter plug, constricting band, tape, 4 inch Tegaderm bandage, alcohol and betadine swabs, 5 cc syringe, a container of IV solution with primed tubing, and DD Form 1380 (U.S. Field Medical Card). You are not in a CBRNE environment.

Standards: Establish a saline lock or convert a saline lock without causing further injury to the patient.

Performance Steps

1. Prepare to establish a saline lock.
    a. Assemble the necessary equipment.
        (1) Two 18 gauge IV catheter/needles.
        (2) One 21 gauge 1 1/4" needle.
        (3) Saline lock adapter plug.
        (4) Constricting band.
        (5) 4 inch Tegaderm bandage.
        (6) Alcohol and betadine swabs.
        (7) 5 cc syringe.
    b. Explain the procedure and the purpose of the saline lock to the patient. Ask about allergies.
    c. Place the patient in a comfortable position with the arms supported.
    d. Select the catheter insertion site.
    e. Prepare the insertion site. Apply a constricting band 2" above the venipuncture site - tight enough to stop venous flow, but not so tight that the radial pulse cannot be felt.
    f. Clean the skin with either an alcohol swab or a betadine swab in a circular motion from the center outward.

2. Insert the saline lock.
    a. Put on gloves.
    b. Perform the venipuncture. Hold the catheter with your dominant hand and remove the protective cover without contaminating the needle. Hold the flash chamber with the thumb and forefinger directly above the vein. Draw skin below the cleansed site downward to hold the skin taut over the site of the venipuncture.
    c. Position the needlepoint, bevel up, parallel to the vein and about 1/2 inch below the venipuncture site. Continue advancing the needle/catheter until the vein is pierced.
    d. When "flash" of blood enters the flash chamber, decrease the angle between the skin and needle until the angle is almost parallel to the skin, and advance further to secure catheter placement in the vein.
    e. Place pressure on the vein above the insertion site by pressing with one finger of the nondominant hand. Release the constricting band.
    f. Remove the needle after advancing the plastic catheter into the vein.
    g. Quickly uncap and insert the male end of the saline lock adapter plug into the hub of the catheter.
    h. Apply a Tegaderm dressing to the site, covering 100% of the site to include insertion site and saline lock adapter plug.

Performance Steps

     i. Flush the IV catheter. Using the 21 gauge needle and 5 cc syringe filled with sterile fluid, penetrate the transparent dressing and insert the needle into the saline lock. Inject 5 cc of sterile fluid into the IV catheter, looking for signs of infiltration.

3. Initiate fluids through a saline lock.
    a. Assemble the necessary equipment - a container of IV solution with primed tubing.
    b. Explain the procedure to the patient. Ask about allergies.
    c. Put on gloves.
    d. Introduce the 18 gauge catheter through the Tegaderm dressing and saline lock until the catheter hub is against the saline lock.
    e. Apply pressure to the vein above the insertion site and withdraw the needle from the catheter and discard it.
    f. Attach the primed IV tubing and initiate the flow of IV fluid through the catheter into the saline lock, checking for signs of infiltration.
    g. Secure the IV tubing to the patient's arm. Unroll approximately 2" of tape from the roll, and place it under the IV tubing with the sticky side up.
    h. Fold the tape back over the tubing and onto itself, and then completely around the patient's arm until it crosses back over the tubing.

4. Document the procedure and observations on the FMC.
    a. Type and size of needle inserted.
    b. Location of the saline lock.
    c. Date and time of insertion.
    d. Date and time an existing IV was converted to a saline lock.
    e. An assessment of the condition of the venipuncture site.
    f. Date and time the saline lock was converted to a continuous infusion IV and the type and amount of IV solution hung.

Performance Measures	GO	NO-GO
1. Prepared to establish a saline lock.	——	——
2. Established a saline lock.	——	——
3. Initiated fluids through saline lock.	——	——
4. Documented all procedures on the FMC.	——	——

Evaluation Guidance: Score each Soldier according to the performance measures. Unless otherwise stated in the task summary, the Soldier must pass all performance measures to be scored GO. If the Soldier fails any steps, show what was done wrong and how to do it correctly.

References

    Required
    DD FORM 1380

    Related
    ISBN 0-07-065351-8

Subject Area 6: Primary Care

PERFORM A PATIENT CARE HANDWASH
081-831-0007

Conditions: You are about to administer patient care or have just had hand contact with a patient or contaminated material. You need to perform a patient care hand-wash. You will need running water or two empty basins, a canteen, a water source, soap, towels (cloth or paper), and a towel receptacle or trash can. You are not in a CBRNE environment.

Standards: Perform a patient care hand-wash without contaminating the hands.

Performance Steps

1. Remove wristwatch and jewelry, if applicable.
NOTE: Rings should not be worn. If rings are worn, they should be of simple design with few crevices for harboring bacteria. Fingernails should be clean, short, and free of nail polish.

2. Roll shirt sleeves to above the elbows, if applicable.

3. Prepare to perform the hand-wash.
   a. If using running water, turn on the warm water.
   b. If running water is not available, set up the basins and open the canteen.

4. Wet your hands, wrists, and forearms.
   a. If using running water, hold your hands, wrists, and forearms under the running water.
   b. If running water is not available, fill one basin with enough water to cover your hands and refill the canteen.

5. Cover your hands, wrists, and forearms with soap.
NOTE: For routine patient care, use regular hand soap. For an invasive procedure such as a catheterization or an injection, use antimicrobial soap.

6. Wash your hands, wrists, and forearms.
   a. Use a circular scrubbing motion, going from the fingertips toward the elbows for at least 15 seconds.
   b. Give particular attention to creases and folds in the skin.
   c. Wash ring(s) if present.

7. Rinse your hands, wrists, and forearms.
   a. If using running water.
      (1) Hold your hands lower than the elbows under the running water until all soap is removed.
      (2) Do not touch any part of the sink or faucet.
   b. If not using running water.
      (1) Use a clean towel to grasp the canteen with one hand.
      (2) Rinse the other hand, wrist, and forearm, letting the water run into the empty basin. Hold your hands lower than the elbows.
      (3) Repeat the procedure for the other arm.
      (4) Do not touch any dirty surfaces while rinsing your hands.

Performance Steps

  8. Dry your hands, wrists, and forearms.
      a. Use a towel to dry one arm from the fingertips to the elbow without retracing the path with the towel.
      b. Dispose of the towel properly without dropping your hand below waist level.
      c. Repeat the process for the other arm using another towel.

  9. Use a towel to turn off the running water, if applicable.

  10. Re-inspect your fingernails and clean them and rewash your hands, if necessary.

Evaluation Preparation:
Setup: None

Brief Soldier: Tell the Soldier to perform a patient care hand-wash. You may specify which method to use. The Soldier need not perform both.

Performance Measures	<u>GO</u>	<u>NO-GO</u>
1. Removed wristwatch and jewelry, if applicable.	——	——
2. Rolled shirt sleeves to above the elbows, if applicable.	——	——
3. Prepared to perform the hand-wash.	——	——
4. Wet the hands, wrists, and forearms.	——	——
5. Covered the hands, wrists, and forearms with soap.	——	——
6. Washed the hands, wrists, and forearms.	——	——
7. Rinsed the hands, wrists, and forearms.	——	——
8. Dried the hands, wrists, and forearms.	——	——
9. Used a towel to turn off the running water, if applicable.	——	——
10. Re-inspected the fingernails and cleaned them and rewashed the hands, if necessary.	——	——

Evaluation Guidance: Score each Soldier according to the performance measures. Unless otherwise stated in the task summary, the Soldier must pass all performance measures to be scored GO. If the Soldier fails any steps, show what was done wrong and how to do it correctly.

References
  Required   **Related**
  None   BASIC NURSING 7

IRRIGATE EYES
081-833-0054

Conditions: You have a patient requiring eye irrigation. You have performed a patient care hand-wash. You will need draping materials, catch basin, light source, gauze or cotton balls, irrigating syringe or similar equipment, gloves, and irrigating solution (normal saline or water). You are not in a CBRNE environment.

Standards: Irrigate the eyes without injuring the eyes.

Performance Steps

1. Identify the patient and explain the procedure.

2. Verify the type, strength, and expiration date of the irrigating solution as appropriate.
CAUTION: Do not irrigate an eye that has an impaled object.

3. Ask the patient to remove contact lenses or glasses, if necessary.

4. Position the patient.
   a. If lying on the back, tilt the head slightly to the side that is to be irrigated.
   b. If seated, tilt the head slightly backward and to the side that is to be irrigated.

5. Position the equipment.
   a. Drape the areas of the patient that may be splashed by the solution.
   b. Place a catch basin next to the face on the affected side.
   c. Position the light so that it does not shine directly into the patient's eyes.

6. Put on gloves.
WARNING: Wear gloves for self-protection against transmission of contaminants whenever handling body fluids.

7. Clean the eyelids with gauze or cotton balls, and rinse debris from the outer eye.

8. Separate the eyelids using the thumb and forefinger, and hold the lids open.
CAUTION: Do not put pressure on the eyeball.

9. Irrigate the eye.
   a. Hold the irrigating tip 1 to 1 1/2 inches away from the patient's eye.
   b. Direct the irrigating solution gently from the inner canthus to the outer canthus.
   c. Use only enough pressure to maintain a steady flow of solution and to dislodge the secretions or foreign bodies.
   d. Instruct the patient to look up to expose the conjunctival sac and lower surface of the eye.
   e. Instruct the patient to look down to expose the upper surface of the eye.

10. Dry the area around the eye by gently patting with gauze sponges.
CAUTION: Do not touch the eye.

11. Remove the gloves, and perform a patient care hand-wash.

12. Record the treatment given on the appropriate form.

Performance Measures	GO	NO-GO
1. Identified the patient and explained the procedure.	——	——
2. Verified the type, strength, and expiration date of the irrigating solution, as appropriate.	——	——
3. Asked the patient to remove contact lenses or glasses, if necessary.	——	——
4. Positioned the patient.	——	——
5. Positioned the equipment.	——	——
6. Put on gloves.	——	——
7. Cleaned the eyelids with gauze or cotton balls, and rinsed debris from the outer eye.	——	——
8. Separated the eyelids using the thumb and forefinger, and held the lids open.	——	——
9. Irrigated the eye.	——	——
10. Dried the area around the eye by gently patting with gauze sponges.	——	——
11. Removed the gloves and performed a patient care hand-wash.	——	——
12. Recorded the treatment given on the appropriate form.	——	——
13. Did not injure the eye.	——	——

Evaluation Guidance: Score each Soldier according to the performance measures. Unless otherwise stated in the task summary, the Soldier must pass all performance measures to be scored GO. If the Soldier fails any steps, show what was done wrong and how to do it correctly.

References

Required	Related
None	BASIC NURSING 7

IRRIGATE AN OBSTRUCTED EAR
081-833-0059

Conditions: You have an order to irrigate an obstructed ear. You will need an irrigating syringe, catch basin, irrigating solution, towels, gauze sponges, and otoscope set. You are not in a CBRNE environment.

Standards: Irrigate the obstructed ear until the obstructing material is removed from the external ear or until the prescribed amount of solution is used. Perform the procedure without causing further injury to the patient.

Performance Steps

1. Gather the irrigation equipment.
NOTE: Common solutions used to irrigate the ear include water, normal saline, hydrogen peroxide and water, and prescribed medication solution. Alcohol may be used to shrink vegetable matter (associated with pediatric patients) and make it easier to expel. Oil or viscous lidocaine may be used for other foreign bodies to make them slippery.

2. Perform a patient care hand-wash (see task 081-831-0007.)

3. Warm and test the solution.
    a. Warm the solution to about body temperature (95 °F to 105 °F) by placing the solution container in a container of warm water.
    b. Test the temperature of the solution by running a small amount of it on the inner wrist.
CAUTION: Cold solutions are not only uncomfortable but may cause dizziness or nausea as a result of stimulation of the equilibrium sensors in the semicircular canals.

4. Identify the patient and explain the procedure.
    a. Tell the patient that some discomfort may be experienced when the solution is instilled.
    b. Emphasize to the patient that he must remain as still as possible.
CAUTION: If the patient moves when the solution is instilled, the syringe may damage the ear canal or tympanic membrane.

5. Insert the otoscope speculum into the external ear canal.
    a. Position the patient to allow a good view into the ear.
    b. Tilt the patient's head toward the shoulder opposite the ear to be irrigated.
    c. Straighten the external ear canal by gently pulling the outer ear upward and backward for an adult or downward and backward for a child.
NOTE: Use the largest speculum that will fit comfortably in the patient's ear.
    d. Turn on the otoscope light and insert the speculum just inside the opening of the ear.
NOTE: To avoid causing pain, the speculum should be inserted gently and not too far into the ear canal.
    e. View the ear canal by looking through the lens of the otoscope.

6. Check for abnormalities.
    a. Check the external ear canal for redness, swelling, drainage, or foreign bodies.
    b. Check the tympanic membrane (TM) for any abnormal conditions.
NOTE: A normal eardrum is slightly cone-shaped, shiny, translucent, and pearly grey.
        (1) A blue, yellow, amber, red, or pink eardrum indicates disease or infection.
        (2) A bulge in the eardrum indicates possible pus or fluid in the middle ear.
        (3) A hole or tear indicates rupture of the TM.

Performance Steps
CAUTION: If an abnormal condition of the TM is suspected, do not irrigate the ear. To do so could cause pain and carry debris or infectious discharge into the middle ear. Report the condition to the supervisor immediately.

7. Position the patient sitting or lying with the head slightly tilted toward the affected side.
NOTE: Do not tilt the head toward the unaffected side, as this interferes with the return of the irrigating solution.

8. Drape the patient's shoulder and upper arm area under the affected ear.

9. Clean the external ear and the entrance to the ear canal with 4 x 4 gauze sponges slightly moistened with the irrigating solution.
WARNING: If a cotton-tipped applicator is used to clean the ear, make sure it does not stick far enough into the ear to rupture the tympanic membrane.

10. Fill the irrigating syringe.

11. Test the flow of solution from the syringe by expelling a small amount back into the solution container.

12. Position the catch basin firmly against the neck just under the affected ear.

13. Straighten the external ear canal by gently pulling the outer ear upward and backward for an adult or downward and backward for a child.

14. Irrigate the patient's ear.
    a. Place the tip of the irrigating syringe just inside the ear, with the tip directed toward the roof of the ear canal.
WARNING: Never allow the syringe to completely block the ear canal. If space is not left around the tip, the solution will not be able to return, and undue pressure will build up in the canal.
    b. Depress the bulb or plunger of the syringe.
        (1) Direct a slow, steady stream of solution against the roof of the ear canal.
        (2) Repeat the procedure until the foreign body is removed, the solution returns free of wax or debris, or the proper amount of solution has been used.

15. Remove the catch basin and dry the external ear with a gauze sponge.

16. Instruct the patient to continue tilting the head toward the affected side for a few minutes to allow any remaining solution to drain from the ear.

17. Remove the drapes from the patient.

18. Dispose of, or clean and store, the equipment.

19. Perform a patient care hand-wash.

20. Document the procedure on the appropriate forms IAW local SOP.
    a. Type and amount of solution used.
    b. Nature of return flow.

Performance Measures	GO	NO-GO
1. Gathered the irrigation equipment.	____	____
2. Performed a patient care hand-wash.	____	____
3. Warmed and tested the solution.	____	____
4. Identified the patient and explained the procedure.	____	____
5. Inserted the otoscope speculum into the external ear canal.	____	____
6. Checked for abnormalities.	____	____
7. Positioned the patient.	____	____
8. Draped the patient's shoulder and upper arm area under the affected ear.	____	____
9. Cleaned the external ear and the entrance to the ear canal.	____	____
10. Filled the irrigating syringe.	____	____
11. Tested the flow of solution.	____	____
12. Positioned the catch basin.	____	____
13. Straightened the external ear canal.	____	____
14. Irrigated the patient's ear.	____	____
15. Removed the catch basin and dried the external ear.	____	____
16. Instructed the patient to continue tilting the head toward the affected side.	____	____
17. Removed the drapes from the patient.	____	____
18. Disposed of, or cleaned and stored, the equipment.	____	____
19. Performed a patient care hand-wash.	____	____
20. Documented the procedure on the appropriate forms IAW local SOP.	____	____
21. Did not cause further injury to the patient.	____	____

Evaluation Guidance: Score each Soldier according to the performance measures. Unless otherwise stated in the task summary, the Soldier must pass all performance measures to be scored GO. If the Soldier fails any steps, show what was done wrong and how to do it correctly.

References

Required	**Related**
None	BASIC NURSING 7

TREAT SKIN DISORDERS
081-833-0125

Conditions: You are evaluating a patient with a skin disorder. No other injuries or anaphylaxis symptoms are present. You will need a standard fully stocked aid bag, oxygen, suction, and ventilation equipment (if available). You are not in a CBRNE environment.

Standards: Perform interventions for the management of skin disorder symptoms.

Performance Steps
1. Provide information on the prevention of skin disorders.
NOTE: All scene size-up, initial assessment, focused history, examination, detailed physical examination, ongoing assessment and transport assessment steps must be taken to ensure that injury(ies) or illness are not overlooked resulting in further injury to the patient.
   a. Maintain healthy skin.
   b. Avoid causative agents.
      (1) Avoid agents that cause skin disorders in most persons (e.g., poison ivy, and excessive sunlight).
      (2) Use sun screen when exposed to excessive sunlight.
   c. Observe skin changes.
   d. Avoid self-treatment.
   e. Inspection of the skin.
      (1) Inspect skin after each mission.
      (2) Always inspect in a well lighted area.
      (3) Look for recent lesions or rashes.
      (4) Ask when and where lesions first appeared.
      (5) Ask how long lesions have been present.
2. Identify and manage viral disorders of the skin.
   a. Herpes simplex.
      (1) Type 1 - causes cold sores, self-limiting, no cure.
      (2) Type 2 - causes lesions in the genital area.
NOTE: Transmission of both types of virus may occur by direct contact with any open lesion. Type 2 mode is primarily sexual. Lesions are present 2 to 3 weeks and are most painful the first week.
      (3) Assessment findings.
         (a) Type 1.
            1) Grouped vesicles on a red base.
            2) Most commonly noted at corner of mouth (cold sore) but can be anywhere.
            3) Vesicle appears, ulcerates, and encrusts.
            4) When vesicle ruptures a burning pain is felt.
         (b) Type 2.
            1) Grouped vesicles on a red base.
            2) Vesicles rupture and encrust, causing ulcerations.
            3) Cervix and labia most common sites on women.
            4) Penis most common site on men.
      (4) Management.
         (a) Abstain from sexual activity until infection is resolved.
         (b) Antiviral medications are indicated- refer to a medical officer for treatment.
         (c) Counsel patient on safe sex practices - condom usage.

Performance Steps
    b. Herpes zoster vesicle (shingles).
        (1) Lesions located along the nerve fibers of spinal ganglia where inflammation occurs.
        (2) Forms an erythematous rash of small vesicles along a spinal nerve pathway.
        (3) Signs and symptoms.
            (a) Rash generally in thoracic region.
                1) Many occur elsewhere.
                2) Follow dermatomal pattern.
            (b) Vesicles rupture and form crust.
            (c) Serous fluid in vesicle may become purulent.
            (d) Last 7 to 28 days.
            (e) Patients report severe burning and/or knife-like pain.
            (f) Does not cross mid line unless the patient is immuno-compromised.
  3. Identify and manage bacterial disorders.
    a. General characteristics.
        (1) Commonly occur in warm, moist locations but may be secondary to local trauma.
        (2) No safe method to kill all skin bacteria.
    b. Impetigo.
        (1) Assessment findings.
            (a) Appears on face, hands, arms, and legs.
            (b) Pustular lesions distributes over involved area.
            (c) Large amount of dried serous exudate (honey colored crust).
            (d) Spread by touching personal articles, linens, and clothing.
        (2) Manage.
            (a) Consider antiseptic soap.
            (b) Consider application of antibiotic cream, ointment, or lotion.
            (c) Refer to a medical officer for treatment.
    c. Folliculitis, furuncles, carbuncles, and felons.
        (1) Assessment findings.
            (a) Edematous, erythematous, and painful.
            (b) Pruritus commonly occurs.
            (c) Infected area becomes shiny, points up, and if furuncle or carbuncle, the center turns yellow.
            (d) Carbuncles can have four to five cores with spontaneous rupture of core.
            (e) Pain stops immediately upon rupture of core.
        (2) Manage. (Felons may spread to fascial planes in the hand and may require surgical exploration and debridement.)
            (a) Isolate patient to prevent spread of infection.
            (b) Refer to a medical officer for treatment.
  4. Identify and manage fungal infections of the skin.
    a. General characteristics.
        (1) Are not part of the normal flora.
        (2) About 20 species produce skin diseases.
    b. Tinea capitis (ringworm of the scalp).
        (1) Spread by contact with infected articles.
        (2) Trauma or irritation breaks in skin facilitates spread.

Performance Steps
        (a) Assessment findings.
          1) Areas of brittle or broken off hairs with some crusting.
          2) Occasionally pruritus.
          3) Non-scarring alopecia occurs at the site.
        (b) Manage.
          1) Consider oral antifungal drugs.
          2) Consider medical officer referral if topical agents are not effective.
    c. Tinea corporis (ringworm of the body). Occurs in parts of body with little or no hair.
      (1) Assessment findings.
        (a) Produces lesions with raised erythemic borders as lesions expand there is
            central clearing (annular lesion).
        (b) May have scale.
        (c) May or may not have pruritus.
      (2) Manage. Consider use of topical or oral antifungal drugs.
    d. Tinea cruris (jock itch). Found in groin area.
      (1) Assessment findings.
        (a) Produces lesions with raised erythemic borders as lesions expand there is
            central clearing (annular or arciform lesions).
        (b) Pruritus and skin excoriation from scratching may be found.
        (c) May spare scrotum.
      (2) Manage.
        (a) Consider methods of drying out area.
          1) Loose clothing (use of boxers or no underwear).
          2) Powder.
        (b) Consider use of topical or oral antifungal drugs.
    e. Tinea pedis (athlete's foot). Normally starts between 4th and 5th toes and then may
    spread.
      (1) Assessment findings.
        (a) Itching and burning.
        (b) Maceration between toes.
        (c) Cracking and peeling of interdigital skin.
        (d) If secondarily infected may have associated discoloration.
      (2) Manage. Consider methods of drying out area.
        (a) Powder.
        (b) Frequent sock changes.
        (c) Rotation of footwear.
      (3) Consider use of topical or oral antifungal drugs.

  5. Identify and manage inflammatory disorders.
NOTE: This disorder is a local or generalized inflammation caused by a number of factors.
    a. General characteristics.
      (1) Can be caused by numerous agents such as drugs, plants, chemicals, metals,
        and food.
      (2) Erythema and edema in acute disorders.
      (3) Skin thickening and chronic pigmentation in chronic disorders.
      (4) Pruritus is almost always present; if present it can cause excoriation due to
        scratching.
    b. Contact dermatitis (irritant and allergic).
      (1) Caused by direct contact with agents who cause irritation or allergic reaction.
      (2) Epidermis becomes inflamed and damaged.

Performance Steps

        (3) Common causes are detergents, soaps, industrial chemicals, medications, hypersensitivity reactions, and plants such as poison ivy.
           (a) Assessment findings.
               1) Lesions appear at point of contact.
               2) Patient feels burning, pain, pruritus, and edema.
               3) Involved area becomes erythematous with papules.
               4) Vesicles appear most often on dorsal surfaces.
           (b) Manage.
               1) Identify cause of hypersensitive reaction.
               2) Symptomatic treatment for inflammation, edema, and pruritus.

6. Identify and manage immersion skin diseases.
    a. Type 1.
      (1) Assessment findings.
         (a) Intermittent exposure to water with dry ground between.
         (b) Confined to the soles of the feet.
         (c) After three days soles become white and wrinkled.
         (d) Creases of feet grow tender on walking.
         (e) After 24 to 48 hours severe pain develops with walking.
         (f) Feet swell slightly.
         (g) Sensation of walking on pieces of rope in boot.
         (h) May not be able to get boot back on due to swelling.
      (2) Manage.
         (a) Dry skin and keep dry.
         (b) Bed rest without boots or socks.
    b. Type 2.
      (1) Assessment findings.
         (a) Continuously standing or wading through water.
         (b) Involves top of feet, ankles, and legs to the top of boots and socks.
         (c) Affects Soldiers in 48 to 60 hours.
         (d) Skin turns red.
         (e) Cellulitis appears.
         (f) Swelling develops.
         (g) Rubbing of boot may cause large deep raw spots or abrasions.
      (2) Manage.
         (a) Dry skin and keep dry.
         (b) Bed rest with head flat and feet elevated.
    c. Type 3.
      (1) Assessment findings.
         (a) When Soldiers wade through water to their waist or neck.
         (b) Clothing stays wet for hours.
         (c) Skin of groin and inner thighs show damage.
         (d) Caused by prolonged wetness or rubbing.
         (e) Becomes very red and painful.
      (2) Management. Permit skin to dry.
    d. Trench foot.
      (1) Assessment findings.
         (a) Prolonged exposure to water ranging from freezing to 50 °F.
         (b) Can appear on feet or hands.
         (c) Lesions may occur from boots rubbing.

Performance Steps
    (2) Manage. Permit skin to dry.
  e. Consider scabies and lice.
    (1) Assessment findings for scabies.
        (a) Intense itching, especially at night, with vesicles, papules, and linear burrows which contain the mites and eggs.
        (b) Males - lesions prominent around finger webs, anterior surfaces of wrists and elbows, armpits, belt line, thighs, and external genitalia.
        (c) Females - lesions prominent on nipples, abdomen, and lower portion of buttocks.
        (d) Infants - head, neck, palms, and soles may be involved, and are generally not seen in older adults.
        (e) Complications generally due to infection of lesions that are broken from scratching.
    (2) Provide management and protective measures for scabies.
        a) Handle patient's underclothing and home bedding observing body substance isolation (BSI), separate bedding from exposed ambulance linen.
        (b) Consider Kwell®, lindane, or other agents.
    (3) Assessment findings for lice.
        (a) Infestation of head lice is of hair, eyebrows, and eyelashes, mustache, and beards.
        (b) Infestation of body lice is of clothing, especially along the seams of the inner surfaces of clothing.
    (4) Provide management and protective measures for lice.
        (a) Personal treatment - use of appropriate body/ hair pediculicide is recommended, repeated 7 to 10 days later.
        (b) Bag linen separately.

Performance Measures	GO	NO-GO
1. Provided information on the prevention of skin disorders.	——	——
2. Identified and managed viral disorders of the skin.	——	——
3. Identified and managed bacterial disorders.	——	——
4. Identified and managed fungal infections of the skin.	——	——
5. Identified and managed inflammatory disorders.	——	——
6. Identified and managed immersion skin diseases	——	——

Evaluation Guidance: Score each Soldier according to the performance measures. Unless otherwise stated in the task summary, the Soldier must pass all performance measures to be scored GO. If the Soldier fails any steps, show what was done wrong and how to do it correctly.

References
  Required
  None

           Related
           EWS NATO HANDBOOK
           HABIF, T. B.

TREAT ABDOMINAL DISORDERS
081-833-0139

Conditions: You have a patient with an abdominal complaint. You will need a stethoscope and the patient's medical record. You are not in a CBRNE environment.

Standards: Solicit a patient history, perform a physical examination, determine the cause of the disease, and administer supportive care IAW guidance from the medical officer.

Performance Steps

1. Solicit a patient history.
   a. History of present illness (HPI).
   b. Chief complaint (usually in patient's own words).
   c. Use onset, provocation/palliation, quality, region/radiation, severity, timing (OPQRST) type questions.
   d. Past medical history (PMH).
   e. Family history (FH).
   f. Social history (SH).
   g. Menstrual history (women of childbearing age).
   h. Travel history.

2. Perform a Physical Exam.
   a. General.
      (1) Use a systematic approach.
      (2) Start with the uppermost part of the gastrointestinal tract (oral cavity).
   b. Abdomen.
      (1) Inspection.
      (2) Auscultation.
      (3) Palpation.

3. Check for red flags of abdominal symptoms (must see a medical officer).
   a. Abdominal pain with guarding or rebound tenderness or progressive severe pain that persists without improvement for over 6 hours.
   b. Recent (< 6 months) abdominal surgery.
   c. Abdominal pain with fever.
   d. Abdominal pain with tachycardia.
   e. Abdominal pain with dehydration.
   f. Abdominal pain in a pregnant patient.

4. Identify and manage gastroesophageal reflux disease (GERD).
   a. Signs and symptoms.
      (1) Heartburn, burping, regurgitation (worse after eating large meal, when lying down, in the middle of night).
NOTE: Cardiac disease must be ruled out before the diagnosis of GERD is made, especially if the symptom is chest pain (consult with medical officer).
      (2) Physical exam is usually normal.
      (3) Red Flags: bloody vomitus, blood in stool, dark tarry stools, significant weight loss.
   b. Treatment.
      (1) Medications: Antacids (Tums, Rolaids), H2 Blockers (Tagamet, Zantac), or Proton Pump Inhibitors (Prilosec, Aciphex).

Performance Steps
        (2) Lifestyle changes: weight loss, avoid alcohol, tobacco, caffeine and large meals; elevate the head of your bed.

5. Identify and manage gastroenteritis.
    a. Signs and symptoms.
        (1) Nausea, vomiting, diarrhea (may be mild or severe).
        (2) Malaise, fever, abdominal cramps.
        (3) May have history of eating or drinking from an unapproved source.
    b. Physical Exam.
        (1) May be normal, but abdomen may be diffusely tender if prolonged vomiting.
        (2) Normal to increased bowel sounds.
    c. Red Flags.
        (1) Vomiting blood, or bloody diarrhea.
        (2) Fever.
        (3) Signs of dehydration.
        (4) Protracted vomiting or diarrhea.
    d. Treatment.
        (1) Medications: check with medical officer for antiemetic if vomiting is severe.
        (2) Correct fluid loss if signs of dehydration.
        (3) Clear liquid diet.
        (4) Bed rest may be indicated (check with medical officer).

6. Identify and manage constipation.
    a. Signs and symptoms.
        (1) History of delayed or difficult bowel movements (BMs) (may be hard and dry).
        (2) Crampy abdominal pain, painful BMs.
    b. Crampy abdominal pain, painful BMs.
        (1) Usually normal (may have tenderness to palpation if severe).
        (2) Bowel sounds variable (may be increased, normal, or decreased).
    c. Treatment.
        (1) Acute constipation: warm prune juice, laxatives, stool softeners.
        (2) Bowel sounds variable (may be increased, normal, or decreased).

7. Identify and manage abdominal pain with peritoneal signs (appendicitis).
    a. Signs and symptoms.
        (1) Anorexia, with pain that is periumbilical which later migrates into the right lower quadrant (RLQ) over 8 hours.
        (2) Nausea, vomiting, diarrhea.
    b. Physical exam.
        (1) Fever.
        (2) Guarding and rebound tenderness present (patient often will point to the RLQ).
    c. Treatment. Appendicitis is a surgical emergency (refer to the medical officer immediately).

8. Identify and manage hemorrhoids.
    a. Signs and symptoms.
        (1) Itching, pain with BM, rectal bleeding.
        (2) Pain with bowel movement.
        (3) Rectal bleeding.
    b. Physical exam.
        (1) Rectal bleeding, with obvious source (external hemorrhoid visible). If no obvious source refer to medical officer.

Performance Steps

        (2) Ensure vital signs do not indicate a severe hemorrhage problem.

    c. Red flag. If bleeding is excessive or vital signs indicate hypovolemia notify the medical officer immediately.

    d. Treatment.

        (1) Increase fiber in diet.

        (2) Avoid straining.

        (3) Sitz bath (sitting in warm water) for 15 min three times a day.

        (4) Stool softener.

        (5) Pain medications, topical (Dibucaine) or oral (Tylenol).

9. Identify and manage cystitis (bladder infection).

    a. Signs and symptoms.

        (1) Urgency, frequency, dysuria (urinary triad).

        (2) Malodorous urine.

        (3) Hematuria (blood in urine).

    b. Physical exam.

        (1) Fever.

        (2) Costal vertebral angle tenderness (flank pain) with percussion.

    c. Treatment.

        (1) Diagnosis confirmed by urinalysis.

        (2) Requires antibiotics (refer to medical officer).

10. Identify and manage diarrheal conditions.

    a. Acute (symptoms are acute in onset and persist for less than 3 weeks).

NOTE: Acute diarrhea is most commonly caused by viral, bacterial, and parasitic infections, frequently resulting from consumption of unpurified water or improperly stored or prepared food. Non-infectious causes include medications, (such as antibiotics), and food allergies.

        (1) Signs and symptoms.

            (a) Frequent loose or watery stools.

            (b) Abdominal pain and cramping.

            (c) History of travel outside the U.S.

        (2) Physical exam.

            (a) Fever.

            (b) Vital signs, orthostatic hypotension- (tilts) A drop in blood pressure when the patient changes position from lying to sitting or from sitting to standing. A drop of 10 millimeters of mercury (mmHg) in systolic pressure or an elevation of the pulse rate by 20 beats per minute (bpm) can indicate a volume deficit.

            (c) Signs of a viral upper respiratory infection.

            (d) Tenderness on palpation of abdomen.

            (e) Bowel sounds may be (normal, increased, or decreased).

        (3) Red flags: dehydration, bloody diarrhea, blood in stool, dark tarry stools, significant weight loss.

        (4) Treatment.

            (a) Diet: clear liquids, progressing to bland diet. Avoid caffeine, dairy products and raw fruit and vegetables.

            (b) Fluid resuscitation: IV or oral fluid and electrolyte replacement.

            (c) Medications: Kaopectate or Loperamide (Imodium). Antibiotics may be required (check with medical officer).

            (d) Consume food and water only from approved sources.

    b. Chronic diarrhea - diarrhea persisting for more than 3 weeks.

Performance Steps
    (1) Chronic diarrhea may be viral, bacterial, or parasitic in nature. Diarrhea not caused by infections could be attributed to a number of malabsorptive, secretory, inflammatory or motility disorders. Sometimes the cause of chronic diarrhea remains unknown.
    (2) Refer to the medical officer.

Performance Measures	GO	NO-GO
1. Solicited a patient history.	——	——
2. Performed a physical examination.	——	——
3. Checked for red flags of abdominal symptoms.	——	——
4. Identified and managed GERD.	——	——
5. Identified and managed gastroenteritis.	——	——
6. Identified and managed constipation.	——	——
7. Identified and managed abdominal pain with peritoneal signs.	——	——
8. Identified and managed hemorrhoids.	——	——
9. Identified and managed cystitis.	——	——
10. Identified and managed diarrheal conditions.	——	——

Evaluation Guidance: Score each Soldier according to the performance measures. Unless otherwise stated in the task summary, the Soldier must pass all performance measures to be scored GO. If the Soldier fails any steps, show what was done wrong and how to do it correctly.

References
    Required
    None

                Related
                EWS NATO HANDBOOK
                ISBN 0-13-084584-1
                ISBN 0-316-12891-0
                ISBN 0-8151-80002-9
                ISBN 08359-5073-5
                ISBN 08359-5089-1 (PBK)
                MED751

DOCUMENT PATIENT CARE USING SUBJECTIVE, OBJECTIVE, ASSESSMENT, PLAN (SOAP) NOTE FORMAT
081-833-0145

Conditions: You are treating a patient and must record the treatment given. You will need medical documentation forms (as specified by local SOP), a black or blue-black ink pen, and SF 600 (Medical Record - Chronological Record of Medical Care). You are not in a CBRNE environment.

Standards: Record patient care accurately using SOAP note documentation format. Make entries legible in non-erasable black or blue-black ink, use only approved medical abbreviations, do not skip lines or leave space between lines, and finish the entry with the required information. Guard confidentiality of all patient information.

Performance Steps
CAUTION: All body fluids should be considered potentially infectious. Always observe body substance isolation (BSI) precautions by wearing gloves and eye protection as a minimal standard of protection.

1. Record the patient's name, rank, SSN, date, and time.
NOTE: An addressograph card can be used on the patient identification block.

2. Write subjective data.
    a. Chief complaint.
    b. The patient's statements regarding the illness or injury history to include Onset, Provokes, Quality, Radiates, Severity, Time (OPQRST).
    c. Usually expressed in the patient's own words.

3. Write objective data.
    a. Observations by the Soldier Medic that support or are related to the subjective data, to include sight, sound, touch, and smell.
    b. Physical assessment data to include the patient's vital signs.
    c. Lab and radiology results.

4. Write the assessment/analysis.
    a. Your interpretation of the patient's problem/condition.
    b. Conclusions reached based upon analysis of the subjective and objective data.

5. Write the plan.
    a. Course of action to resolve the problem.
        (1) Treatments made.
        (2) Profiles.
        (3) Medications.
        (4) Patient education.
    b. Follow-up appointment or referral.
    c. Each item in your plan should be numbered.

6. Correct recording errors, if applicable.
    a. Draw a single line through the error.
    b. Write the word error above it.
    c. Initial next to the error.
    d. Record the note correctly.

Performance Steps

7. Finish the entry with your signature, printed name, rank, and title.
NOTE: Medical confidentiality of all patient information must be guarded. Unauthorized disclosure of medical information is grounds for UCMJ action against the informant.

Evaluation Preparation:

Setup: For training and evaluation, construct a written scenario that has all of the elements necessary for the Soldier to develop and write a concise record of patient care using the SOAP note documentation format.

Brief Soldier: Tell the Soldier to record patient care using the SOAP note documentation format.

Performance Measures	GO	NO-GO
1. Recorded required patient information.	——	——
2. Wrote subjective data.	——	——
3. Wrote objective data.	——	——
4. Wrote the assessment/analysis.	——	——
5. Wrote the plan.	——	——
6. Corrected errors, if applicable.	——	——
7. Finished the entry with required information.	——	——

Evaluation Guidance: Score each Soldier according to the performance measures. Unless otherwise stated in the task summary, the Soldier must pass all performance measures to be scored GO. If the Soldier fails any steps, show what was done wrong and how to do it correctly.

References
    Required
    SF 600

                     Related
                     BASIC NURSING 7

PERFORM PATIENT HYGIENE
081-833-0165

Conditions: A patient requires assistance with personal hygiene. You will need washcloths, towels, bath blanket or cover sheet, toiletry items, clean hospital gown, gloves, wash basin, toothbrush or toothettes, emesis basin, suction equipment, water soluble lubricant, brush, comb, shampoo, razor, shaving cream, orange stick nail file, sheets, and waterproof pads. You are not in a CBRNE environment.

Standards: Perform patient hygiene without causing further injury to the patient.

Performance Steps

1. Verify the activity with doctor's orders or nursing care plan.

2. Explain the procedure to the patient.
NOTE: Some of the following steps may be omitted based upon the patient's condition. Patients should be encouraged to participate in self care to the extent that they are able.

3. Provide a bed bath.
   a. Provide privacy.
     (1) Close the door and draw a curtain around the patient.
     (2) Expose only the areas being bathed.
   b. Raise the entire bed to comfortable working height.
   c. Place the bath blanket or sheet over the patient and remove top covers without exposing the patient.
   d. Remove the patient's gown.
     (1) If the patient has an IV, remove the gown from the arm without the IV first. Move the IV bag and the tubing through the sleeve and re-hang the bag.
NOTE: If an IV pump is used, turn off the pump, clamp the tube, and then remove it as described above. Unclamp the tube, reinsert it into the pump, turn on the pump, and adjust the rate.
     (2) If the patient has an injured extremity, remove the sleeve from the unaffected side first.
   e. Place a towel under the patient's head.
   f. Wash the face.
     (1) Wash the patient's eyes from inner to outer canthus, using a clean part of the cloth for each eye.
NOTE: If the patient is unconscious, clean the eyes as above. Instill prescribed eye drops or ointment, if applicable (See task 081-835-3022.) If the patient does not have a blink reflex, keep the eyelids closed and cover with a patch. Do not tape the eyelid.
     (2) Wash, rinse, and dry the forehead, cheeks, ears, nose, and neck with plain warm water.
NOTES: Soap tends to dry the face. Men may want to be shaved (see step 5b).
   g. Wash the upper body.
     (1) Remove the bath blanket from over the arm. Place a towel under the arm.
     (2) Bathe the arm using long firm strokes from distal to proximal end.
     (3) Lift the arm above the head if possible and wash and dry the axilla completely.
     (4) Repeat steps 3f (1) through 3f (3) on other arm.
     (5) Apply powder or deodorant to the axilla if applicable.
     (6) Bathe and dry the chest.

Performance Steps

NOTE: Take special care to wash the skin under a female's breasts. Lift the breasts upward if necessary. Clean and dry thoroughly.

    h. Wash the lower body.

        (1) Place a bath towel over the chest and abdomen. Fold the blanket down to just above the patient's pubic region.

        (2) Wash, rinse, and dry the abdomen paying attention to the umbilicus and the skin folds of the abdomen and groin.

        (3) Wash and dry the leg nearest you.

            (a) Place a towel under the leg.

            (b) Support the leg at the knee and place the foot flat on the bed.

NOTE: The patient's foot may be placed in the basin to soak while the leg is being washed. However, soaking feet is NOT recommended for patients with diabetes mellitus or peripheral vascular disease.

            (c) Wash and dry the leg using long firm strokes. Wash from ankle to knee and then from knee to thigh.

CAUTION: Avoid massaging the legs when the patient is at risk for thrombosis or emboli.

            (d) Wash and dry the foot completely.

            (e) Move to the opposite side and repeat steps 3h (3)(a) through 3g (3)(d) for the other leg.

NOTE: Always raise the side rail for safety.

    i. Change bath water and gloves.

    j. Wash the perineum.

        (1) Place the patient in a side lying position and keep the patient covered with a bath blanket as much as possible.

        (2) Wash the buttocks and anus from front to back.

NOTE: If feces is present, wrap it in an underpad fold and remove as much as possible with disposable wipes first. Use as many wash cloths as necessary to clean completely. Ensure to cleanse the gluteal folds.

        (3) Dry the area and replace the underpad with a clean one.

    k. Wash the genitals.

        (1) Female.

            (a) Position the patient supine with a waterproof pad beneath the buttocks. Drape the patient with a bath blanket to maintain privacy.

            (b) Wash the labia majora and then gently pull back the labia majora to wash the groin from perineum to rectum.

            (c) Clean the pubic area from front to back.

NOTE: Clean around an indwelling catheter if applicable without pulling tension on it. Ensure the catheter is secured to the upper thigh or positioned over the thigh (not under it).

        (2) Male.

            (a) Gently grasp the penis. Retract the foreskin if uncircumcised.

            (b) Wash the tip of the penis and urinary meatus cleansing away from the meatus. Use a circular motion.

            (c) Clean the penile shaft, scrotum, and underlying folds.

            (d) Rinse and dry.

    l. Change bath water and gloves.

    m. Wash the back.

        (1) Place the patient on his side.

        (2) Clean and dry the back from neck to buttocks using long firm strokes.

    n. Apply lotion to the skin if needed.

    o. Replace the gown.

Performance Steps

    4. Provide oral care.
       a. Place the casualty in a side-lying position with a towel under the chin. Have an emesis basin available.
       b. Separate the upper and lower teeth.
NOTE: Oral suction must be available, especially if the patient has no gag reflex.
       c. Clean the mouth using a toothbrush, moistened 4 x 4 gauze, or toothette with water. Ensure the tongue, roof of mouth, inside cheeks, and tooth surfaces have been cleaned.
NOTE: The toothbrush should be soft bristled. Angle the brush at 45 degrees to clean the teeth. Avoid using glycerin or lemon swabs.
       d. Rinse with a clean toothette and water.
NOTE: Use as little water as possible to avoid aspiration.
       e. Suction the oral cavity as secretions accumulate if the patient is unable to remove them.
       f. Apply lip balm or water-soluble jelly to the lips.

    5. Provide hair care.
       a. Shampoo the hair.
         (1) Place a towel and waterproof pad under the head.
         (2) Comb or brush the patient's hair to release any tangles.
         (3) Position the patient supine with a plastic trough under the head.
         (4) Pour warm water over the head until completely wet.
NOTES: Protect the patient's face and eyes by placing a towel or washcloth over them. If hair is matted with blood, apply hydrogen peroxide to dissolve it, and then rinse with saline or water.
         (5) Apply shampoo and lather.
         (6) Massage gently starting at the hairline and working toward the back of the scalp.
         (7) Rinse the hair.
         (8) Apply conditioner if needed.
         (9) Dry the hair.
         (10) Complete styling of the hair as necessary.
NOTE: Braids may be helpful to prevent tangling of long hair.
       b. Shave the beard.
         (1) Position the patient into a sitting position if possible. Place a towel over the chest.
         (2) Place a moist, warm washcloth over the patient's face.
         (3) Apply shaving cream.
         (4) While pulling the skin taut, angle the razor to 45 degrees. Shave in the direction of hair growth.
NOTE: Ask the patient to direct you on his usual technique.
         (5) Rinse and dry the face. Apply after shave if patient desires.

    6. Perform foot and nail care.
       a. Using an orange stick, gently clean under the patient's nails.
       b. Clip the nails straight and even with the digits. File the nails to shape and smooth rough edges.
CAUTION: Never cut the toenails. A patient with diabetes or hypertrophy should be referred to a podiatrist.
       c. Push the cuticle back gently with an orange stick.
       d. Apply lotion.

Performance Steps

7. Change the patient's linen (make an occupied bed).
   a. Raise the entire bed to a comfortable working height.
   b. Lower the head of the bed, if tolerated by the patient.
   c. Remove the bedspread or blanket. Leave a sheet covering the patient.
   d. Roll the patient to a side-lying position on the far side of the bed.
NOTE: Make sure any tubing is not pulled.
   e. Roll the bottom sheet, draw sheet, and underpad toward the patient as far as possible.
   f. Place a clean bottom sheet on the bed.
      (1) The sheet may be fitted.
      (2) Flat sheet. Center the sheet on the bed, and pull the bottom hem toward the foot of the bed. Open the sheet toward the patient. Tuck and miter the top under the head of the bed.
   g. Place draw sheets or waterproof pads on the center of the bed. Fan-fold toward the patient.
   h. Cover the unoccupied side of bed with the linen. Tuck the draw sheet under the mattress.
   i. Assist the patient to logroll over all the linen toward the other side of the bed.
   j. Raise the bed rail on the side facing the patient. Go to the other side and lower the bed rail.
   k. Remove soiled linens. Place them on the floor or in the hamper.
CAUTION: Never leave the patient alone with the side rails down.
   l. Pull clean linen toward you. Straighten the linen out.
   m. Tuck and miter the corners.
   n. Tuck in the draw sheet.
   o. Straighten the waterproof pads.
   p. Assist the patient to a supine position.
   q. Place a clean top sheet and blanket over the patient.
   r. Remove the original cover sheet.
   s. Tuck the bottom of the covers under the mattress making a modified miter. Loosen the linen at the feet for comfort.
   t. Change the patient's pillowcase.

8. Assist the patient to a position of comfort and place needed items within reach.

9. Raise the side rails and lower the bed.

10. Remove soiled supplies.

11. Document what was performed and the patient's response. Inability to tolerate a procedure should be documented.

Performance Measures	<u>GO</u>	<u>NO-GO</u>
1. Verified the activity with doctor's orders or nursing care plan.	——	——
2. Explained the procedure to the patient.	——	——
3. Provided a bed bath.	——	——
4. Provided oral care.	——	——

Performance Measures	<u>GO</u>	<u>NO- GO</u>
5. Provided hair care.	___	___
6. Performed foot and nail care.	___	___
7. Changed the patient's linen.	___	___
8. Assisted the patient to a position of comfort and placed needed items within reach.	___	___
9. Raised the side rails and lowered the bed.	___	___
10. Removed soiled supplies IAW local SOP.	___	___
11. Documented the procedures and the patient's response.	___	___

Evaluation Guidance: Score each Soldier according to the performance measures in the evaluation guide. Unless otherwise stated in the task summary, the Soldier must pass all performance measures to be scored a GO. If the Soldier fails any step, show what was done wrong and how to do it correctly.

References
 Required   **Related**
 None   BASIC NURSING 7

PERFORM VISUAL ACUITY TESTING
081-833-0193

Conditions: You have a patient who requires evaluation of visual acuity. You will need an opaque eye card and a Snellen chart. You are not in a CBRNE environment.

Standards: Administer the visual acuity test.

Performance Steps

1. Identify the patient and provide privacy.

2. Wash your hands. (See task 081-831-0007.)

3. Explain the procedure to the patient/family.

4. Position the patient 20 feet from the eye chart.

5. Instruct the patient to leave corrective lenses on, if worn, except for reading glasses.

6. Cover one eye with an opaque card and instruct the patient to read through the chart to the smallest line possible. Have the patient read from top to bottom, left to right.

7. Repeat the procedure for the other eye.

8. Repeat the exam with both eyes open.
NOTE: Certain situations, such as physical exams, may require this exam to be performed with and without corrective lenses. Review the facility standing operating procedure for guidance. For pediatric patients who do not read, use symbols and picture charts.

9. If in field environment determine gross visual acuity.
    a. Have the patient read any printed material available.
    b. If the patient is unable to read printed material--
        (1) Have patient determine the number of fingers held up.
        (2) Distinguish between light and dark.
NOTE: Record the results as a fraction with the number 20 (= distance) as the numerator and the number of the last line read (= acuity) as the denominator. Also indicate the number of letters missed. Example: 20/30 -2.

10. Record results of the test.

Performance Measures	GO	NO-GO
1. Identified the patient and provided privacy.	——	——
2. Washed hands.	——	——
3. Explained the procedure to the patient/family.	——	——
4. Positioned the patient 20 feet from the eye chart.	——	——
5. Instructed the patient to leave corrective lenses on during exam.	——	——
6. Tested one eye.	——	——

Performance Measures	GO	NO-GO
7. Repeated testing on the other eye.	____	____
8. Repeated testing with both eyes open.	____	____
9. Performed gross visual acuity in field environment, if applicable.	____	____
10. Recorded the results.	____	____

Evaluation Guidance: Score each Soldier according to the performance measures. Unless otherwise stated in the task summary, the Soldier must pass all performance measures to be scored GO. If the Soldier fails any steps, show what was done wrong and how to do it correctly.

References

Required	Related
BATES, B.	None

REMOVE A PATIENT'S RING
081-833-0195

Conditions: You have a patient who requires a ring to be removed. You will need a Penrose drain, water-soluble lubricant, and a 25 inch length of umbilical tape or string, or thick silk suture. You are not in a CBRNE environment.

Standards: Remove the patient's ring without causing further injury to the patient.

Performance Steps

1. Lubricate the digit with a water-soluble lubricant and apply traction on the ring while turning in a circular motion.
NOTE: Frequently a ring must be removed to prevent laceration of tissue or vascular compromise.

2. Attempt to remove the ring using the string-wrap method.
    a. Wrap the penrose drain circumferentially around the finger in a distal to proximal direction to reduce soft tissue swelling. For maximal effect, the wrap should stay in place for a few minutes.
    b. A 20 to 25 inch piece of string, umbilical tape or thick silk suture is first passed between the ring and the finger. If there is marked soft tissue swelling, the tip of a hemostat may be passed under the ring to grasp the string and pull it through the ring.
    c. The distal string is wrapped clockwise around the swollen finger (proximal to distal) to include the proximal interphalangeal (PIP) joint and the entire swollen finger.
        (1) The wrapping is begun next to the ring and should be snug enough to compress the swollen tissue.
        (2) The successive loops of the wrap are placed next to each other to keep any swollen tissue from bulging between the strands.
    d. When the wrapping has been completed, the proximal end of the string is carefully unwound in the same clockwise direction, forcing the ring over that portion of the finger that has been compressed by the wrap.

3. If unsuccessful, a ring cutter should be used if there is excessive swelling.
    a. The ring cutter has a small hook that fits under the ring and serves as a guide for a saw-toothed wheel that cuts the metal.
    b. The cut ends of the ring are spread using large hemostats and the ring is removed.

4. Do not cause further injury to the patient.

Performance Measures	GO	NO-GO
1. Lubricated the digit with a water-soluble lubricant and applied traction on the ring while turning in a circular motion.	——	——
2. Attempted to remove the ring using the string-wrap method.	——	——
3. Attempted to remove the ring using a ring cutter.	——	——
4. Did not cause further injury to the patient.	——	——

Evaluation Guidance: Score each Soldier according to the performance measures. Unless otherwise stated in the task summary, the Soldier must pass all performance measures to be scored GO. If the Soldier fails any steps, show what was done wrong and how to do it correctly.

References
    Required
    None

    Related
    BASIC NURSING 7

TREAT COMMON EYE, EAR, NOSE, AND THROAT (EENT) DISORDERS
081-833-0203

Conditions: You encounter a patient with symptoms of an eye, ear, nose, and throat (EENT) disorder. No other injuries are present. You will need printed material, Snellen chart, penlight, otoscope, and the patient's medical records. You are not in a CBRNE environment.

Standards: Perform interventions for the management of EENT disorders.

Performance Steps

1. Solicit patient history, perform a physical exam of the eyes and provide treatment.
    a. Solicit patient history.
        (1) Mechanism of injury (if there is a history of injury)?
        (2) Does the patient wear glasses or contact lenses?
        (3) History of eye disease or previous eye trauma/surgery?
        (4) Is there eye pain or loss of vision? If there is vision loss, is it in one eye or both?
    b. Perform a physical exam.
        (1) Includes adequate lighting and avoiding pressure on the globe.
        (2) Visual acuity is the most important step in evaluating ocular problems. Have the patient:
            (a) Read any available printed material.
            (b) Count fingers.
            (c) Distinguish between light and dark.
            (d) Read the snellen chart (eye chart). (See task 081-833-0193.)
        (3) If a patient has corrective lenses, test without glasses first, and then test with glasses on.
        (4) Note any drainage or bleeding from the eye.
        (5) Eyelids - inspect lids for edema, discoloration, and foreign bodies.
        (6) Conjunctiva and sclera.
            (a) Inspect for erythema or exudate.
            (b) Note color of sclera.
        (7) Cornea- should be clear and avascular.
        (8) Pupil-note any irregularity in the shape of the pupils.
NOTE: Unequal size of pupils (anisocoria) may be congenital (approximately 20% of normal people have minor or noticeable differences in pupil size, but pupillary reaction is normal.)
        (9) Test pupillary reaction to light both directly and consensually.
            (a) Dim the lights in the room so that the pupils dilate.
NOTE: Do not shine into both eyes simultaneously.
            (b) Shine a penlight directly into one eye and observe the pupil constrict.
            (c) Note consensual reaction response of the opposite pupil constricting simultaneously with the tested pupil.
    c. Provide treatment.
        (1) Medical officer consultation is required in all ocular complaints.
        (2) Eye pain with decreased visual acuity should be considered an emergency and be evaluated on an urgent basis.

2. Solicit patient history, perform a physical exam of the ear and provide treatment.
    a. Solicit patient history.
        (1) Ask patient about symptoms associated with ear problems, pain, swelling, redness, drainage.

Performance Steps

          (2) Ask patient about history of recent illness, upper respiratory infection (URI), sore throat, or recent swimming.

          (3) Ask patient about any hearing loss.

      b. Perform a physical exam.

          (1) Examine external pinna for erythema, swelling, deformity, trauma, or drainage from canal.

          (2) Using otoscope observe canal and tympanic membrane for signs of inflammation, infection, or foreign body material.

NOTE: Stabilize examining hand against patients' head to prevent injury. If the TM is obscured by cerumen, the canal can be cleaned by warm water irrigation. If you suspect perforation of the TM, or if the auditory canal is filled with blood or discharge, it should never be irrigated. Consult with the medical officer prior to ear irrigation

          (3) TM should have no perforations. Look for air bubbles, air/fluid levels and scarring. A healthy TM may be pearly gray to amber in color. Redness indicates infection.

          (4) Inner ear - tested by evaluating hearing. Gross hearing testing begins when the patient responds or fails to respond to your questions. Complete hearing evaluations are performed through Audiology services

      c. Provide treatment - refer to medical officer.

   3. Solicit patient history, perform a physical exam of the nose and sinuses and provide treatment.

      a. Solicit patient history.

          (1) Ask about recent upper respiratory infection symptoms, drainage, bleeding, congestion, trauma, pressure, recent dental problems, and pain.

          (2) Ask about any medications, prior history or chronic illnesses.

          (3) Ask about duration of symptoms.

          (4) Ask about any allergies or family history of allergies.

NOTE: If the patient has a history of multiple nosebleeds, ask about family history of bleeding problems.

      b. Perform a physical exam.

          (1) Examine the external structure of the nose, inspect and palpate.

          (2) Examine the paranasal sinuses, inspect and palpate the frontal and maxillary sinuses.

      c. Provide treatment.

          (1) Epistaxis - have the patient sit up and lean forward. Tip the head downward and pinch the entire nose firmly for 10-15 minutes. If this does not control the bleeding, then refer to the medical officer.

          (2) Sinuses - refer to medical officer for decongestants or antihistamines.

   4. Solicit patient history, perform a physical exam of the mouth and oropharynx and provide treatment.

      a. Solicit patient history.

          (1) Ask patient about recent symptoms, bad breath, sore throat, hoarseness, difficulty swallowing, and inability to open mouth.

          (2) Ask patient about duration of symptoms, smoking habits and drooling.

      b. Perform a physical exam.

          (1) Lips - inspect and palpate.

Performance Steps

          (2) Buccal mucosa, teeth, and gums- have the patient remove any dental appliances and open mouth partially. Use a tongue blade and bright light to inspect. Inspect and count the teeth, noting cavities, ulcerations, lesions or missing teeth.

          (3) Tongue- inspect.

          (4) Oropharynx- inspect.

       c. Provide treatment - refer to medical officer.

   5. Record all treatment in the patient's medical record.

   6. Seek the advice and assistance of a higher medical authority whenever possible.

Performance Measures	GO	NO-GO
1. Solicited patient history, performed a physical exam and provided treatment of the eyes.	——	——
2. Solicited patient history, performed a physical exam and provided treatment of the ear.	——	——
3. Solicited patient history, performed a physical exam and provided treatment of the nose and sinuses.	——	——
4. Solicited patient history, performed a physical exam and provided treatment of the mouth and oropharynx.	——	——
5. Recorded all treatment in the patient's medical record.	——	——
6. Sought the advice and assistance of a higher medical authority whenever possible.	——	——

Evaluation Guidance: Score each Soldier according to the performance measures. Unless otherwise stated in the task summary, the Soldier must pass all performance measures to be scored GO. If the Soldier fails any steps, show what was done wrong and how to do it correctly.

References
    Required
    None

                   Related
                   ISBN 0-07-065351-8

TREAT SUBUNGUAL HEMATOMA
081-833-0208

Conditions: You have a patient requiring treatment for a subungual hematoma. You will need a #11 scalpel blade or a paper clip and a lighter. You are not in a CBRNE environment.

Standards: Solicit the patient history and treat a subungual hematoma without causing further injury to the patient.

Performance Steps

1. Obtain a history of the patient's complaint.

2. Gather the materials for the procedure.

3. Perform a patient care hand-wash. (See task 081-831-0007.)

4. Explain the procedure to the patient.

5. Treat the hematoma.
   a. Scalpel blade. Place the tip of the scalpel blade on the nail and twist until blood drains.
   b. Paper clip.
      (1) Heat the paper clip until the tip is red-hot.
      (2) Applying gentle pressure, puncture the nail with the hot paper clip and drain the blood.

6. Soak the affected finger or toe in antibacterial soap and water twice a day for 2 to 3 days.

7. Record all treatment given.

Evaluation Preparation: This task is best evaluated by verbalization of the steps. Give the Soldier a simulated patient and a scenario in which he must treat a patient's subungual hematoma.

Performance Measures	GO	NO-GO
1. Obtained a history of the patient's complaint.	——	——
2. Gathered the materials for the procedure.	——	——
3. Performed a patient care hand-wash.	——	——
4. Explained the procedure to the patient.	——	——
5. Treated the hematoma.	——	——
6. Soaked the affected finger or toe in antibacterial soap and water twice a day for 2 to 3 days.	——	——
7. Recorded all treatment given.	——	——

Evaluation Guidance: Score each Soldier according to the performance measures. Unless otherwise stated in the task summary, the Soldier must pass all performance measures to be scored GO. If the Soldier fails any steps, show what was done wrong and how to do it correctly.

References
> **Required**
> None

> **Related**
> ISBN 0-07-065351-8

MAINTAIN A HEALTH RECORD
081-833-0220

Conditions: You need to maintain a health record. You have access to medical records and local standing operating procedures (SOP). You will need a terminal digit file system (if available). You are not in a CBRNE environment.

Standards: Identify and correct discrepancies on the record jacket and within the record, and file the record in the terminal digit file.

Performance Steps

1. Ensure that records are properly filed in the terminal digit file.
    a. Medical records are filed on the basis of the last four digits of the patient's social security number.
    b. Records are filed sequentially in numerical order.
    c. If files are removed, they are replaced with a charge out file.
NOTE: Medical forms accumulated during the absence of the medical record are placed in the charge out file until the medical record is returned.
        (1) The charge out file should be marked with the patient's name, rank, and social security number.
        (2) The charge out file should also be marked with the destination of the medical record or the reason for its removal.

2. Ensure that record jackets are serviceable and properly marked. Initiate a new record jacket if needed. Refer to Figure 3-13.

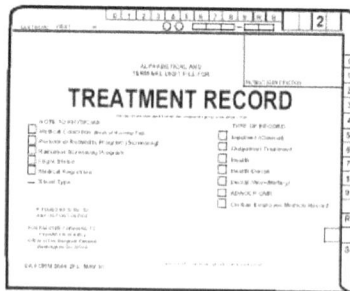

Figure 3-13. Medical record jacket

3. Ensure that forms within the medical record are properly filed.
    a. Check the form for the required patient identification data.
    b. Determine on which side of the folder to file the form.
    c. File the form in the proper sequence on its respective side.

Evaluation Preparation: This task is best evaluated by performance of the steps. Give the Soldier a real or training medical record which he must maintain.

Performance Measures	GO	NO-GO
1. Ensured that records were properly filed in the terminal digit file.	——	——
2. Ensured that record jackets were serviceable and properly marked. Initiated a new record jacket if needed.	——	——
3. Ensured that forms within the medical record were properly filed.	——	——

Evaluation Guidance: Score each Soldier according to the performance measures. Unless otherwise stated in the task summary, the Soldier must pass all performance measures to be scored GO. If the Soldier fails any steps, show what was done wrong and how to do it correctly.

References

    Required
    None

    Related
    AR 40-66

TREAT COMMON MUSCULOSKELETAL DISORDERS
081-833-0222

Conditions: You have a patient who presents with a musculoskeletal complaint. You will need a shoulder sling, splint, ace wrap, crutches anti-inflammatory drugs, ice packs, and patient's medical record. You are not in a CBRNE environment.

Standards: Treat common musculoskeletal or foot complaints without causing further harm to the patient.

Performance Steps

1. Review the patient's medical record, if available.
2. Obtain a history.
   a. O = onset. When did it start? What were you doing when it started? What was the position of the foot (inverted, supinated)?
   b. P = provocative and palliative factors. What makes it better? What makes it worse? Is there any pain without weight bearing?
   c. Q = quality. Is it sharp, dull, aching, pounding, constant, or intermittent? What is the character of the pain?
   d. R = region and radiation. Where exactly is the pain? Does it seem to spread anywhere or does it stay right there? Is there any involvement of other joints?
   e. S = severity. How bad is the pain? Is it incapacitating? Does it cause you to change your activity?
   f. T = time, temporal characteristics (duration). When does it hurt? How long does it last? Have you had prior episodes? Any history of trauma or prior surgery?
3. Manage cervical pain.
   a. Cervical strain.
      (1) A strain happens when a muscle-tendon unit is overloaded or stretched.
      (2) Motion of the neck becomes painful.
      (3) Peaks after several hours or the next day.
      (4) Treat with nonsteroidal anti-inflammatory drugs (NSAIDs), heat, massage, and other therapeutic modalities.
   b. Cervical sprain.
      (1) Movement is limited.
      (2) Ligamentous disruption may be extensive enough to result in instability with associated neurologic involvement.
      (3) Routine cervical spine radiographs are indicated.
      (4) Treatment of a cervical sprain consists of immobilization, rest, support, and NSAIDs.
      (5) Return to participation is permitted when motion and muscle strength normalize.
   c. Cervical fracture. Any patient suspected of cervical fracture or having any neurologic deficit as a result of a cervical injury requires x-rays and must be evaluated by a medical officer.
4. Manage low back pain.
   a. Lumbosacral strain (mild to moderate).
      (1) Signs and symptoms.
         (a) Usually have reduced range of motion.
         (b) Discomfort which is localized to the lumbar-sacral area.
         (c) Palpable muscle tenderness/spasm.

Performance Steps

        (d) Negative straight leg raise (SLR).

    (2) Treatment. Decrease activity and ice massages. Medications, if required, usually consist of anti-inflammatory drugs and/or muscle relaxants. Often obesity is a factor in low back pain and patients should be encouraged to lose weight.

5. Manage shoulder pain.
   a. Rotator cuff tear.
     (1) Usually presents with shoulder pain/tenderness.
     (2) History of trauma.
     (3) Patient is unable to abduct the arm or hold it abducted against gravity.
     (4) Treat initially with a shoulder sling and oral anti-inflammatory drugs (ASA, Motrin). Any shoulder complaint with a history of trauma must be referred to a medical officer.
   b. Impingement Syndrome (shoulder pain).
     (1) Most common cause of shoulder pain and refers to mechanical compression and/or wear of the rotator cuff tendons.
     (2) Any process which compromises this normal gliding function may lead to mechanical impingement.
     (3) Most commonly seen in tennis players, pitchers and swimmers.
     (4) The first step in treating shoulder impingement is to eliminate any identifiable cause or contributing factor.
     (5) Non-steroidal anti-inflammatory medication may be used.
     (6) The mainstay of treatment involves exercises to restore normal flexibility and strength to the shoulder girdle.
   c. Acute bursitis.
     (1) Usually produces pain with movement.
     (2) Follows overuse in most instances.
     (3) Most frequently tender to palpation over subdeltoid bursa.
     (4) Treated with anti-inflammatory drugs and progressive shoulder exercises. There should be a reduction of certain physical activities including lifting, pushups and pulling for 7 days.
   d. Septic arthritis.
     (1) Should be considered if the patient has a fever or other signs and symptoms of inflammation.
     (2) Emergent referral to a medical officer is indicated.
   e. Dislocation.
     (1) Usually follows a history of trauma but may occur spontaneously in some people.
     (2) Sudden onset of pain with gross deformity of shoulder joint.
     (3) Severe limitation of motion.
     (4) X-ray should be done to rule out (R/O) associated fracture if a history of trauma. Often deferred until after reduction in order not to delay.
     (5) Splint and assess distal pulses.
     (6) Prompt referral to a medical officer.
     (7) Pain medication and/or muscle relaxant may be used to relieve anxiety, pain and muscle spasm prior to reducing.

6. Manage knee pain.
   a. Septic knee joint.
     (1) Hot, tender knee with or without swelling.
     (2) Orthopedic emergency requiring referral to a medical officer.
   b. Sprain/strain.

Performance Steps

        (1) Tenderness over medial collateral ligament (MCL)or lateral collateral ligament (LCL) without laxity may indicate grade I sprain or strain.

        (2) If mild laxity and tenderness of MCL/LCL is present, possible grade II sprain.

        (3) If ecchymosis, effusion present with laxity, possible grade III sprain (torn ligament).

        (4) Initial treatment consists of ice packs, ace wrap and elevation for the first 24 hours. Crutches may be indicated for comfort. Anti-inflammatory agents are used as required.

    c. Patellar dislocation. Gross instability of the patella indicates that injury to the soft tissues of the medial aspect of the knee has been extensive.

        (1) When dislocation of the patella occurs alone, it may be caused by a direct force or activity of the quadriceps, and the direction of dislocation of the patella is usually lateral.

        (2) Spontaneous reduction may occur if the knee joint is extended.

        (3) Initially treat with rest, ice, compression, elevation (RICE), NSAID, profile, and crutches if unstable. Will need ortho referral to evaluate for arthroscopic surgery.

    d. Retropatellar (patellofemoral) pain syndrome.

        (1)) The symptoms probably represent the majority of knee pain complaints in athletes.

        (2) Vague knee pain, which is usually after several hours of exercise.

        (3) Walking downhill or downstairs, bending at the knees, and kneeling exacerbates pain.

        (4) Initially treat with RICE, NSAID, stretches and exercises to strengthen quadriceps. Physical therapy consult for prolonged cases.

7. Manage foot pain.

    a. Perform a physical examination (PE).

        (1) Inspect the problem area.

        (2) Determine the range of motion.

        (3) Palpate the problem area.

        (4) Check muscle strength.

    b. Refer to a medical officer (MO) for x-rays of problem area, if appropriate and available.

    c. Formulate assessment based upon history, PE and/or x-rays.

8. Manage ankle injuries.

    a. Grade I ankle sprain.

        (1) Antalgic gait.

        (2) Able to bear weight.

        (3) Minimal edema.

        (4) Mild tenderness of malleolar area.

        (5) Negative drawer sign.

        (6) Initially treated with ice, compression, and elevation for 24-48 hrs. Crutches are indicated for up to 48 hrs in Grade I sprains. Anti-inflammatory agents (Motrin) and ace wrap protection are indicated for 5-7 days; with gradually increased exercises.

    b. Grade II ankle sprain.

        (1) Unable to bear weight.

        (2) Edema.

        (3) Possible ecchymosis.

        (4) Acute tenderness.

Performance Steps
        (5) Negative drawer sign.
        (6) Neurovascular status intact.
        (7) Range of motion reduced.
        (8) An x-ray should be done to rule out an associated fracture.
        (9) May require posterior or "U" splinting for 3-5 days with ice, elevation, crutches and
            analgesics (Motrin). An ace wrap is indicated with gradual increase of activity
            after 72 to 96 hours.
    c. Grade III ankle sprain.
        (1) Unable to bear weight.
        (2) Edema.
        (3) Ecchymosis present.
        (4) Acute tenderness.
        (5) Positive drawer sign.
        (6) Neurovascular status may be compromised.
        (7) Range of motion markedly reduced.
        (8) Should be referred to a MO for x-rays to be done to rule out an associated
            fracture.
        (9) Immobilization using either a splint or non-weight bearing cast. Initially, ice,
            compression, and elevation are used to reduce edema and pain. Crutches,
            without weight bearing, and follow-up with podiatry or orthopedics is usually
            indicated. Nonsteroidal anti-inflammatory drugs or a mild narcotic will often be
            needed for pain relief. In all sprains, physical activity must be reduced
            appropriately and will vary in length from 72 hours to several weeks.

  9. Manage achilles tendonitis.
NOTE: A frequent complaint in runners.
    a. Pain, swelling, tenderness along tendon.
    b. Treat Achilles tendonitis with RICE, NSAIDs, ice for twenty minutes after activity, heel
       lift, and crutches if severe.

 10. Manage metatarsalgia.
    a. Pain under the metatarsals that is exacerbated with functional activities and may
       present as burning.
    b. Commonly seen in women and in the second metatarsal.
    c. Important to rule out stress fracture, neuroma, and avascular necrosis of the
       metatarsal head.
    d. Conservative management is directed at relieving the pressure beneath the area of
       maximum pain.
    e. The patient should obtain a shoe of appropriate style and adequate size to allow an
       orthotic device to be inserted.

 11. Manage bunion.
    a. Excessive bony growth (exostosis) on the head of the first metatarsal.
    b. Callous formation and bursal inflammation.
    c. The patient should be encouraged to wear shoes of adequate size and shape.
    d. Pads may be placed in the first web space or over the median eminence to help take
       pressure off of a painful median eminence.
    e. Pads are also may be placed underneath the metatarsal heads to take pressure off
       painful calluses or sesamoids.
     f. Podiatric surgical intervention may be considered.

Performance Steps

12. Manage plantar fasciitis.
    a. Inflammation of plantar aponeurosis.
    b. Tenderness along plantar fascia.
    c. Treatment of pain 1-2 weeks in duration is with NSAIDs, rest, and stretches.
    d. The patient should perform ice massages with a cold bottle under the arch after activity.
    e. Over the counter insole arch support may help alleviate tension on the arch.
    f. Chronic pain may require a podiatry consult.

13. Record all treatment in the patient's medical record.

Performance Measures	**GO**	**NO-GO**
1. Reviewed the patient's medical record if available.	——	——
2. Obtained history.	——	——
3. Managed cervical pain.	——	——
4. Managed low back pain.	——	——
5. Managed shoulder pain.	——	——
6. Managed knee pain.	——	——
7. Managed foot pain.	——	——
8. Managed ankle injuries.	——	——
9. Managed Achilles tendonitis.	——	——
10. Managed Metatarsalgia.	——	——
11. Managed Bunion.	——	——
12. Managed plantar fasciitis.	——	——
13. Recorded all treatment in the patient's medical record.	——	——

Evaluation Guidance: Score each Soldier according to the performance measures. Unless otherwise stated in the task summary, the Soldier must pass all performance measures to be scored GO. If the Soldier fails any steps, show what was done wrong and how to do it correctly.

References
Required	Related
None	BERHOW, R (16)

TREAT COMMON RESPIRATORY DISORDERS
081-833-0223

Conditions: You have a patient with a respiratory complaint. You must treat the respiratory disorder. You have a stethoscope, sphygmomanometer, oto-ophthalmoscope, tongue depressors, and patient's medical records. You are not in a CBRNE environment.

Standards: Identify the signs and symptoms of a common respiratory disorder and treat without causing any further injury.

Performance Steps

1. Identify the signs and symptoms of pneumonia.
NOTE: When trying to determine the likely causes of pneumonia, you must consider where and how patients acquired it and what other medical problems they have. These considerations are important because of the different types of infectious agents that cause pneumonia.
   a. Solicit a patient history.
      (1) Cough frequently productive of purulent sputum (green, brown, rusty colored).
      (2) Chest pain, pleuritic (worse with cough or deep breath).
      (3) Shortness of breath at rest.
      (4) Malaise, lethargy.
      (5) Poor appetite.
   b. Perform a physical exam and identify findings of pneumonia.
      (1) Fever occasionally with shaking chills.
      (2) Tachycardia.
      (3) Tachypnea.
      (4) Respiratory distress (retractions).
      (5) Abnormal breath sounds: rhonchi, rales, wheezing.
      (6) Abnormal pulse oximetry < 95%.
NOTE: Possible red flags (warning) for pneumonia: fever > 101 degrees and shortness of breath.
   c. Consult with medical officer for treatment as applicable.
   d. Manage the respiratory disorder and provide treatment.
      (1) Motrin or tylenol for fever.
      (2) Decongestant: sudafed, entex (do not give antihistamines).
      (3) Cough suppressants if trouble sleeping at night.
      (4) Increase fluid intake.
      (5) Antibiotics are the mainstay of therapy (consult medical officer).
      (6) Bronchodilators: albuterol inhaler.
      (7) Consider bed rest/ profile, evacuate if in field environment.
   e. Record all treatment in the patient's medical record.
   f. Patients with a suspected pneumonia will be referred to a medical officer.

2. Identify the signs and symptoms of asthma.
   a. Solicit a patient history (may vary widely from mild to life threatening).
      (1) Shortness of breath after exercise or upon awakening.
      (2) History of wheezing.
      (3) Chronic cough (usually non productive).
      (4) Nocturnal attacks.
      (5) Triggers.
         (a) Emotional upsets.

Performance Steps

        (b) Physical exertion.

        (c) Cold weather.

        (d) Upper respiratory infection (URI).

        (e) Allergic components: pollen, mold, house dust, animal dander, smoke, medications, etc.

  b. Perform a physical exam and identify findings of asthma.

      (1) Dyspnea.

      (2) Wheezing.

      (3) Cough.

      (4) Tachycardia.

      (5) Decreased blood oxygenation.

NOTE: Possible red flags (warning) for pneumonia to be aware of are: shortness of breath, severe wheezing and accessory muscle use.

  c. Consult with medical officer for treatment as applicable.

  d. Manage the respiratory disorder and provide treatment (acute attacks only).

      (1) Inhaled bronchial dilators either MDI or nebulizers.

      (2) IV hydration.

      (3) Oxygen.

      (4) Refer to medical officer.

      (5) Evacuate immediately if in field environment.

  e. Record all treatment in the patient's medical record.

  f. Seek the advice and assistance of a higher medical authority whenever possible.

3. Identify the signs and symptoms of a viral upper respiratory infection.

  a. Solicit a patient history.

      (1) Nasal congestion.

      (2) Sore throat.

      (3) Cough (productive or non-productive).

      (4) Hoarseness.

      (5) Malaise.

      (6) Fatigue.

      (7) Headache.

      (8) Sinus pressure.

  b. Perform a physical exam and identify findings of a viral upper respiratory infection.

      (1) Eyes; conjunctiva injected, increased lacrimation.

      (2) Ears; tympanic membrane may be injected, moves poorly with Valsalva maneuver.

      (3) Nose: mucoid or purulent nasal discharge, swollen mucus membranes, decreased air movement.

      (4) Throat: oropharynx injected, tonsillar pillars may be swollen with or without exudate.

      (5) Neck: supple, tender to palpation with shoddy, lymph nodes usually in the anterior chain.

      (6) Chest: lungs may be clear or have scattered rhonchi or mild wheezing, usually no retractions or accessory muscle use.

      (7) Vital signs: temperature, normal to low grade 100-101.

NOTE: Red flags are: fever > 101, shortness of breath, productive cough with chest pain, tonsillar swelling with exudate, pain when touching chin to chest, and difficulty swallowing saliva.

  c. Consult with medical officer for treatment as applicable.

Performance Steps
    d. Manage the viral upper respiratory infection and provide treatment.
       (1) Treatment is symptomatic: decongestants, throat lozenges, cough syrup, Tylenol, increased fluids, rest, etc.
    e. Record all treatment in the patient's medical record.
    f. Seek the advice and assistance of a higher medical authority whenever possible.

Evaluation Preparation: This task is best evaluated by verbalization of the steps. Give the Soldier a scenario in which he must manage respiratory disorders.

Performance Measures	**GO**	**NO-GO**
1. Identified the signs and symptoms of pneumonia. Treated the pneumonia.	——	——
2. Identified the signs and symptoms of asthma. Treated the asthma.	——	——
3. Identified the signs and symptoms of a viral upper respiratory infection. Treated the viral upper respiratory infection	——	——

Evaluation Guidance: Score the Soldier go if all steps are passed. Score the Soldier no-go if any steps are failed. If the Soldier fails any step, show what was done wrong and how to do it correctly.

References
    Required
    None

                      Related
                      ISBN 0-07-065351-9

Subject Area 7: Musculosketeal

APPLY A SAM SPLINT
081-831-1052

Conditions: You have a casualty with a suspected fracture to an arm or leg. All other more serious injuries have been assessed and treated. You need to apply a SAM splint. You will need a SAM splint and four muslin bandages. You are not in a CBRNE environment.

Standards: Splint the limb with the suspected fracture so it does not move and circulation is not impaired.

Performance Steps

    1. Prepare the casualty for application of the splint.
        a. Have the casualty lie on his back with the affected limb exposed.
        b. Check for a distal pulse below the break.
WARNING: If a pulse is not found, apply gentle manual traction in line with the long axis of the limb. This maneuver may restore the pulse. If a pulse does not return after one attempt, splint the limb in the most comfortable position for the casualty and evacuate the casualty as soon as possible.

    2. Prepare the SAM splint.
        a. Unroll and flatten the splint.
        b. Fold the splint. Refer to Figure 3-14.

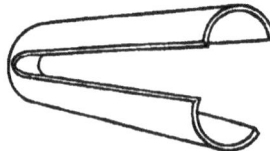

Figure 3-14. Fold SAM splint

        (1) If you will apply the splint to a forearm, wrist, lower leg, or ankle, fold it in half so it is a tall V-shape.
        (2) If you will apply the splint to an upper arm, fold it into an irregular (uneven) V-shape so one side of the V is about 4 to 6 inches shorter than the other.
        c. Bend the edges of the splint in until the shape of the splint generally conforms to the curve and shape of the limb being splinted.

    3. Prepare cravats from muslin bandages.
NOTE: If muslin bandages are not available, strips of cloth from a blanket or clothing or tape from an aid bag can be used.

    4. Apply the splint with the limb in a position of function.
        a. If you are applying the splint to a forearm or wrist, position it so the bend is at the elbow and the fracture is between the two sides of the splint. Refer to Figure 3-15.

Performance Steps

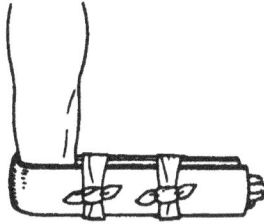

Figure 3-15. Splint with the limb

b. If you are applying the splint to a lower leg or ankle, position it so the bend is on the bottom of the footgear or foot. Refer to Figure 3-16.

Figure 3-16. Splint to a lower leg or ankle

c. If you are applying the splint to an upper arm, position it so the short side is in the casualty's armpit (but not pressing on the armpit), the long side extends to the shoulder, and the upper arm is between the two sides to the splint. Refer to Figure 3-17.

Figure 3-17. Splint to an upper arm

d. Adjust the shape of the splint to conform to the limb, if necessary.

5. Secure the splint using at least two cravats.
    a. Position the cravats above and below the fracture site.
WARNING: Do not apply a cravat directly over the fracture site.
    b. Tie the tails of the cravat in a nonslip knot on the outside of the splint.
    c. Tuck the ends of the tails into the cravat to prevent accidental entanglement when the casualty is moved.

Performance Steps

6. Recheck the casualty's pulse below the most distal cravat. Loosen the cravats and reapply the splint, if needed.

NOTE: The distal pulse should be checked periodically to ensure that swelling has not compromised the pulse.

7. Apply a sling and swathes to further immobilize a fractured arm.

8. Evacuate the casualty.

Evaluation Preparation:

Setup: Provide a Soldier to act as a casualty. Have the SAM splint and materials for securing and immobilizing the splint available.

Brief Soldier: Tell the Soldier that the casualty has a suspected closed fracture and where it is located. Tell the Soldier to splint the suspected fracture. Do not evaluate performance measure 6 in the simulated mode.

Performance Measures

	GO	NO-GO
1. Prepared the casualty for application of the splint.	____	____
2. Prepared the SAM splint.	____	____
3. Prepared cravats from muslin bandages.	____	____
4. Applied the splint with the limb in a position of function.	____	____
5. Secured the splint using at least two cravats.	____	____
6. Rechecked the casualty's pulse below the most distal cravat. Loosened the cravats and reapplied the splint, if needed.	____	____
7. Applied a sling and swathes to further immobilize a fractured arm.	____	____
8. Evacuated the casualty.	____	____

Evaluation Guidance: Score each Soldier according to the performance measures. Unless otherwise stated in the task summary, the Soldier must pass all performance measures to be scored GO. If the Soldier fails any steps, show what was done wrong and how to do it correctly.

References

Required	Related
None	ISBN 0-7637-4406-9

APPLY AN ELASTIC BANDAGE
081-833-0060

Conditions: You have a patient needing an elastic bandage applied. You will need roller bandages, adhesive tape, and scissors. You are not in a CBRNE environment.

Standards: Select and apply an appropriate bandage and wrap without causing further injury to the patient.

Performance Steps
CAUTION: All body fluids should be considered potentially infectious. Always observe body substance isolation (BSI) precautions by wearing gloves and eye protection as a minimal standard of protection.

   1. Select the appropriate bandaging material for the injury.
NOTE: The width of the bandage to use is determined by the size of the part to be covered. As a general rule, the larger the part or area, the wider the bandage.
      a. Use gauze or a flex roller for bleeding injuries of the forearm, upper arm, thigh, and lower leg.
      b. Use a flexible roller gauze bandage for bleeding injuries of the hand, wrist, elbow, shoulder, groin, knee, ankle, and foot.
      c. Use an elastic roller bandage for amputations, arterial bleeding, sprains, and torn muscles.
         (1) Hand - 2 inch bandage.
         (2) Lower arm, lower leg, and foot - 3 inch bandage.
         (3) Thigh and chest - 4 to 6 inch bandage.
NOTE: Elastic roller bandages may be used wherever pressure support or restriction of movement is needed. They should not be used to secure dressings.

   2. Prepare the patient for bandaging.
      a. Position the body part to be bandaged in a normal resting position (position of function).
NOTE: Bending a bandaged joint changes the pressure of the bandage in places of stress (elbow, knee, and ankle).
      b. Ensure that the body part to be bandaged is clean and dry.
      c. Place pads over bony places or between skin surfaces to be bandaged (such as fingers and armpits).

   3. Apply the anchor wrap.
CAUTION: Do not wrap too tightly. The roller bandage may act as a tourniquet on an injured limb, causing further damage.
      a. Lay the bandage end at an angle across the area to be bandaged.
      b. Bring the bandage under the area, back to the starting point, and make a second turn.
      c. Fold the uncovered triangle of the bandage end back over the second turn.
      d. Cover the triangle with a third turn, completing the anchor.

   4. Apply the bandage wrap to the injury.
      a. Use a circular wrap to end other bandage patterns, such as a pressure bandage, or to cover small dressings.
      b. Use a spiral wrap for a large cylindrical area such as a forearm, upper arm, calf, or thigh. The spiral wrap is used to cover an area larger than a circular wrap can cover.

Performance Steps

c. Use a spiral reverse wrap to cover small to large conical areas, for example, from ankle to knee.

d. Use a figure eight wrap to support or limit joint movement at the hand, elbow, knee, ankle, or foot.

e. Use a spica wrap (same as the figure eight wrap) to cover a much larger area such as the hip or shoulder.

f. Use a recurrent wrap for anchoring a dressing on fingers, the head, or on a stump.

NOTE: Bandage width depends on the site: 1 inch wide for fingers and 3, 4, or 6 inches wide for the stump or head.

5. Check circulation after application of the bandage.

a. Check the pulse distal to the injury.

b. Check for capillary refill (<2 seconds is normal), if applicable.

c. Inspect the skin below the bandaging for discoloration.

d. Ask the patient if any numbness, coldness, or tingling sensations are felt in the bandaged part.

e. Remove and reapply the bandage, if necessary.

6. Check for irritation.

a. Ask the patient if the bandage rubs.

b. Check for bandage wrinkles near the skin surface.

c. Check for red skin or sores (ulcers) when the bandage is removed.

d. Remove and reapply the bandage, if necessary.

7. Elevate injured extremities to reduce swelling (edema) and control bleeding, if appropriate.

8. Record the treatment given on the appropriate medical form.

Evaluation Preparation:

Setup: For training and evaluation, have another Soldier act as the patient. Tell the patient not to assist in any way. Have roller bandages of various widths, scissors, and adhesive tape available. Designate the part the Soldier should apply the roller bandage to.

Brief Soldier: Tell the Soldier to apply a roller bandage to the designated part.

Performance Measures	GO	NO-GO
1. Selected the appropriate bandaging material.	——	——
2. Prepared the patient for bandaging.	——	——
3. Applied the anchor wrap.	——	——
4. Applied the bandage wrap.	——	——
5. Checked circulation after the bandage was applied.	——	——
6. Checked for irritation.	——	——
7. Elevated the injured extremity, if appropriate.	——	——
8. Recorded the treatment given.	——	——

Performance Measures	<u>GO</u>	<u>NO-</u> <u>GO</u>
9. Did not cause further injury to the patient.	——	——

Evaluation Guidance: Score each Soldier according to the performance measures. Unless otherwise stated in the task summary, the Soldier must pass all performance measures to be scored GO. If the Soldier fails any steps, show what was done wrong and how to do it correctly.

References
    Required   **Related**
    None   EMERG CARE AND TRANS 9

IMMOBILIZE A SUSPECTED FRACTURE OF THE ARM OR DISLOCATED SHOULDER
081-833-0062

Conditions: You have a casualty with a suspected fracture of the arm or dislocated shoulder. You need to immobilize the injury. You will need a DD Form 1380 (U.S. Field Medical Card), wire ladder splint, cravat bandages, basswood splint, and materials for improvising a splint. You are not in a CBRNE environment.

Standards: Complete all the steps necessary to immobilize a suspected fracture of the arm or dislocated shoulder without causing additional injury.

Performance Steps

1. Check the casualty's radial pulse.
   a. If a radial pulse is impaired or absent, gently attempt to reposition the extremity towards a normal position.
   b. If one or two attempts do not restore circulation, splint the limb as it lies and rapidly transport the casualty to the nearest appropriate facility.

2. Position the injury.
   a. Position a fractured arm by having the casualty support it with the uninjured arm and hand in the least painful position, if possible.
CAUTION: Do not try to reduce or set the fracture. Splint it where it lies unless a severe deformity makes it necessary to reposition the limb to keep it within the confines of the litter and/or evacuation vehicle.
   b. Position the arm for shoulder dislocations.
CAUTION: Do not use force when moving the limb.
      (1) Posterior. Position the forearm across the midsection of the casualty's body with the hand or wrist slightly higher than the elbow.
      (2) Anterior. Maintain the arm in a fixed, locked position away from the body.
      (3) Turn the palm of the hand in towards the body, if possible.

3. Immobilize the injury.
   a. Use an arm sling to immobilize a dislocated shoulder.
   b. Use a basswood or an improvised splint for a fractured forearm.
      (1) Pad the splint.
      (2) Place the padded splint under the casualty's forearm so that it extends from the elbow to beyond the fingertips.
      (3) Place a rolled cravat or similar material in the palm of the cupped hand.
      (4) Apply the cravats in the following order and recheck the radial pulse after each cravat is applied.
         (a) Above the fracture site near the elbow.
         (b) Below the fracture site near the wrist.
         (c) Over the hand and tied in an "X" around the splint.
      (5) Apply an arm sling and swathe.
NOTE: Ensure that the fingernails are left exposed so that capillary refill can be assessed.
   c. Use a wire ladder splint for a fractured humerus and for multiple fractures of an arm or a forearm when the elbow is bent.
      (1) Prepare the splint using the uninjured arm for measurements.
         (a) Bend the prong ends of the splint away from the smooth side, about 1 1/2 inches down on the outside of the splint.

Performance Steps

        (b) With the smooth side against the elbow, place one end of the splint even with the top of the uninjured shoulder.

        (c) Select a point slightly below the elbow.

        (d) Remove the splint from the arm and bend the splint at the measured point to form an "L".

        (e) Pad the splint.

NOTE: If padding is unavailable, apply the splint anyway.

    (2) Position the splint on the outside of the injured arm, extending from the shoulder to beyond the fingertips.

NOTE: Extend the "L" angle of the splint beyond but do not touch the elbow of the injured arm. Extend the leg of the angle touching the forearm beyond the ends of the fingers. If the splint is too short, extend it with a basswood splint. If possible, have the casualty support the splint.

    (3) Place a rolled cravat or similar material in the palm of the cupped hand.

    (4) Check the radial pulse. Make a note on the FMC if the pulse is absent or if the pulse was lost after treatment.

    (5) Apply the cravats in the following order and recheck the radial pulse after each cravat is applied.

        (a) On the humerus above any fracture site.

        (b) On the humerus below any fracture site.

        (c) On the forearm above any fracture of the forearm.

        (d) On the forearm below any fracture site.

        (e) Around the hand and splint.

    (6) Tie each cravat on the outside edge of the splint.

NOTE: If the pulse is weaker or absent after tying the cravat, loosen and retie the cravat.

    (7) Apply an arm sling and swathe.

  d. Use a wire ladder splint for a fractured or dislocated humerus, elbow, or forearm when the elbow is straight.

    (1) Prepare the splint as in step 3c(1) but bend it only enough to fit the injured arm.

    (2) Position the splint on the outside of the arm against the back of the hand.

    (3) Apply the cravats in the following order and recheck the radial pulse after each cravat is applied.

        (a) Above the injury.

        (b) Below the injury.

        (c) High on the humerus, above the first cravat.

        (d) Around the hand and wrist.

    (4) Tie each cravat on the outside of the splint.

NOTE: If the pulse is weaker or absent after tying the cravat, loosen and retie the cravat.

    (5) Apply swathes.

        (a) Place the arm toward the midline in front of the body. Bind the forearm to the pelvic area with a cravat. Tie the knot on the uninjured side.

        (b) Apply an additional cravat above the elbow. Secure it on the uninjured side at breast pocket level.

4. Record the treatment given on the FMC.

5. Evacuate the casualty.

Performance Measures	<u>GO</u>	<u>NO-</u> <u>GO</u>
1. Checked the radial pulse first and after each intervention.	____	____
2. Positioned the injury.	____	____
3. Immobilized the injury.	____	____
4. Recorded the treatment on the FMC.	____	____
5. Evacuated the casualty.	____	____
6. Did not cause further injury to the casualty.	____	____

Evaluation Guidance: Score each Soldier according to the performance measures. Unless otherwise stated in the task summary, the Soldier must pass all performance measures to be scored GO. If the Soldier fails any steps, show what was done wrong and how to do it correctly.

References

Required	**Related**
DD FORM 1380	PHTLS

IMMOBILIZE THE HIP
081-833-0064

Conditions: You have a casualty with a suspected dislocated or fractured hip. You need to immobilize the hip. Three other Soldiers are available to assist you. You will need litter, splints, cravats or commercial straps, padding material, spine board or other rigid object, a traction splint, and pneumatic anti-shock garment (PASG). You are not in a CBRNE environment.

Standards: Immobilize a suspected dislocated or fractured hip without impairing circulation or causing further injury to the casualty.

Performance Steps

1. Check for the signs and symptoms of a hip injury.
CAUTION: Both a dislocated and a fractured hip are accompanied by considerable pain. The casualty will resist any movement because of pain. It is essential that medical personnel take all possible precautions, using the best available materials at hand while preparing the casualty to be immediately evacuated.
    a. Anterior dislocation.
NOTE: Anterior dislocation is very rare and is caused by the legs suddenly being forced widely apart and locked in this position.
        (1) Hip pain.
        (2) Severe deformity of the affected leg.
            (a) The knee is turned outward.
            (b) The affected leg is shortened.
            (c) The hip is drawn away from the midline of the body.
            (d) The leg has rotated away from the midline of the body.
        (3) Impaired circulation in the affected extremity.
            (a) Loss of pulse distal to the injury.
            (b) Coolness and/or cyanosis.
            (c) Swelling due to internal blood loss.
            (d) Hypovolemic shock.
WARNING: Significant blood loss may occur before swelling is evident. Take the casualty's vital signs as soon as possible and monitor them during stabilization and transport.
        (4) Impaired sensation in the affected extremity.
            (a) Tingling or other abnormal sensations (paresthesia).
            (b) Loss of sensation.
    b. Posterior dislocation.
NOTE: Posterior dislocation is the most common type of hip dislocation.
        (1) Hip pain.
        (2) Severe deformity of the affected leg.
            (a) The hip joint is flexed with the knee drawn up.
            (b) The hip is drawn toward the midline of the body.
            (c) The leg has rotated toward the midline of the body.
        (3) Impaired circulation in the affected extremity.
            (a) Loss of pulse distal to the injury.
            (b) Coolness and/or cyanosis.
            (c) Swelling due to internal blood loss.

Performance Steps

      (4) Impaired sensation in the affected extremity.

         (a) Paresthesia.

         (b) Loss of sensation.

NOTE: Weakness of muscles that raise the foot may occur. This condition, known as "foot drop," may be a sign of damage to the sciatic nerve.

      c. Fracture.

NOTE: Some of the most common fractures are those that occur at the proximal (upper) end of the femur. These have been called "hip fractures" even though the hip joint is rarely involved.

      (1) Hip pain.

      (2) The casualty is unable to walk on or move the affected leg.

      (3) Deformity.

         (a) The affected leg has rotated toward the midline of the body.

         (b) The affected leg will usually be shorter than the uninjured one.

NOTE: Fractures of the femur are often open. Whether closed or open, they are always associated with a loss of large amounts of blood. Therefore, you should treat the casualty with high-flow oxygen and monitor vital signs frequently, watching for signs of shock.

      (4) Impaired circulation in the affected extremity.

         (a) Loss of pulse in the femoral or popliteal arteries distal to the injury.

         (b) Coolness and/or cyanosis.

         (c) Swelling due to internal blood loss.

      (5) Impaired sensation in the affected extremity.

         (a) Paresthesia.

         (b) Loss of sensation.

2. Check for circulation in the affected leg by checking the femoral and popliteal pulses and observing for swelling or cyanosis.

3. Check for impaired sensation by asking the casualty if he has tingling, abnormal sensations, or loss of sensation in the affected limb.

4. Immobilize the injury.

CAUTION: Do not log roll a casualty with a hip injury onto the injured side. If available, place the casualty on a spine board using a scoop litter.

      a. Hip dislocations.

      (1) Place the casualty on a firm surface, such as a spine board. See task 081-833-0092

      (2) Support the leg in its abnormal position using pillows, blankets, or similar material.

      (3) Secure the support material with cravats.

      b. Hip fracture.

      (1) Place the casualty on a firm surface.

      (2) Place support material under the buttocks to reduce abdominal pain only if there are no other major fractures in the lower extremities.

      (3) Place bulky support material between the casualty's legs and strap them together.

      (4) Bring the casualty's knees up.

      (5) Place bulky support material underneath the knees.

5. Check for complications.

      a. Impaired circulation in the affected limb.

      b. Neurological deficit.

      c. Hypovolemic shock.

Performance Steps

6. Record the treatment given.

WARNING: Spontaneous reduction of dislocation may occur during any movement. This may be accompanied by additional damage to nerves and blood vessels. The receiving facility must be informed if this occurs.

7. Evacuate the casualty.
   a. Position the casualty and spine board on a litter.
   b. Position the casualty resting slightly on the uninjured side.
   c. Support the injured side with padding material.
   d. Secure the casualty and spine board to the litter.

WARNING: Avoid any bumping or jerking during transport. Excessive movement of a fracture or dislocation can increase blood loss and pain. Hip and leg injuries allow for a greater area of pooling of blood that is not evident early on, and may result in the casualty going in to hypovolemic shock.

Performance Measures	GO	NO-GO
1. Checked for the signs and symptoms of a hip injury.	——	——
2. Checked for circulation in the affected leg.	——	——
3. Checked for impaired sensation.	——	——
4. Immobilized the injury.	——	——
5. Checked for complications.	——	——
6. Recorded the treatment given.	——	——
7. Evacuated the casualty.	——	——
8. Did not cause further injury to the casualty.	——	——

Evaluation Guidance: Score each Soldier according to the performance measures. Unless otherwise stated in the task summary, the Soldier must pass all performance measures to be scored GO. If the Soldier fails any steps, show what was done wrong and how to do it correctly.

References

Required	Related
None	EMERG CARE AND TRANS 9

APPLY A TRACTION SPLINT
081-833-0141

Conditions: You encounter a casualty with a suspected femur fracture. All other more serious injuries have been assessed and treated. You must apply a traction splint, another Soldier is available to assist you. You will need a traction splint, long spine board, securing devices, padding material, and DD Form 1380 (U.S. Field Medical Card). You are not in a CBRNE environment.

Standards: Apply the traction splint without restricting circulation. Immobilize the fracture and maintain traction throughout the procedure, minimizing the effects of the injury.

Performance Steps
CAUTION: All body fluids should be considered potentially infectious. Always observe body substance isolation (BSI) precautions by wearing gloves and eye protection as a minimal standard of protection.

1. Check for signs of a femur fracture. With unresponsive casualties, you should attempt to identify the mechanism of injury (MOI).
    a. Inspect the extremity for deformities, contusions, abrasions, punctures or penetrations, burns, tenderness, lacerations, and swelling (DCAP-BTLS).
    b. Palpate for tenderness, instability, or crepitus (TIC) in the extremity.
WARNING: Do not use a traction splint for an injury close to or involving the knee, injury of the hip or pelvis, partial amputations or avulsions with bone separation, or lower leg or ankle injury.

2. Cut the casualty's trouser leg, or otherwise expose the injured leg.

3. Assess the pulse, motor, and sensory function (PMS).

4. Direct the assistant to manually support and stabilize the injured leg.

5. Place the traction splint beside the casualty's uninjured leg. Adjust the splint to the proper length.
    a. Place the ring at the ischial tuberosity (next to the casualty's iliac crest).
    b. Loosen the locking sleeve.
    c. Extend the splint 8 to 12 inches beyond the casualty's foot.
    d. Tighten the locking sleeve.
    e. Open and adjust the four Velcro support straps, which should be positioned at the mid-thigh, above the knee, below the knee, and above the ankle.

6. Fasten the ankle hitch about the casualty's ankle and foot. Normally, the boot is removed for this procedure.
    a. Thread the ankle hitch under the casualty's ankle at the void created by the heel.
    b. Place the lower edge of the ankle hitch even with the bottom of the heel.
    c. Crisscross the side straps high on the instep.
    d. Bring the crisscrossed straps down to meet the center strap and hold them in place.

7. While you support the leg at the site of the suspected injury (one hand above the site and one hand below the site) to support the fracture as traction is pulled and the leg is lifted, direct the assistant to manually apply gentle longitudinal (in-line) traction to the ankle hitch and foot.

Performance Steps

CAUTION: Apply only enough traction to align the limb to fit into the splint. Do not attempt to align the fracture fragments anatomically. Once manual traction has been applied, it must remain constant until the traction splint has been put in place and is providing traction.

NOTE: While applying gentle traction, the assistant may lift the casualty's leg far enough to fit the splint into place.

8. Slide the splint into position under the injured leg.
   a. Pull the release ring on the ratchet and release the traction strap.
   b. Move the splint between the assistant's legs so that it is aligned with the casualty's injured leg.
   c. Move one hand from the fracture site and pull the splint from between the assistant's legs.
   d. Slide the splint under the leg until the ischial ring is at the buttock.

NOTE: Make sure the splint is aligned with the leg.
   e. When the splint is in place, position the hand back under the fracture site for stabilization only.
   f. On the assistant's signal, lower the leg into the cradle of the splint while maintaining manual traction.
   g. Extend and position the heel stand after the splint is in position under the leg.

9. Pad the groin area (cravat, Kerlix, etc.) and fasten the ischial strap.

10. Apply mechanical traction.
   a. Attach the rings from the ankle hitch to the "S" hook from the splint.
   b. Tighten the ratchet mechanism by turning it clockwise.
   c. Direct the assistant to alert you when mechanical traction is equal to their manual traction.

NOTE: Adequate traction has been applied when the injured leg is the same length as the other leg or the casualty feels relief.

11. Secure the Velcro support straps. Direct the assistant to maintain manual stabilization until all four support straps are secure.

12. Reevaluate the ischial strap and ankle hitch.

13. Reassess the PMS.

14. Place the casualty securely on a long spine board to immobilize the hip.

15. Secure the splint to the long spine board to prevent movement of the splint (strapping, tape, etc.).

16. Document the procedure on a FMC.

Evaluation Preparation:

Setup: For training and evaluation, have another Soldier act as the casualty. You will need another Soldier to act as an assistant. Describe a scenario that involves a femur fracture (parachute or vehicle accident, etc.). Tell the assisting Soldier to only perform those actions the Soldier being evaluated directs.

CAUTION: Do not allow the Soldiers to apply full traction to the casualty.

Brief Soldier: To test step 1, tell the Soldier to state the signs of a femur fracture. Tell the Soldier that the casualty has a femur fracture. Tell the Soldier to, using the assistant, apply a traction splint to the fractured femur and prepare the casualty for transport.

Performance Measures	GO	NO-GO
1. Checked for signs of a femur fracture.	____	____
2. Exposed the injured leg.	____	____
3. Assessed the PMS.	____	____
4. Directed the assistant to manually support and stabilize the injured leg.	____	____
5. Measured the traction splint on the other extremity and adjusted the splint to the proper length.	____	____
6. Applied the ankle hitch about the casualty's ankle and foot.	____	____
7. Applied manual, in-line traction to the ankle hitch and foot.	____	____
8. Placed the splint into position under the injured leg.	____	____
9. Padded the groin area and fastened the ischial strap.	____	____
10. Applied mechanical traction.	____	____
11. Secured the Velcro support straps.	____	____
12. Reevaluated the ischial strap and ankle hitch.	____	____
13. Reassessed the PMS.	____	____
14. Placed the casualty on a long spine board.	____	____
15. Secured the splint to the long spine board.	____	____
16. Documented the procedure on the FMC.	____	____

Evaluation Guidance: Score each Soldier according to the performance measures. Unless otherwise stated in the task summary, the Soldier must pass all performance measures to be scored GO. If the Soldier fails any steps, show what was done wrong and how to do it correctly.

References
Required
DD FORM 1380

Related
EMERG CARE AND TRANS 9

PROVIDE BASIC EMERGENCY TREATMENT FOR A PAINFUL, SWOLLEN, DEFORMED EXTREMITY
081-833-0154

Conditions: You have encountered a casualty who presents with a musculoskeletal injury. All other more serious injuries have been assessed and treated. You will need cravats, dressings, splinting materials, and DD Form 1380 (U.S. Field Medical Card). You are not in a CBRNE environment.

Standards: Immobilize the injured extremity without causing further injury to the casualty.

Performance Steps
CAUTION: All body fluids should be considered potentially infectious. Always observe body substance isolation (BSI) precautions by wearing gloves and eye protection as a minimal standard of protection.

1. Identify the signs and symptoms of a musculoskeletal injury.
    a. Pain and tenderness, especially when the injured part is touched or moved.
    b. Deformity or angulation.
NOTE: When in doubt, look at the uninjured side and compare it to the injured one.
    c. Crepitus.
    d. Edema (swelling).
    e. Ecchymosis (bruising).
    f. Exposed bone.
    g. Joints locked into position.
    h. Impaired circulation, motor function, and sensation.

2. Splint the extremity (see tasks 081-833-0141, 081-833-0062, and 081-833-0064).
NOTES: 1. In order for any splint to be effective, it must immobilize the adjacent joints and bone ends. 2. If the casualty is unstable, immobilize on a long spine board and transport immediately.
    a. Assess pulse, motor function, and sensation (PMS).
       (1) Check for a pulse distal to the injury.
       (2) Ask if the casualty can feel your touch distal to the injury.
       (3) Ask the casualty to wiggle his fingers or toes, grasp your fingers, or push his feet against your hands.
NOTE: If the fracture is open, apply a dressing to control bleeding and to protect the area.
    b. Manually stabilize the injury site. This can be done by you, your assistant, or the casualty.
NOTE: Maintain manual stabilization or traction during positioning and until the splinting process is complete.
    c. Attempt to realign an angulated fracture once, if necessary.
NOTE: Attempt to realign only if there is impaired circulation (loss of distal pulses, cold to the touch) or the extremity is so grossly angulated that splinting would not be effective.
       (1) Gently grasp the distal extremity while your assistant places one hand above and one hand below the injury site.
       (2) Gently pull manual traction in the direction of the long axis of the bone.
       (3) If resistance is felt or it appears the bone ends will come through the skin, stop and splint the extremity in the position found. Evacuate the casualty as soon as possible.

Performance Steps

        (4) If no resistance is felt, maintain gentle traction until the extremity is properly splinted.

    d. Measure or adjust the splint. Apply padding.

    e. Apply and secure the splint to immobilize adjacent bones.

    f. Splint the extremity in the position of function.

    g. Reassess PMS distal to the injury.

3. Treat for shock. (See task 081-833-0047.)

4. Consider administration of pain medication.

5. Document all care given on the FMC. (See tasks 081-833-0145 and 081-831-0033.)

6. Transport to the nearest medical treatment facility.

Evaluation Preparation:

Setup: For training and evaluation, have one Soldier act as a casualty with a musculoskeletal injury. Brief the casualty on the location and complaints of a musculoskeletal injury. Use moulage if available. Tell the casualty not to assist the Soldier in any way.

Brief Soldier: Ask the Soldier for the signs and symptoms of a musculoskeletal injury and have him perform the appropriate treatment.

Performance Measures

	GO	NO-GO
1. Identified the signs and symptoms of a musculoskeletal injury.	——	——
2. Splinted the extremity.	——	——
3. Treated for shock.	——	——
4. Considered administration of pain medication.	——	——
5. Documented care given on the FMC.	——	——
6. Transported to the nearest medical treatment facility.	——	——

Evaluation Guidance: Score each Soldier according to the performance measures. Unless otherwise stated in the task summary, the Soldier must pass all performance measures to be scored GO. If the Soldier fails any steps, show what was done wrong and how to do it correctly.

References

Required	Related
DD FORM 1380	EMERG CARE AND TRANS 9

TREAT A CASUALTY WITH A SUSPECTED SPINAL INJURY
081-833-0176

Conditions: You encounter a casualty with a suspected spinal injury. All other more serious injuries have been assessed and treated. Three other Soldiers are available for assistance. You will need straps, cravats, long and short spine boards, immobilization vest-type device, cervical collar or materials to improvise a cervical collar, head supports and DD Form 1380 (U.S. Field Medical Card). You are not in a CBRNE environment.

Standards: Complete all of the steps necessary to immobilize a casualty with a suspected spinal injury and prepare him for transport without causing additional injury to the casualty.

Performance Steps
CAUTION: All body fluids should be considered potentially infectious. Always observe body substance isolation (BSI) precautions by wearing gloves and eye protection as a minimal standard of protection.

  1. Check for the signs and symptoms of a spinal injury. With unresponsive casualties, you should attempt to identify the mechanism of injury (MOI).
WARNING: If you suspect the casualty has a spinal injury, you must treat him as though he has a spinal injury; when in doubt, immobilize.
    a. If possible, inspect the spine for deformities, contusions, abrasions, punctures or penetrations, burns, tenderness, lacerations, and swelling (DCAP-BTLS).
    b. Lacerations and/or contusions in the spinal region indicate severe trauma and usually accompany a spinal injury.
    c. Palpate for tenderness, instability, or crepitus (TIC) in the spinal region.
NOTE: The ability to walk, move the extremities, feel sensation, and the absence of pain does not necessarily rule out a spinal cord injury.
      (1) Carefully insert your hand under the casualty's neck and palpate along the cervical spine as far as can be done without moving the casualty.
      (2) Carefully insert your hand into the area of the small of the back and palpate along the thoracic spine and down the lumbar spine as far as possible without moving the casualty.
    d. Check for weakness, loss of sensation, paresthesia (tingling), and/or paralysis.
      (1) A cervical spine injury may cause numbness or paralysis in all four extremities.
      (2) A waist level (lumbar) spinal injury may cause numbness or paralysis below the waist.
      (3) Ask the casualty to try to move his fingers and toes to check for motor function.

  2. Immobilize a sitting casualty using a short spine board.
    a. Have your assistant carefully move the casualty's head into a proper, neutral in-line position. Continue manual stabilization until the casualty is secured to a long spine board. Whenever possible, kneel behind the casualty and place hands around the base of the skull on either side. Careful movement of the head and neck into a neutral position must be stopped if movement results in any of the following:
      (1) Neck muscle spasm.
      (2) Increased pain.
      (3) Increase in numbness, tingling, or loss of motor ability.
      (4) Compromise of the airway or ventilation.
    b. In these situations, the casualty's head must be immobilized in the position in which it was initially found.

Performance Steps

    c. Stabilize the head and neck.

      (1) Place your hands on both sides of the casualty's skull, with the palms above the ears.

      (2) Support the jaw (mandible) with the fingers.

      (3) Maintain manual stabilization until directed to release the stabilization.

    d. Assess the pulse, motor, and sensory function (PMS).

    e. Apply a rigid cervical collar, if available, or improvise one.

NOTE: Measure the rigid cervical collar according to the manufacturer's specifications. (See task 081-833-0177.) An improperly sized device has a potential for further injury.

    f. Push the board as far into the area behind the casualty as possible.

    g. Tilt the upper end of the board toward the head.

    h. Direct the assistant to position the back of the casualty's head against the board, maintaining manual in-line stabilization, by moving the head and neck as one unit.

NOTE: If the cervical collar or improvised collar does not fit flush with the spine board, place a roll in the hollow space between the neck and board. The roll should only be large enough to fill the gap, not to exert pressure on the neck.

    i. Secure the short spine board to the casualty's torso.

      (1) Place the buckle of the first strap in the casualty's lap.

      (2) Pass the other end of the strap through the lower hole in the board, up the back of the board, through the top hole, under the armpit, over the shoulder, and across the back of the board at the neck.

      (3) Buckle the second strap to the first strap and place the buckle on the side of the board at the neck.

      (4) Pass the other end over the shoulder, under the armpit, through the top hole in the board, down the back of the board, through the lower hole, and across the lap. Secure it by buckling it to the first strap.

    j. Secure the casualty's head and head supports to the board with straps or cravats.

WARNING: Ensure that the cravats or head straps are firmly in place before the assistant releases stabilization.

      (1) Apply head supports.

      (2) Use two rolled towels, blankets, sandbags, or similar material.

      (3) Place one close to each side of the head.

      (4) Using a cravat-like material across the forehead, make the supports and head one unit by tying to the board.

    k. Reassess PMS.

    l. Tie the casualty's hands together and place them in his lap.

NOTE: When positioning a casualty who is secured to a short spine board, on a long spine board, line up the hand grip holes of the short spine board with the holes of the long spine board, if possible, and secure the two boards together with straps.

  3. Immobilize a sitting casualty using a vest-type device, such as a Kendrick Extrication Device (KED).

    a. Stabilize and assess the sitting casualty as in steps 2a through 2e.

NOTE: Before placing the vest-type device behind the casualty, the two long straps (groin straps) are unfastened and placed behind the device.

    b. Position the immobilization device behind the casualty. The side flaps are placed around the casualty's torso and moved until they are in contact with the casualty's armpits.

    c. Secure the vest-type device to the casualty's torso.

Performance Steps

  (1) Immobilize the torso, beginning with the middle strap, followed by the lower strap and finally the upper strap. Tighten each strap after attachment.

  (2) Position and tighten each groin strap; ensure you pad the groin area.

 d. Secure the casualty's head to the vest-type device.

  (1) Pad behind the casualty's head as necessary.

  (2) Place the first strap or cravat across the chin angling upward toward the ear. Attach to the head flaps on either side of the head. Ensure the strap/cravat does not interfere with the airway.

  (3) Place the second strap or cravat across the forehead angling downward toward the base of the head. Attach to the head flaps on either side of the head.

 e. Evaluate and adjust the straps as needed. They must be tight enough so the device does not move excessively up, down, left, or right, but not so tight as to restrict the casualty's breathing.

NOTE: The pelvic straps must be released after being placed on a long spine board in order to place the casualty in a supine position.

 4. Place the casualty on a long spine board.

NOTE: If a long spine board is not available, utilize a standard litter or improvised litter made from a board or door. A hard surface is preferable to one that gives with the casualty's weight.

 a. The log-roll technique.

  (1) Position the long spine board next to, and parallel with, the casualty.

  (2) Maintain manual stabilization of the casualty's head and neck. This individual will direct all movements while maintaining in-line support of the head and neck.

   (a) Place your hands on both sides of the casualty's skull, with the palms above the ears.

   (b) Support the jaw (mandible) with the fingers.

   (c) Maintain manual stabilization until the casualty has been placed on the spine board.

  (3) Apply a cervical collar, if available, or improvise one.

  (4) Brief each of the three assistants on their duties and instruct them to kneel on the same side of the casualty, with the long spine board on the opposite side of the casualty.

   (a) First assistant. Place the near hand on the shoulder and the far hand on the waist.

   (b) Second assistant. Place the near hand on the hip and the far hand on the thigh.

   (c) Third assistant. Place the near hand on the knee and the far hand on the ankle.

  (5) On the command of the team leader stabilizing the casualty's head, and in unison, the assistants roll the casualty slightly toward them. The head and neck must be maintained in-line with the casualty's spine during all movements.

  (6) Instruct the assistants to reach across the casualty with one hand, grasp the spine board at its closest edge, and slide it against the casualty. Instruct the number two assistant to reach across the board to the far edge and hold it in place to prevent board movement.

  (7) Instruct the assistants to slowly roll the casualty back onto the board, keeping the head and spine in a straight line.

Performance Steps

(8) Reassess PMS.

NOTE: If the cervical collar or improvised collar does not fit flush with the spine board, place a roll in the hollow space between the neck and board. The roll should only be large enough to fill the gap, not to exert pressure on the neck.

b. The straddle-slide technique.

NOTE: Use this method when limited space makes it impossible to use the log roll technique.

(1) Stand (team leader) at the head of the casualty with your feet wide apart.

(2) Apply stabilization to the casualty's head and apply a cervical collar, if available, or improvise one.

(3) Instruct the first assistant to stand behind you (facing your back), to line up the spine board, and to gently slide the spine board under the casualty at your command.

(4) Instruct the second assistant to straddle the casualty while facing you and gently elevate the shoulders so that the spine board can be slid under them.

(5) Instruct the third assistant (facing you) to carefully elevate the hips while the spine board is being slid under the casualty.

(6) Instruct the fourth assistant (facing you) to carefully elevate the legs and ankles while the board is being slid into place under the casualty.

WARNING: Complete all movements simultaneously, keeping the head and spine in a straight line.

NOTE: If the cervical collar or improvised collar does not fit flush with the spine board, place a roll in the hollow space between the neck and board. The roll should only be large enough to fill the gap, not to exert pressure on the neck.

5. Secure the casualty to the long spine board.

a. While maintaining manual stabilization, secure the torso to the long spine board by applying straps across chest, pelvis, and legs. Adjust these straps as needed.

b. While continuing to maintain manual stabilization, apply the head supports to each side of the casualty's head.

NOTE: If commercial head supports are not available, use two rolled towels, blankets, sandbags, socks filled with sand, or similar material.

WARNING: Do not release manual stabilization until the cravats or head straps are firmly in place.

c. Fasten a strap or cravat-like material tightly over the head supports and the lower forehead. A second strap/cravat is placed over the pads and the rigid cervical collar and is fastened securely to the long board.

WARNING: Ensure the head straps do not interfere with the casualty's airway.

d. Reassess PMS.

6. Record the treatment on the Field Medical Card FMC.

7. Evacuate the casualty.

Evaluation Preparation:

Setup: For training and evaluation, have another Soldier act as the casualty. You will need three Soldiers to act as the assistants. The Soldier being tested is to act as the team leader and direct the actions of the assistants. The casualty may be placed in a vehicle or other scenario, depending on available resources and the technique you are testing. Tell the casualty not to assist the Soldier in any way. Tell the assisting Soldiers to only perform those actions the Soldier being evaluated directs.

Brief Soldier: To test step 1, tell the Soldier to state the signs and symptoms of a spinal injury. Tell the Soldier that the casualty has a suspected spinal injury. Then tell the Soldier to position the casualty on a spine board and to direct the actions of the assistants.

Performance Measures	GO	NO-GO
1. Checked for signs and symptoms of a spinal injury.	——	——
2. Secured the casualty on a short spine board or vest-type device (KED), if appropriate.	——	——
3. Placed the casualty on the long spine board.	——	——
4. Secured the casualty on the long spine board.	——	——
5. Recorded the treatment on the FMC.	——	——
6. Evacuated the casualty.	——	——
7. Did not cause further injury to the casualty.	——	——

Evaluation Guidance: Score each Soldier according to the performance measures. Unless otherwise stated in the task summary, the Soldier must pass all performance measures to be scored GO. If the Soldier fails any steps, show what was done wrong and how to do it correctly.

References
    Required
    DD FORM 1380

    Related
    EMERG CARE AND TRANS 9

APPLY A CERVICAL COLLAR
081-833-0177

Conditions: A casualty has received traumatic injuries and now complains of neck pain. All other more serious injuries have been assessed and treated. You need to apply a cervical collar. Another Soldier is available to assist you. You will need straps, cravats, a long spine board, a rigid cervical collar device, and head supports. You are not in a CBRNE environment.

Standards: Complete all the steps necessary to immobilize and transport a casualty with a suspected cervical spine injury without causing additional injury to the casualty.

Performance Steps
CAUTION: All body fluids should be considered potentially infectious. Always observe body substance isolation (BSI) precautions by wearing gloves and eye protection as a minimal standard of protection.

1. While your assistant is stabilizing the casualty's cervical spine, complete an initial assessment and care for all life-threatening injuries before applying the cervical collar.

2. Use the mechanism of injury, level of responsiveness, and location of injuries to determine the need for cervical immobilization.
WARNING: If you suspect the casualty has a spinal injury, treat him as though he has a spinal injury.

3. While maintaining manual cervical spine stabilization and neutral neck alignment, assess the casualty's neck prior to placing the collar. Once the collar is in place, you will not be able to assess or palpate the back of the neck.

4. Reassure the casualty and explain the procedure to him.

5. Determine the size of collar to apply.
   a. The front height of the collar should fit between the point of the chin and the chest at the suprasternal notch.
   b. Once in place, the collar should rest on the shoulder girdle and provide firm support under both sides of the mandible without obstructing the airway or any ventilation efforts.
   c. If the collar is too large, the casualty's neck may be placed in hyperextension.
   d. If the collar is too small, the casualty's neck may be placed in hyperflexion.

6. Apply the collar to a seated casualty, if applicable.
   a. Have the other Soldier apply in-line stabilization of the head and neck from behind the casualty.
   b. Place the chin support first.
   c. Wrap the collar around the neck.
   d. Secure the Velcro strap in place.
   e. Maintain manual stabilization of the head and neck until the casualty is immobilized on a long spine board.
NOTE: Cervical collars do not fully immobilize the cervical spine; therefore, you must maintain manual stabilization of the casualty's neck until the casualty is fully immobilized on a long spine board. (See task 081-833-0181.)

Performance Steps

   7. Apply the collar to a supine casualty, if applicable.
      a. Have the other Soldier kneel at the casualty's head and manually apply in-line stabilization of the head and neck.
      b. Set the collar in place around the neck.
      c. Secure the Velcro strap in place.
      d. Maintain manual stabilization of the head and neck until the casualty is immobilized on a long spine board.

Evaluation Preparation:
Setup: For training and evaluation, have another Soldier act as the casualty. You will need another Soldier to act as an assistant. The Soldier being tested is to act as the team leader and direct the actions of the assistant. The casualty may be placed in a vehicle or other scenario, depending on available resources and the technique you are testing. Tell the casualty not to assist the Soldiers in any way. Tell the assisting Soldier to only perform those actions the Soldier being evaluated directs.

Brief Soldier: Tell the Soldier to state the signs and symptoms of a spinal injury. Tell the Soldier that the casualty has a suspected cervical spinal injury. Then tell the Soldier to apply a cervical collar and to direct the actions of the assistant.

Performance Measures	GO	NO-GO
1. Completed the initial assessment.	——	——
2. Used the mechanism of injury, level of responsiveness, and location of injuries to determine the need for cervical immobilization.	——	——
3. Maintained manual stabilization and assessed the casualty's neck prior to placing the collar.	——	——
4. Reassured the casualty.	——	——
5. Determined the size of collar to apply.	——	——
6. Applied the collar to a seated casualty, if applicable.	——	——
7. Applied the collar to a supine casualty, if applicable.	——	——

Evaluation Guidance: Score each Soldier according to the performance measures. Unless otherwise stated in the task summary, the Soldier must pass all performance measures to be scored GO. If the Soldier fails any steps, show what was done wrong and how to do it correctly.

References
    Required   **Related**
    None   EMERG CARE AND TRANS 9

APPLY A KENDRICK EXTRICATION DEVICE
081-833-0178

Conditions: You encounter a casualty with a suspected spinal injury. The casualty is in a sitting position. All other more serious injuries have been assessed and treated. Another Soldier is available to assist you. You will need a Kendrick Extrication Device (KED), a rigid cervical collar, straps, cravats, a long spine board and DD Form 1380 (U.S. Field Medical Card). You are not in a CBRNE environment.

Standards: Apply a KED to the casualty with a suspected spinal injury without causing further injury.

Performance Steps
CAUTION: All body fluids should be considered potentially infectious. Always observe body substance isolation (BSI) precautions by wearing gloves and eye protection as a minimal standard of protection.

1. Check for signs and symptoms of a spinal injury. With unresponsive casualties, you should attempt to identify the mechanism of injury (MOI).
WARNING: If you suspect the casualty has a spinal injury, you must treat him as though he has a spinal injury; when in doubt, immobilize.
   a. If possible, inspect the spine for deformities, contusions, abrasions, punctures or penetrations, burns, tenderness, lacerations, and swelling (DCAP-BTLS).
   b. Lacerations and/or contusions in the spinal region indicate severe trauma and usually accompany a spinal injury.
   c. Palpate for tenderness, instability, or crepitus (TIC) in the spinal region.

2. Stabilize the casualty's spine.
   a. Have your assistant carefully move the casualty's head into a proper, neutral in-line position. Continue manual stabilization until the casualty is secured to a long spine board. Careful movement of the head and neck into a neutral position must be stopped if movement results in any of the following:
      (1) Neck muscle spasm.
      (2) Increased pain.
      (3) Increase in numbness, tingling, or loss of motor ability.
      (4) Compromise of the airway or ventilation.
   b. Assess the casualty's pulse, motor function and sensory function in each extremity.
   c. The casualty's head must be immobilized in the position in which it was initially found. This is best accomplished with large amounts of padding placed in the spaces between the casualty's head and the head flaps.

3. Apply a rigid cervical collar. (See task 081-833-0177.)

4. Position the KED.
   a. Continue to manually stabilize the casualty's head in a neutral, in-line position.
   b. Assess distal pulse, motor, and sensory function (PMS).
NOTE: Before placing the KED behind the casualty, the two long straps (groin straps) are unfastened and placed behind the device.
   c. Position the KED between the casualty's upper back and the seat back.
   d. Lift the KED so that the side flaps are in contact with the casualty's armpits.
   e. Secure the KED to the casualty's torso in the following order:
      (1) Middle strap.

Performance Steps

      (2) Lower strap.

      (3) Upper strap.

  f. Evaluate and pad behind the casualty's head as necessary. Secure the casualty's head to the KED using forehead and chin straps or cravats. Ensure the strap/cravat does not interfere with the airway.

  g. Evaluate and adjust the torso straps. They must be tight enough to prevent the device from moving up, down, and to the side excessively, but not so tight as to restrict the casualty's ability to breathe.

  h. Ensure the groin straps are well positioned and secure them.

  i. Secure the casualty's wrists together using a cravat.

  j. Reassess PMS.

5. Transfer the casualty to a long spine board. (See task 081-833-0176.)

  a. Unbuckle the groin straps before stretching out the legs.

  b. Complete the spinal immobilization as usual.

NOTE: Manual stabilization is maintained at all times until the casualty is fully immobilized on the long spine board.

6. Record the treatment on the FMC.

Evaluation Preparation:

Setup: For training and evaluation, have another Soldier to act as the casualty. The Soldier being tested is to act as the team leader and direct the actions of the assistant. The casualty may be placed in a vehicle or other confined space. Tell the casualty not to assist the Soldier in any way. Tell the assistant to only perform those actions the Soldier being evaluated directs.

Brief Soldier: To test step 1, tell the Soldier to state the signs and symptoms of a spinal injury. Tell the Soldier that the casualty has a suspected spinal injury. Then tell the Soldier to apply the KED and position the casualty on a long spine board for transport.

Performance Measures	GO	NO-GO
1. Checked for signs and symptoms of a spinal injury.	——	——
2. Stabilized the casualty's spine.	——	——
3. Applied a rigid cervical collar.	——	——
4. Applied the KED.	——	——
5. Transferred the casualty to a long spine board.	——	——
6. Recorded the treatment on the FMC.	——	——
7. Caused no further harm to the casualty.	——	——

Evaluation Guidance: Score each Soldier according to the performance measures. Unless otherwise stated in the task summary, the Soldier must pass all performance measures to be scored GO. If the Soldier fails any steps, show what was done wrong and how to do it correctly.

References
    Required
    DD FORM 1380

    Related
    EMERG CARE AND TRANS 9

APPLY A KENDRICK TRACTION DEVICE
081-833-0180

Conditions: You encounter a casualty with a suspected femur fracture. All other more serious injuries have been assessed and treated. You need to apply a Kendrick Traction Device (KTD). One other Soldier is available for assistance. You will need a KTD, and a long spine board. You are not in a CBRNE environment.

Standards: Complete all steps necessary to apply a KTD without causing additional injury to the casualty.

Performance Steps

1. Take body substance isolation (BSI) precautions.

2. Direct the assistant to manually stabilize the injured leg.

3. Direct the application of manual traction.
NOTE: If the KTD is used without elevating the leg, application of manual traction is not necessary. If the leg is elevated at all, manual traction must be applied before elevating the leg. The ankle hitch may be applied before elevating the leg and used to provide manual traction.

4. Assess motor, sensory, and distal circulation of the injured leg.

5. Adjust the splint to the proper length.
   a. Apply the ankle hitch snugly just above the ankle.
   b. Apply the thigh strap with traction pole receptacle at belt line or pelvic crest.
   c. Size the pole so that one section extends beyond the casualty's foot.
NOTE: The distal end of the splint should be 8 to 12 inches beyond the foot.
   d. Insert the pole ends into the traction pole receptacle.
   e. Secure the elastic strap around the knee.
   f. Apply traction.
   g. Apply elastic straps over the thigh and lower leg.

6. Assess the distal pulse, motor function, and sensation of the injured leg.

7. Secure the torso to the long board to immobilize the hip. Reassess pulse, motor, and sensory functions.

8. Secure the splint to the long board to prevent movement of the splint. Reassess pulse, motor, and sensory functions.

9. Record the treatment.

10. Do not cause further injury to the casualty.

Evaluation Preparation: Setup: For training and evaluation, have another Soldier act as the casualty. Tell the casualty not to assist in any way.

Brief Soldier: Tell the Soldier to apply a traction device to the designated part.

Performance Measures	GO	NO-GO
1. Took BSI precautions.	____	____
2. Directed the assistant to manually stabilize the injured leg.	____	____
3. Directed the application of manual traction, as needed.	____	____
4. Assessed motor, sensory, and distal circulation of the injured leg.	____	____
5. Adjusted the splint to the proper length.	____	____

    a. Applied the ankle hitch snugly just above the ankle.
    b. Applied the thigh strap with traction pole receptacle at belt line or pelvic crest.
    c. Sized the pole so that one section extended beyond the casualty's foot.
    d. Inserted the pole ends into the traction pole receptacle.
    e. Secured the elastic strap around the knee.
    f. Applied traction.
    g. Applied elastic straps over the thigh and lower leg.

6. Assessed the distal pulse, motor function, and sensation of the injured leg.	____	____
7. Secured the torso to the long board to immobilize the hip.	____	____
8. Secured the splint to the long board to prevent movement of the splint.	____	____
9. Recorded the treatment given.	____	____
10. Did not cause further injury to the casualty.	____	____

Evaluation Guidance: Score each Soldier according to the performance measures. Unless otherwise stated in the task summary, the Soldier must pass all performance measures to be scored GO. If the Soldier fails any steps, show what was done wrong and how to do it correctly.

References

    Required
    None

    Related
    0-7637-4406-9

APPLY A LONG SPINE BOARD
081-833-0181

Conditions: You have determined that your casualty needs a long spine board. All other more serious injuries or conditions have been treated. Three or four Soldiers are available for assistance. You will need straps, cravats, towels, long spine boards, safety pins, materials to improvise a cervical collar and head supports, and DD Form 1380 (U.S. Field Medical Card). You are not in a CBRNE environment.

Standards: Complete all the steps necessary to immobilize and transport a casualty with a suspected spine injury on a long spine board, without causing additional injury to the casualty.

Performance Steps

1. Check for the signs and symptoms of a spinal injury.
WARNING: If you suspect that the casualty has a spinal injury, treat him as though he does have a spinal injury.
    a. Spinal deformity. Its presence indicates a severe spinal injury, but its absence does not rule one out.
    b. Tenderness and/or pain in the spinal region.
       (1) Detect it by palpation or ask the casualty.
       (2) The presence of any pain is sufficient cause to suspect the presence of a spinal injury.
    c. Lacerations and/or contusions in the spinal region indicate severe trauma and usually accompany a spinal injury.
NOTE: The absence of lacerations and/or contusions does not rule out a spinal injury.
    d. Weakness, loss of sensation, and/or paralysis.
       (1) A neck level (cervical) spine injury may cause numbness or paralysis in all four extremities.
       (2) A waist level spinal injury may cause numbness or paralysis below the waist.
       (3) Ask the casualty to try to move the fingers and toes to check for paralysis.

2. Place the casualty on a long spine board.
NOTE: If a spine board is not available, utilize a standard litter or improvised litter made from a board or door. A hard surface is preferable to one that gives with the casualty's weight.
    a. The log roll technique.
       (1) Place the spine board next to, and parallel with, the casualty.
       (2) Immobilize the casualty's head and neck using manual stabilization.
          (a) Place your hands on both sides of the casualty's head, cradling the skull with your hands.
          (b) Maintain manual stabilization until the casualty has been placed on the spine board.
       (3) Apply a cervical collar, if available, or improvise one. (See task 081-833-0177.)
       (4) Brief each of the three assistants on their duties and instruct them to kneel on the same side of the casualty, with the spine board on the opposite side of the casualty.
          (a) First assistant. Place the near hand on the shoulder and the far hand on the waist.
          (b) Second assistant. Place the near hand on the hip and the far hand on the thigh.

Performance Steps

        (c) Third assistant. Place the near hand on the knee and the far hand on the ankle.

    (5) On your command, and in unison, the assistants roll the casualty slightly toward them. Turn the casualty's head, keeping it in a straight line with the spine.

    (6) Instruct the assistants to reach across the casualty with one hand, grasp the spine board at its closest edge, and slide it against the casualty. Instruct the second assistant to reach across the board to the far edge and hold it in place to prevent board movement.

    (7) Instruct the assistants to slowly roll the casualty back onto the board. Keep the head and spine in a straight line.

    (8) Place the casualty's wrists together at the waist and tie them together loosely.

NOTE: If the cervical collar or improvised collar does not fit flush with the spine board, place a roll in the hollow space between the neck and board. The roll should only be large enough to fill the gap, not to exert pressure on the neck

    b. The straddle-slide technique.

NOTE: Use this method when limited space makes it impossible to use the log roll technique (such as fractured pelvis).

    (1) Stand at the head of the casualty with your feet wide apart.

    (2) Apply stabilization to the casualty's head and apply a cervical collar.

    (3) Instruct the first assistant to stand behind you (facing your back), to line up the spine board, and to gently push the spine board under the casualty at your command.

    (4) Instruct the second assistant to straddle the casualty while facing you and gently elevate the shoulders so that the spine board can be slid under them.

    (5) Instruct the third assistant (facing you) to carefully elevate the hips while the spine board is being slid under the casualty.

    (6) Instruct the fourth assistant (facing you) to carefully elevate the legs and ankles while the board is being slid into place under the casualty.

WARNING: Complete all movements simultaneously, keeping the head and spine in a straight line.

NOTE: If the cervical collar or improvised collar does not fit flush with the spine board, place a roll in the hollow space between the neck and board. The roll should only be large enough to fill the gap, not to exert pressure on the neck.

  3. Secure the casualty to the long spine board

    a. Secure the casualty's torso and lower extremities with straps across the chest, hips, thighs, and lower legs

NOTE: Include the arms if the straps are long enough. If the spine board is not provided with straps and fasteners, use cravats or other long strips of cloth.

    b. Secure the casualty's head and head supports to the board with straps or cravats.

WARNING: Do not release manual stabilization until the cravats or head straps are firmly in place.

    (1) Apply head supports

    (2) Use two rolled towels, blankets, boots, or similar material. (Do not use sandbags.)

    (3) Place one close to each side of the head.

    (4) Using a cravat-like material across the forehead, make the supports and head one unit by tying to the board.

Performance Steps

   4. Record the treatment on the FMC.

   5. Evacuate the casualty.

Evaluation Preparation:
Setup: For training and evaluation, have another Soldier act as the casualty. You will need three or four Soldiers to act as the assistants. The Soldier being tested is to act as the team leader and direct the actions of the assistants. The casualty may be placed in a vehicle or other scenario, depending on available resources and the technique you are testing. Tell the casualty not to assist the Soldiers in any way.

Brief Soldier: To test step 1, tell the Soldier to state the signs and symptoms of a spinal injury. Tell the Soldier that the casualty has a suspected spinal injury. Then tell the Soldier to position the casualty on a spine board and to direct the actions of the assistants.

Performance Measures	GO	NO-GO
1. Checked for signs and symptoms of a spinal injury.	——	——
2. Placed the casualty on the long spine board.	——	——
3. Secured the casualty on the long spine board.	——	——
4. Recorded the treatment on the FMC.	——	——
5. Evacuated the casualty.	——	——
6. Did not cause further injury to the casualty.	——	——

Evaluation Guidance: Score each Soldier according to the performance measures in the evaluation guide. Unless otherwise stated in the task summary, the Soldier must pass all performance measures to be scored GO. If the Soldier fails any step, show what was done wrong and how to do it correctly.

References
   Required   **Related**
   DD FORM 1380   EMERG CARE AND TRANS 9

APPLY A REEL SPLINT
081-833-0182

Conditions: You and an assistant have encountered a casualty. You have done your initial assessment and you suspect a femur fracture. You need to apply a reel splint. You will need a reel splint, long spine board, securing devices, and padding material. You are not in a CBRNE environment.

Standards: Apply the splint without restricting circulation. Immobilize the fracture and maintain traction throughout the procedure, minimizing the effect of the injury.

Performance Steps

1. Take body substance isolation (BSI) precautions.

2. Direct the assistant to manually stabilize the injured leg.

3. Assess motor, sensory, and distal circulation of the injured leg.

4. Place the splint next to the uninjured leg, ischial pad next to the iliac crest.

5. Adjust the splint to the proper length.
   a. Loosen the sleeve locking device.
   b. Place the splint next to the uninjured leg so that the ischial pad of the splint is next to the casualty's iliac crest.
   c. Extend the splint until the bend in the splint is level with the casualty's heel.
   d. Lock the sleeve.
   e. Open the support straps.
   f. Place the straps under the splint.
   g. Release the ischial strap.
   h. Pull the release ring on the ratchet and release the traction strap.
   i. Extend and position the heel stand after the splint is in position under the casualty.

6. Apply the ankle hitch.

7. Apply manual traction while the assistant positions the splint.

8. Maintain manual traction and lower the limb onto the cradles of the splint.

9. Have the assistant secure the ischial strap.

10. Have the assistant connect the ankle hitch to the windlass and tighten the ratchet to equal manual traction.

11. Have the assistant secure the cradle straps.

12. Assess the distal pulse, motor function, and sensation of the injured leg.

13. Secure the torso to the long board to immobilize the hip.

14. Secure the splint to the long board to prevent movement of the splint.

Performance Measures	<u>GO</u>	<u>NO-GO</u>
1. Took BSI precautions.	——	——
2. Directed the assistant to manually stabilize the injured leg.	——	——
3. Assessed motor, sensory, and distal circulation of the injured leg.	——	——
4. Placed the splint next to the uninjured leg, ischial pad next to the iliac crest.	——	——
5. Adjusted the splint to the proper length.	——	——
6. Applied the ankle hitch.	——	——
7. Applied manual traction while the assistant positioned the splint.	——	——
8. Maintained manual traction and lowered the limb onto the cradles of the splint.	——	——
9. Had the assistant secure the ischial strap.	——	——
10. Had the assistant connect the ankle hitch to the windlass and tighten the ratchet to equal manual traction.	——	——
11. Had the assistant secure the cradle straps.	——	——
12. Assessed the distal pulse, motor function, and sensation of the injured leg.	——	——
13. Secured the torso to the long board to immobilize the hip.	——	——
14. Secured the splint to the long board to prevent movement of the splint.	——	——

Evaluation Guidance: Score each Soldier according to the performance measures in the evaluation guide. Unless otherwise stated in the task summary, the Soldier must pass all performance measures to be scored GO. The Soldier will be retrained if a NO-GO is received in any of the following areas: failure to maintain traction after it has been assumed; failure to reassess the distal pulse, motor function, and sensation before and after splinting; failure to secure the ischial strap before taking traction; failure to apply mechanical traction before securing the leg straps; or if final immobilization fails to support the femur or prevent rotation of the injured leg.

References
    Required
    None

    Related
    ISBN 0-7637-4406-9

Subject Area 8: Chemical Agent Injuries

TREAT A NERVE AGENT CASUALTY IN THE FIELD
081-833-0083

Conditions: You are in a chemical environment and have a casualty who is lying on the ground wearing a chemical protective overgarment and mask carrier. The casualty is displaying the signs and symptoms of nerve agent poisoning. You need to treat the casualty. You are wearing MOPP level 4. You will need a medical aid bag, MARK I nerve agent antidote kits (NAAK) or antidote treatment, nerve agent, autoinjectors (ATNAA) and convulsant antidote for nerve agents (CANA) autoinjectors, a DD Form 1380 (U.S. Field Medical Card), and decontaminable litter.

Standards: Complete all steps necessary to treat a nerve agent casualty in the field without causing further injury to the casualty.

Performance Steps

1. Assess the casualty for the signs and symptoms of nerve agent poisoning.
NOTE: If the casualty has been exposed to vapor or aerosol, the pupils will become pinpointed immediately. However, if the nerve agent is absorbed through the skin only or by ingesting contaminated food or water, the pinpointing of the pupils may be delayed or absent.
    a. Vapor exposure.
NOTE: Effects from vapor exposure will occur within seconds to minutes after being exposed and will not normally worsen after being removed from the exposure for 15 to 20 minutes.
        (1) Mild.
NOTE: Exposure to small amounts of vapor for a brief period usually causes effects in the eyes, nose, and lungs.
            (a) Unexplained runny nose.
            (b) Unexplained sudden headache.
            (c) Sudden and excessive salivation (drooling).
            (d) Difficulty in seeing (dimness of vision and miosis).
            (e) Tightness in the chest or difficulty in breathing.
            (f) Stomach cramps.
            (g) Nausea with or without vomiting.
            (h) Tachycardia or bradycardia.
        (2) Moderate.
            (a) All or most of the mild symptoms.
            (b) Fatigue.
            (c) Weakness.
            (d) Muscular twitching.
        (3) Severe.
NOTE: Effects may occur after one breath but normally take place within several seconds of a large vapor exposure.
            (a) All or most of the mild and moderate symptoms.
            (b) Strange or confused behavior.
            (c) Wheezing, dyspnea, and coughing.
            (d) Severely pinpointed pupils.
            (e) Red eyes with tearing.
            (f) Vomiting.

Performance Steps
        (g) Severe muscular twitching and general weakness.
        (h) Involuntary urination and defecation.
        (i) Convulsions.
        (j) Unconsciousness.
        (k) Respiratory failure.
        (l) Bradycardia.
        (m) Paralysis.
  b. Skin (percutaneous) exposure.
NOTES: 1. It is difficult to separate this type of exposure into categories due to the continued absorption of nerve agent into skin layers. Due to continued absorption, the effects from the nerve agent may be progressive in nature. They may occur from minutes up to 18 hours after exposure and continue even after the skin has been decontaminated. 2. The greater the amount of exposure to nerve agent, the shorter the onset time of symptoms with increased severity.
     (1) Mild exposure.
       (a) Localized sweating at the exposure site.
       (b) Muscular twitching at the exposure site.
       (c) Stomach cramps and nausea.
     (2) Moderate exposure.
       (a) Fatigue.
       (b) Weakness.
       (c) Muscular twitching.
     (3) Severe exposure.
       (a) Sudden loss of consciousness.
       (b) Vomiting.
       (c) Convulsions.
       (d) Severe muscular twitching and general weakness.
       (e) Difficulty breathing or cessation of respirations.
NOTE: Death would be the result of complete respiratory system failure.

  2. Mask the casualty.
    a. Instruct the casualty to mask self if they are able.
    b. Position the casualty face up and mask the casualty. Do not fasten the hood at this time.
CAUTION: Do not kneel or come into unnecessary contact with the chemically-contaminated ground.

  3. Check the casualty's pocket flaps and the area around the casualty for expended autoinjectors.

  4. Administer the antidote.
    a. Mild symptoms. Instruct the casualty to administer one MARK I NAAK or one ATNAA. (See STP 21-1-SMCT, task 081-831-1044.)
    b. Severe symptoms. Administer three MARK I NAAKs or three ATNAA autoinjectors and one CANA autoinjector to the casualty. (See STP 21-1-SMCT, task 081-831-1044.)
NOTE: Removal of any liquid nerve agent on the skin, on clothing, or in the eyes should be accomplished as soon as possible after administration of the antidote. Decontamination should be performed by the casualty, if able, or by a buddy.

  5. Check the casualty for signs of effectiveness of treatment.

Performance Steps
    a. Atropinization.
        (1) Heart rate above 90 beats per minute.
        (2) Reduced bronchial secretions.
        (3) Reduced salivation.
    b. Cessation of convulsions.

  6. Administer additional atropine or CANA, if needed.
    a. Administer additional atropine at approximately 15 minute intervals until atropinization is achieved.
    b. Administer additional atropine at intervals of 30 minutes to 4 hours to maintain atropinization or until the casualty is evacuated to a medical treatment facility (MTF).
    c. Administer a second and, if needed, a third CANA at 5 to 10 minute intervals to casualties suffering convulsions.
CAUTION: Do not give more than two additional CANA injections for a total of three.

NOTE: Additional atropine and the two additional CANA injections can be administered by a Combat Lifesaver (CLS), the Soldier Medic, or other medical personnel.

NOTE: Ensure all expended autoinjectors are secured to the casualty's left upper pocket of the battle dress overgarment (BDO) or the left pocket on the sleeve of the joint service lightweight integrated suit technology (JSLIST) overgarment (which has no pockets on the upper torso portion of the garment).

  7. Provide assisted ventilation for severely poisoned casualties, if equipment is available.
NOTE: The resuscitation device, individual chemical (RDIC) is a hand-powered ventilator equipped with a CBRNE filter. When the casualty reaches an MTF where oxygen and a positive pressure ventilator are available, these should be employed continuously until adequate spontaneous respiration is resumed.

  8. Record the number of injections given and all other treatment provided on the FMC.

  9. Evacuate the casualty.

Performance Measures

	GO	NO-GO
1. Assessed the casualty for the signs and symptoms of nerve agent poisoning.	——	——
2. Masked the casualty.	——	——
3. Checked the casualty's pocket flaps and the area around the casualty for expended autoinjectors.	——	——
4. Administered the antidote (MARK I NAAK or ATNAA).	——	——
5. Checked the casualty for signs of effectiveness of treatment.	——	——
6. Administered additional atropine or CANA, if needed.	——	——
7. Provided assisted ventilation for severely poisoned casualties, if equipment was available.	——	——

Performance Measures	<u>GO</u>	<u>NO- GO</u>
8. Recorded the number of injections given and all other treatment given on the FMC.	——	——
9. Evacuated the casualty.	——	——
10. Did not cause further injury to the casualty.	——	——

Evaluation Guidance: Score each Soldier according to the performance measures. Unless otherwise stated in the task summary, the Soldier must pass all performance measures to be scored GO. If the Soldier fails any steps, show what was done wrong and how to do it correctly.

References

Required	**Related**
DD FORM 1380	FM 4-02.285 (FM 8-285)
	STP 21-1-SMCT

TREAT A BLOOD AGENT (HYDROGEN CYANIDE) CASUALTY IN THE FIELD
081-833-0084

Conditions: You are in a chemical environment and have a casualty who is lying on the ground wearing protective overgarments, overboots, and mask carrier. You are wearing MOPP level 4. The casualty is displaying the signs and symptoms of blood agent poisoning. You must treat the casualty. You will need ventilation equipment (if available), and DD Form 1380 (U.S. Field Medical Card).

Standards: Complete all the steps necessary to treat a blood agent casualty in the field, without causing further injury to the casualty.

Performance Steps
CAUTION: Blood agent (hydrogen cyanide) causes symptoms ranging from convulsions to coma. After inhaling a high concentration of blood agent, a person may become unconscious and die within minutes. Blood agents in high concentration act quickly and death may result in 15 seconds. These agents release an odor of bitter almonds or peach kernels. Anyone smelling the odors should mask immediately.

1. Check for signs and symptoms of blood agent poisoning.
    a. Vertigo.
    b. Nausea.
    c. Increased respirations.
    d. Headache.
    e. Pink color of the skin.
    f. Violent convulsions.
    g. Coma.
    h. Respiratory arrest.
    i. Cardiac arrest.

2. Mask the casualty immediately.

3. Administer positive pressure ventilation, if available.
CAUTION: No device currently exists that can provide medical assistance in a contaminated environment.

4. Record the treatment given on the FMC.

5. Evacuate the casualty.

Evaluation Preparation:
Setup: For training and evaluation, have another Soldier act as the casualty and exhibit symptoms, such as hyperventilation. Tell the Soldier that the casualty is exhibiting symptoms such as slow pulse rate. You may decide whether the casualty is already masked or not.

Brief Soldier: Tell the Soldier to state the signs and symptoms of blood agent poisoning, and then treat the casualty.

Performance Measures	GO	NO-GO
1. Checked for the signs and symptoms of blood agent poisoning.	——	——
2. Masked the casualty immediately.	——	——
3. Administered positive pressure ventilation, if available.	——	——
4. Recorded the treatment given on the FMC.	——	——
5. Evacuated the casualty.	——	——

Evaluation Guidance: Score each Soldier according to the performance measures. Unless otherwise stated in the task summary, the Soldier must pass all performance measures to be scored GO. If the Soldier fails any steps, show what was done wrong and how to do it correctly.

References

    Required
    DD FORM 1380

    Related
    FM 4-02.285 (FM 8-285)

TREAT A CHOKING AGENT CASUALTY IN THE FIELD
081-833-0085

Conditions: You are in a chemical environment and have a casualty who is lying on the ground wearing protective overgarments, overboots, and mask carrier. You are wearing MOPP level 4. The casualty is displaying the signs and symptoms of a chocking agent poisoning. You must treat the casualty. You will need ventilation equipment (if available), and DD Form 1380 (U.S. Field Medical Card).

Standards: Complete all the steps necessary to treat a choking agent casualty in the field, without causing further injury to the casualty.

Performance Steps
NOTE: The treatment available for the choking agent casualty in the field is limited. It is essential that the casualty be masked and evacuated to increase the possibility of survival.

   1. Check for the signs and symptoms of choking agent poisoning.
      a. Immediate signs and symptoms.
NOTE: Although heavy concentrations of poison bring on these symptoms very quickly, small doses may take up to 2 to 6 hours before there is any sign of poisoning.
         (1) Watery eyes.
         (2) Coughing.
         (3) Choking.
         (4) Tightness in the chest.
         (5) Nausea.
         (6) Vomiting.
         (7) Headache.
         (8) Transient blindness.
         (9) Increased salivation.
        (10) Tingling burning sensation on the skin.
      b. Delayed signs and symptoms.
         (1) Rapid shallow breathing.
         (2) Cyanosis.
         (3) Apprehension.
         (4) Severe coughing, producing frothy fluid.
         (5) Weak and rapid pulse.
         (6) Chest wall retractions.
         (7) Pulmonary edema.
      c. Asymptomatic. The casualty has been exposed, but shows no signs or symptoms.

   2. Mask the casualty, but do not fasten the hood.

   3. Position the casualty.
      a. Supine.
      b. Seated.

   4. Treat the casualty.
      a. Asymptomatic.
         (1) Restrict the casualty's activities to light duties to avoid stress to the respiratory system.
         (2) Monitor the casualty for the onset of symptoms.
      b. Symptomatic.

Performance Steps
  (1) Keep the casualty at rest seated.
  (2) Provide intermittent positive pressure ventilation, if equipment is available.
  (3) Keep the casualty warm.

  5. Record the treatment given on the FMC.

  6. Evacuate the casualty.

Evaluation Preparation:
Setup: For training and evaluation, have another Soldier act as the casualty and exhibit signs such as choking or coughing (coach the casualty on how to answer the Soldier's questions on symptoms such as headache). Tell the medic the casualty is exhibiting symptoms such as cyanosis. You may decide whether the casualty is already masked or not.

Brief Soldier: Tell the Soldier to state the signs and symptoms of a choking agent casualty, and then treat the casualty.

Performance Measures	GO	NO-GO
1. Checked for signs and symptoms of choking agent poisoning.	——	——
2. Masked the casualty but did not fasten the hood.	——	——
3. Positioned the casualty.	——	——
4. Treated the casualty.	——	——
5. Recorded the treatment given on the FMC.	——	——
6. Evacuated the casualty.	——	——

Evaluation Guidance: Score each Soldier according to the performance measures. Unless otherwise stated in the task summary, the Soldier must pass all performance measures to be scored GO. If the Soldier fails any steps, show what was done wrong and how to do it correctly.

References
  Required   **Related**
  DD FORM 1380   FM 4-02.285 (FM 8-285)

TREAT A BLISTER AGENT CASUALTY (MUSTARD, LEWISITE, PHOSGENE OXIME) IN THE FIELD
081-833-0086

Conditions: You are in a chemical environment and are treating a casualty who is lying on the ground wearing MOPP level 4. You are wearing MOPP level 4. The casualty is displaying the signs and symptoms of blister agent poisoning. You must treat the casualty. You will need the casualty's canteen and personal decontamination kit, and DD Form 1380 (U.S. Field Medical Card).

Standards: Complete all the steps necessary to treat a blister agent casualty in the field, without causing further injury to the casualty. Do not kneel when providing treatment.

Performance Steps

1. Check for the signs and symptoms of blister agent poisoning.
NOTE: Moist areas of the body are highly susceptible to blister agents. Therefore, during hot weather, blister agents can cause a greater number of casualties.
    a. Skin.
        (1) Itching.
        (2) Redness.
        (3) Blisters.
        (4) Pain.
            (a) Intense and immediate if contaminated by lewisite (L) (arsenical) or phosgene oxime.
            (b) Delayed from 1 hour to days if contaminated by mustard (HD).
    b. Eyes (L--immediate, HD--1 hour).
        (1) Extremely sensitive to light.
        (2) Gritty feeling.
        (3) Painful.
        (4) Watery.
        (5) Involuntary spasms of the eyelids.
        (6) Swelling and blistering of eyelids.
        (7) Corneal lesion.
        (8) Permanent blindness (direct contact).
        (9) Redness.
    c. Respiratory tract (L--immediate, HD--4 to 6 hours).
        (1) Coughing.
        (2) Sore throat.
        (3) Frothy sputum.
        (4) Phlegm.
        (5) Nasal secretions.
        (6) Adema.
    d. Systemic.
        (1) Malaise.
        (2) Headache.
        (3) Nausea and vomiting.
        (4) Severe skin burns.
        (5) Bloody diarrhea.
CAUTION: Seek overhead protection, or heavy foliage if available.

Performance Steps

2. Tell the casualty to take a deep breath, hold it, and close the eyes.
CAUTION: While the eyes are being irrigated, the breath should be held and the mouth kept closed to prevent contamination and absorption through mucous membranes.

3. Lift the casualty's mask.

4. Irrigate the casualty's eyes.
    a. Use water from the casualty's canteen.
NOTE: If the casualty's water has been contaminated, use sterile water or sterile normal saline from the aid bag.
    b. Tilt the casualty's head to one side.
    c. Tell the casualty to open the eyes as much as possible.
    d. Pour water slowly into one eye.
    e. To avoid spreading contamination, let the water run off the side of the face.
    f. Repeat steps 4a through 4e for the other eye.
NOTE: It may be necessary for the casualty to re-mask and take additional breaths if unable to hold the breath until both eyes are irrigated.

5. Use the casualty's personal decontamination kit on both the face and the portion of the mask in contact with the face. (See STP 21-1-SMCT, task 031-503-1013.)

6. Replace the casualty's mask.

7. Tell the casualty to clear and check the mask.

8. Tell the casualty to breathe normally.
NOTE: Further decontamination procedures will be performed by the casualty (self-aid) or buddy aid.

9. Record the treatment given on the FMC.

10. Evacuate the casualty, if necessary.

Evaluation Preparation:

Setup: For training and evaluation, have another Soldier act as the casualty and exhibit signs such as coughing. Coach the casualty on how to answer the Soldier's questions on symptoms such as headache. Tell the Soldier that the casualty is exhibiting signs such as blisters. Training decontamination kits must be used.

Brief Soldier: Tell the Soldier to state the signs and symptoms of blister agent poisoning, and then treat the casualty. For step 4, have the Soldier tell you what should be done.

Performance Measures	GO	NO-GO
1. Checked for signs and symptoms of blister agent poisoning.	——	——
2. Told the casualty to take a deep breath, hold it, and close the eyes.	——	——
3. Lifted the casualty's mask.	——	——
4. Irrigated the casualty's eyes.	——	——

Performance Measures	GO	NO-GO
5. Used the casualty's personal decontamination kit on both the face and the portion of the mask in contact with the face.	——	——
6. Replaced the casualty's mask.	——	——
7. Told the casualty to clear and check the mask.	——	——
8. Told the casualty to breathe normally.	——	——
9. Recorded the treatment given on the FMC.	——	——
10. Evacuated the casualty, if necessary.	——	——
11. Did not kneel at any time.	——	——

Evaluation Guidance: Score each Soldier according to the performance measures. Unless otherwise stated in the task summary, the Soldier must pass all performance measures to be scored GO. If the Soldier fails any steps, show what was done wrong and how to do it correctly.

References

Required	**Related**
DD FORM 1380	STP 21-1-SMCT

DECONTAMINATE A CASUALTY
081-833-0095

Conditions: You are supervising the contaminated side of an established chemical decontamination station. Medical personnel and nonmedical augmentees are in MOPP level 4. Chemically contaminated casualties have been triaged by the senior medic and have been routed to your area for decontamination. You will need a M258A1 or M291 decontamination kit, 5% chlorine solution, 0.5% chlorine solution, butyl rubber aprons, butyl rubber gloves, stainless steel buckets, cellulose sponges, water source, plastic bags, litters, litter stands, bandage scissors, M8 chemical detection paper, chemical agent monitor (CAM), contaminated disposal containers, bandages, gauze, DD Form 1380 (U.S. Field Medical Card), and tourniquets.

Standards: Remove the casualty's clothing without further contaminating the casualty or contaminating decontamination team personnel. Remove dressings, replace tourniquets, and decontaminate splints. Effectively decontaminate and transfer the casualty across the shuffle pit without contaminating the clean side of the hot line.

Performance Steps
NOTES: 1. The supported unit must provide a minimum of 8 nonmedical personnel to augment the decontamination station as the decontamination team. Although casualty decontamination is routinely performed by these nonmedical personnel, the supervision of and final determination as to the completeness of the decontamination rests with medical personnel. 2. Steps 1 through 17 will be performed by personnel in the clothing removal area. At the clothing removal area two to four persons will be working together as a team, one or two on either side of the casualty.

   1. Decontaminate the casualty's hood.
      a. Cover the mask air inlets with your hand. Instruct the casualty to do this if he is able.
      b. Wipe off the front, sides, and top of the hood with a cellulose sponge soaked with 5% calcium hypochlorite solution or use the M258A1 or M291 skin decontaminating kit.
NOTE: The medical equipment set (MES) for chemical agent casualty decontamination contains powdered calcium hypochlorite (high test hypochlorite or HTH). It is mixed with water to make the 5% and 0.5% decontaminating solutions. Liquid chlorine bleach (household bleach), a 5% solution of sodium hypochlorite, may also be used.
      c. Uncover the mask air inlets.

   2. Cut off the casualty's hood.

      a. Dip scissors in the 5% solution.
CAUTION: Dip and scrub the scissors in the 5% solution after each separate cutting procedure and rinse your gloves in the same solution in order to reduce the spread of contamination.
      b. Cut the neck cord.
      c. Cut away the drawstring below the voicemitter.
      d. Release or cut the hood shoulder straps.
      e. Unzip the hood zipper.
      f. Begin cutting at the zipper, below the voicemitter.
      g. Proceed cutting upward, close to the filter inlet covers and eye lens outserts.
      h. Cut upward to the top of the eye lens outserts.
      i. Cut across the forehead to the outer edge of the next eye outsert.
      j. Cut downward toward the casualty's shoulder, staying close to the eye lens outserts and filter inlet covers.

Performance Steps

  k. Cut across the lower part of the voicemitter to the zipper.
  l. Dip the scissors and rinse your gloves in the 5% solution.
  m. Cut from the center of the forehead, over the top of the head.
  n. Fold the left and right sides of the hood to the sides of the casualty's head, laying the sides of the hood on the litter.

3. Decontaminate the casualty's mask and exposed skin.
  a. Use the M258A1 or M291 skin decontamination kit or 0.5% solution.
  b. Cover the mask air inlets as in step 1a.

CAUTION: Use only the 0.5% solution to decontaminate the skin and the parts of the mask that touch the face. The 5% solution is corrosive and may burn the skin.

  c. Decontaminate the exterior of the mask.
  d. Wipe down all the exposed skin areas, to include the neck and behind the ears.
  e. Uncover the mask air inlets.

4. Remove the casualty's FMC.
  a. Cut the FMC tie-wire, allowing the FMC to fall into a plastic bag. If possible, do not allow any of the tie-wire to remain attached to the card. This will prevent the wire from poking a hole in the bag.
  b. Seal the plastic bag and rinse the plastic bag with the 0.5% solution.
  c. Place the plastic bag under the protective mask head straps.

5. Remove gross contamination on the overgarment by wiping all visible contamination spots with a sponge soaked in 5% solution.

6. Remove the casualty's protective overgarment jacket.

CAUTION: Dip and scrub the scissors in the 5% solution before doing each cutting procedure to avoid contaminating the inner garment or the casualty's skin.

  a. Cut the sleeves from the cuff up to the shoulder of the jacket, and then through the collar. Keep the cuts close to the inside of the arms so that most of the sleeve material can be folded outward.

CAUTION: Medical items are not removed at the clothing removal area. Cut around medical items such as dressings, splints, and tourniquets.

  b. Unzip the jacket (or cut alongside the jacket's zipper).
  c. Roll the chest sections to the respective sides, with the inner black liner outward. Carefully tuck the cut jacket between the arm and the chest.
  d. Roll the cut sleeves away from the arms, exposing the black liner.

7. Remove the casualty's protective overgarment trousers.

CAUTION: Dip and scrub the scissors in the 5% solution before doing each cutting procedure to avoid contaminating the inner garment or the casualty.

  a. Cut the trouser legs from the ankle to the waist. Keep the cuts near the insides of the legs, along the inseam, to the crotch.
    (1) Cut up the right leg and across the crotch of the trousers.
    (2) Cut up the left leg, cross over the crotch cut, and continue to cut up through the waistband.

NOTE: Avoid cutting through the pockets.

  b. Fold the cut trouser halves onto the litter with the contaminated sides away from the casualty. Make sure the outer side of the protective overgarment does not touch the skin or undergarments of the casualty.
  c. Roll the inner leg portion under and between the legs.

Performance Steps

8. Remove the casualty's butyl rubber gloves.
   a. Decontaminate your butyl rubber gloves in the 5% solution.
   b. Lift the casualty's arm up and out of the cutaway sleeve unless contraindicated by the casualty's condition.
   c. Pull the butyl rubber gloves off by rolling the cuff over the fingers, turning the glove inside out. Do not remove the white glove liners at this time.
   d. Lower the casualty's arms and fold them across the chest.

CAUTION: Do not allow the arms to come into contact with the exterior of the protective overgarments.

   e. Place the gloves in a contaminated disposal container.
   f. Decontaminate your butyl rubber gloves in the 5% solution.

9. Remove the casualty's protective overboots.
   a. Stand at the foot of the litter facing the casualty.
   b. Cut the protective overboot laces.
   c. Grasp the heel of the protective overboot with one hand and the toe of the protective overboot with the other hand.
   d. Pull the heel downward, and then toward you until the overboot is removed.

NOTE: While you and another team member hold the casualty's raised feet, have a third member wipe down the end of the litter with the 5% solution before lowering the feet to the litter.

   e. Place the overboots in a contaminated disposal container.

10. Remove and secure the casualty's personal effects.
   a. Remove the casualty's personal articles from the overgarment and ACU pockets.
   b. Place the articles in plastic bags.
   c. Label the bags with the casualty's name and SSN. (Print the information on a piece of paper and place the paper in the plastic bag.)
   d. Seal the plastic bags.
   e. If the articles are not contaminated, return them to the casualty. If the articles may be contaminated, place the bags in the contaminated holding area until they can be decontaminated. The articles will then be returned to the casualty.

11. Remove the combat boots following the same procedures as for removing the protective overboots.

NOTE: Remove the boots without touching the casualty's inner clothing or exposed skin.

12. Cut off the casualty's army combat uniform (ACU).

CAUTION: Decontaminate your butyl rubber gloves in the 5% solution before you touch the casualty's garments or exposed skin.

   a. Cut off the ACU shirt.
      (1) Uncross the casualty's arms.
      (2) Cut the ACU shirt using the same procedure as for the protective overgarment jacket.
      (3) Recross the casualty's arms over the chest.
   b. Unbuckle or cut the belt material.
   c. Cut off the ACU trousers following the same procedure as for the protective overgarment trousers.

13. Cut off the casualty's undergarments.

CAUTION: Decontaminate your butyl rubber gloves in 5% solution before you touch the casualty's garments or exposed skin.

Performance Steps

    a. Cut off the underpants.

    b. Cut off the T-shirt.

    c. Cut off the brassiere, if necessary.

       (1) Lift the casualty's arm off the chest.

       (2) Cut between the cups.

       (3) Cut both shoulder straps where they attach to the cup.

       (4) Lay the cups away from the casualty onto the litter.

       (5) Lay shoulder straps up and over the shoulders onto the litter.

NOTE: At this point, the white glove inner liners for a female may be removed while the casualty's arms are lifted off her chest.

14. Remove the casualty's glove inner liners.

    a. Remove the glove liners using the same procedure as for removing butyl rubber gloves.

    b. Cross the casualty's arms over the chest.

15. Remove the casualty's socks.

    a. Decontaminate your butyl rubber gloves in 5% solution.

    b. Position yourself at the foot of the litter.

    c. Remove each sock by rolling it down over the foot, turning it inside out or by cutting the sock off.

    d. Place the socks into a contaminated disposal container.

16. Decontaminate the casualty's ID tags.

    a. Decontaminate your butyl rubber gloves in the 5% solution.

    b. Wipe the ID tags with the 0.5% solution.

17. Move the casualty to the skin decontamination area.

CAUTION: Observe proper body mechanics to avoid injury to your back. Use your legs instead of your back to lift the casualty.

    a. Decontaminate your butyl rubber aprons and gloves in the 5% solution.

    b. Lift the casualty out of the cutaway garments, using a three person arms carry.

       (1) Lifter #1 slides his arms (palms turned upward) under the casualty's head/neck and shoulders.

       (2) Lifter #2 slides his arms (palms turned upward) under the casualty's back and buttocks.

       (3) Lifter #3 slides his arms (palms turned upward) under the casualty's thighs and calves.

       (4) On the command of Lifter, bearer #1, lift the casualty. (PREPARE TO LIFT: LIFT.)

    c. Once the casualty has been lifted off the litter, all three lifters stand upright and turn the casualty in against their chests.

NOTE: At this point, the casualty has nothing on his body except the protective mask and medical items (dressings, splints, tourniquets).

    d. While the casualty is being held, another team member quickly removes the contaminated litter and replaces it with a clean litter. A decontaminatable mesh litter should be positioned, if available.

    e. Lower the casualty onto the clean litter, in a supine position, on the command of lifter #1.

    f. Carry the litter to the skin decontamination area, and then return to the clothing removal area.

Performance Steps
      g. Dispose of all contaminated material at the clothing removal area.
         (1) The casualty's contaminated clothing is placed in a bag and put in a contaminated disposal container.
         (2) The dirty litter is rinsed with the 5% decontamination solution and placed in a dirty litter storage area.

CAUTION: Before obtaining another casualty, the clothing removal team should rinse their gloves and aprons in the 5% decontaminating solution and drink enough water to compensate for the heat and workload.

NOTE: Steps 18 through 23 are performed by personnel in the skin decontamination area. At the skin decontamination area, two to four persons will be working together as a team, one or two on either side of the casualty.

  18. Perform spot skin decontamination.
      a. Spot decontaminate potential areas of chemical contamination with the M258A1 or M291 Skin Decontaminating Kit or the 0.5% solution.
      b. Pay particular attention to areas where gaps exist in the MOPP gear, such as the neck, lower part of the face, waistline, wrists, and ankles.

  19. Remove field dressings and bandages.
NOTE: This step must be performed by medical personnel.
      a. Carefully cut off dressings and bandages.
      b. Cut off any remaining clothing that was covered by the dressings and bandages.
      c. Decontaminate the exposed areas of skin with the 0.5% solution.
      d. Irrigate the wound with the 0.5% solution if the wound is suspected to be contaminated.
NOTE: Bandages are not replaced unless there is a critical medical need (for example, to control bleeding). Bandages are replaced when the casualty is in the clean (uncontaminated) treatment area.
      e. Place removed dressings and clothing in a contaminated disposal container.

  20. Replace any tourniquets.
NOTE: Medical personnel must perform this step.
      a. Decontaminate an area above the existing tourniquet.
      b. Place a new tourniquet 1/2 to 1 inch above the old tourniquet.
      c. Remove the old tourniquet.
      d. Remove any remaining clothing or dressings covered by the old tourniquet.
      e. Decontaminate the newly exposed areas.
      f. Place the removed tourniquet, dressings, and clothing in a contaminated disposal container.

  21. Decontaminate any splints.
NOTE: Splints are only removed by medical personnel.
      a. Stabilize the splinted extremity.
      b. Decontaminate the splint and the extremity by liberally flushing them with the 0.5% solution.
CAUTION: Do not remove any part of a traction splint from a femoral fracture.

  22. Check the casualty for contamination.
      a. Use M8 chemical agent detector paper or the CAM.
      b. Decontaminate any areas of detected contamination, as necessary.

Performance Steps

CAUTION: Under no circumstances should a casualty who has not been entirely decontaminated be moved across the hot line. If a wound or splinted area cannot be entirely decontaminated, inform the senior medic. Do not move the casualty across the hot line. He must be treated on the contaminated side of the casualty decontamination station.

23. Transfer the casualty to the shuffle pit.
    a. Personnel decontaminate themselves by rinsing their butyl rubber gloves and apron with the 5% solution.
    b. Carry the casualty to the shuffle pit on the skin decontamination litter.
    c. Place the litter on the litter stand located in the shuffle pit.
    d. Lift the casualty from the decontamination litter using the same technique described in step 17.
    e. Remove the decontamination litter from the stand and a medic from the clean side will replace it with a clean litter.
    f. Lower the casualty onto the clean litter and move back from the hot line.

NOTE: Do not step across the hot line. Personnel from the clean side of the hot line will take the casualty to the clean treatment station.

Performance Measures	GO	NO-GO
1. Decontaminated the casualty's hood.	——	——
2. Cut off the casualty's hood.	——	——
3. Decontaminated the casualty's mask and exposed skin.	——	——
4. Removed the casualty's FMC.	——	——
5. Removed gross contamination.	——	——
6. Removed the casualty's protective overgarment jacket.	——	——
7. Removed the casualty's protective overgarment trousers.	——	——
8. Removed the casualty's butyl rubber gloves.	——	——
9. Removed the casualty's protective overboots.	——	——
10. Removed and secured the casualty's personal effects.	——	——
11. Removed the casualty's combat boots.	——	——
12. Removed the casualty's ACU.	——	——
13. Cut off the casualty's undergarments.	——	——
14. Removed the casualty's glove inner liners.	——	——
15. Removed the casualty's socks.	——	——
16. Decontaminated the casualty's ID tags.	——	——
17. Moved the casualty to the skin decontamination area.	——	——
18. Performed spot skin decontamination.	——	——

Performance Measures	<u>GO</u>	<u>NO-</u> <u>GO</u>
19. Removed field dressings and bandages.	——	——
20. Replaced any tourniquets.	——	——
21. Decontaminated any splints.	——	——
22. Checked the casualty for contamination.	——	——
23. Transferred the casualty to the shuffle pit.	——	——

Evaluation Guidance: Score each Soldier according to the performance measures. Unless otherwise stated in the task summary, the Soldier must pass all performance measures to be scored GO. If the Soldier fails any steps, show what was done wrong and how to do it correctly.

References
    Required   **Related**
    DD FORM 1380   STP 21-1-SMCT

TREAT A BIOLOGICAL EXPOSED CASUALTY
081-833-0137

Conditions: You encounter a casualty with symptoms consistent with biological agent exposure. All other life-threatening injuries have been treated. You will need a fully stocked aid bag, intravenous (IV) administration equipment and fluids, oxygen, suction and ventilation equipment (if available), selected medications, documentation forms, and chemical personal protective equipment.

Standards: Perform appropriate identification and treatment for a casualty exposed to a biological agent.

Performance Steps

1. Determine the biological warfare agent.
NOTE: Ensure personal protective equipment is available for yourself and the casualty.

NOTE: All scene size-up, initial assessment, focused history, examination, detailed physical examination, on-going assessment, and transport assessment steps must be taken to ensure that injuries or illnesses are not overlooked resulting in further injury to the casualty.
    a. Pneumonia-like agents.
        (1) Anthrax.
        (2) Tularemia.
        (3) Plague.
        (4) Q fever.
    b. Encephalitis-like agents.
        (1) Smallpox.
        (2) Venezuelan equine encephalitis.
    c. Biological toxins.
        (1) Botulinium.
        (2) Staphylococcal enterotoxin B.
        (3) Ricin.
        (4) Mycotoxins.
    d. Other agents.
        (1) Cholera.
        (2) Viral hemorrhagic fevers.
        (3) Brucellosis.

2. Recognize findings.
    a. Pneumonia-like agents.
        (1) Anthrax.
            (a) Fever.
            (b) Malaise.
            (c) Cough and mild chest discomfort, followed by severe respiratory distress.
            (d) Diaphoresis.
            (e) Lesions and scabs, if skin exposed.
            (f) Shock.
        (2) Plague.
            (a) High fever.
            (b) Chills.
            (c) Headache.

Performance Steps
          (d) Cough with hemoptysis.
          (e) Malaise.
          (f) Rapid progression of pneumonia.
          (g) Black, necrotic skin lesions.
          (h) Cyanosis.
          (i) Shock.
       (3) Tularemia.
          (a) Fever.
          (b) Substernal chest discomfort.
          (c) Nonproductive cough.
       (4) Q fever.
          (a) Fever.
          (b) Myalgias.
          (c) Headache.
          (d) Malaise.
          (e) Cough.
          (f) Chest pain.
          (g) Pleuritic chest pain.
   b. Encephalitis-like agents.
       (1) Smallpox.
          (a) Malaise.
          (b) Fever.
          (c) Rigors.
          (d) Vomiting.
          (e) Headache.
          (f) Backache.
          (g) Lesions that eventually progress to pustular vesicles.
          (h) Rash.
       (2) Venezuelan equine encephalitis.
          (a) Fever.
          (b) Chills.
          (c) Sever headache.
          (d) Photophobia.
          (e) Myalgia.
          (f) Nausea.
          (g) Vomiting.
          (h) Sore throat.
   c. Biological toxins.
       (1) Botulinum.
          (a) Generalized weakness.
          (b) Generalized paralysis.
          (c) Ptosis.
          (d) Diplopia.
          (e) Difficulty breathing.
          (f) Respiratory failure.
       (2) Staphylococcal enterotoxin B.
          (a) Fever.
          (b) Chills.
          (c) Headache.
          (d) Myalgia.

Performance Steps
        (e) Cough.
        (f) Shock.
    (3) Ricin.
        (a) Weakness.
        (b) Fever.
        (c) Cough.
        (d) Hypotension.
    (4) T-2 mycotoxins.
        (a) Pain.
        (b) Itching.
        (c) Redness.
        (d) Lesions on exposed skin.
        (e) Nose and throat pain.
        (f) Runny nose.
        (g) Sneezing.
        (h) Shock.
  d. Other agents.
    (1) Cholera.
        (a) Abdominal cramping.
        (b) Profuse watery stools.
        (c) Vomiting.
        (d) Malaise.
    (2) Brucellosis.
        (a) Fever.
        (b) Malaise.
        (c) Myalgia.
        (d) Sweats.
        (e) Arthralgias.
        (f) Cough.
    (3) Viral hemorrhagic fevers.
        (a) Easy bleeding.
        (b) Petechiae.
        (c) Hypotension.
        (d) Flushing of face and chest.
        (e) Edema.
        (f) Vomiting and diarrhea.
        (g) Headache.
    (4) Glanders.
        (a) Fever.
        (b) Rigors.
        (c) Sweats.
        (d) Myalgia.
        (e) Headache.
        (f) Pleuritic chest pain.
        (g) Cervical adenopathy.
        (h) Splenomegaly.
        (i) Generalized papular/pustular eruptions.

3. Terminate exposure. Physically remove the casualty from the contaminated environment.

Performance Steps

4. Provide generalized emergency care.
    a. Recognition and identification.
        (1) Differentiate between chemical and biological weapons.
        (2) Call for additional resources.
            (a) Medical officers will be needed to prescribe proper antibiotic coverage and/or antitoxin treatment.
            (b) Laboratory support will be needed for positive identification of the agent.
    b. Isolation of selected cases.
        (1) Smallpox, plague, and ebola are highly transmissible.
        (2) Isolate biological casualties from unaffected individuals.
        (3) Enforce protective measures.
    c. Supportive care.
        (1) Secure and maintain the airway. (See task 081-833-0107.)
        (2) Initiate IV fluid or saline locks.
    d. Arrange for antibiotic or antitoxin therapy.
NOTE: This requires the expertise of a medical officer.
        (1) Ciprofloxacin.
        (2) Doxycycline.
        (3) Gentamycin.
        (4) Oral tetracycline.
        (5) Erythromycin.

5. Treat for specific exposure.
    a. Pneumonia-like agents.
        (1) Anthrax.
            (a) Aggressive respiratory and cardiovascular support.
            (b) Intravenous fluids to counteract septic shock.
            (c) Intravenous or oral ciprofloxacin or oral doxycycline.
            (d) Contact isolation is required.
        (2) Plague.
            (a) Aggressive fluid resuscitation.
            (b) Oral streptomycin, tetracycline, or sulfamethoxazole-trimethoprim
            (c) Contact isolation is required.
        (3) Tularemia.
            (a) Respiratory and fluid support as needed.
            (b) Gentamicin.
            (c) Contact isolation is required.
        (4) Q Fever -oral tetracycline or doxycycline.
    b. Encephalitis-like agents.
        (1) Smallpox.
            (a) Respiratory and contact isolation.
            (b) Consider immediate vaccination.
        (2) Venezuelan equine encephalitis.
            (a) Provide supportive treatment.
            (b) Blood and body fluid precautions required.
            (c) Investigational live, attenuated vaccine available.
    c. Biological toxins.
        (1) Botulinum.
            (a) Provide respiratory support.
            (b) Intravenous administration of antitoxin.

Performance Steps
    (2) Staphylococcal Enterotoxin B.
      (a) Provide respiratory support.
      (b) Initiate saline lock for vascular access.
      (c) Hypochlorite solution effectively inactivates toxin when applied to most nonporous surfaces.
    (3) Ricin.
      (a) Provide respiratory support.
      (b) Consider initiation of intravenous fluids.
    (4) Mycotoxins.
      (a) Thorough decontamination with hypochlorite solution.
      (b) Provide respiratory support.
      (c) Consider intravenous fluid.
  d. Other agents.
    (1) Cholera.
      (a) Oral hydration with "World Health Organization oral rehydration solution".
      (b) Intravenous normal saline.
      (c) Antibiotic treatment with tetracycline or doxycycline.
      (d) Consider ciprofloxacin or erythromycin for resistant strains of cholera.
      (e) Treat children with tetracycline, erythromycin, or trimethoprim-sulfamethoxazole.
    (2) Brucellosis.
      (a) Intravenous fluid initiation.
      (b) Oral doxycycline combined with streptomycin.
      (c) Endocarditis or other serious complication may require triple antibiotic coverage.
    (3) Viral hemorrhagic fevers.
      (a) Judicious use of intravenous fluids.
      (b) Consider antiviral drug ribavirin.
      (c) Consider immediate vaccination.
      (d) Consider antiviral therapy with ribavirin.
 6. Triage casualties based upon level of exposure. (See task 081-833-0140.)
  a. Minimal: all ambulatory casualties.
  b. Delayed: moderate to severe symptoms.
  c. Immediate.
    (1) Respiratory failure.
    (2) Decompensated shock.
  d. Expectant.
    (1) Pulseless.
    (2) Persistent decompensated shock despite adequate IV fluids.
 7. Provide protection for biological agents.
  a. Recognition and identification of agent.
  b. Personal protective equipment.
    (1) Protective masks.
      (a) Medical high-efficiency particulate air (high-efficiency particulate air (HEPA), occupational safety and health administration (OSHA) N95 mask).
      (b) Self-contained breathing apparatus.
    (2) Protective overgarment.
      (a) Hood.
      (b) Gloves.

Performance Steps

        (c) Boots.

    c. Immunization and prophylaxis.

       (1) Immunizations.

          (a) Anthrax.

          (b) Plague.

          (c) Q fever (experimental).

          (d) Tularemia.

          (e) Smallpox.

          (f) Venezuelan equine encephalitis (experimental).

          (g) Viral hemorrhagic fevers (experimental).

          (h) Botulinum.

       (2) Prophylaxis.

          (a) Anthrax: ciprofloxacin/doxycycline.

          (b) Plague: tetracycline.

          (c) Q fever: tetracycline.

          (d) Brucellosis: doxycycline and rifampin prophylaxis.

          (e) Tularemia: tetracycline.

Performance Measures	GO	NO-GO
1. Determined the biological warfare agent.	——	——
a. Considered pneumonia-like agents.		
b. Considered encephalitis-like agents.		
c. Considered biological toxins.		
d. Considered other agents.		
2. Recognized findings.	——	——
3. Terminated exposure.	——	——
4. Provided generalized emergency care.	——	——
a. Recognized and identified agent.		
b. Isolated case.		
c. Supportive care provided.		
d. Antibiotic and/or antitoxin therapy arranged.		
5. Treated for specific emergency care.	——	——
a. Treated for pneumonia-like agents.		
b. Treated for encephalitis-like agents.		
c. Treated for biological toxins.		
d. Treated for other agents.		
6. Triaged casualties.	——	——
7. Provided protection for biological agents.		

Evaluation Guidance: Score each Soldier according to the performance measures. Unless otherwise stated in the task summary, the Soldier must pass all performance measures to be scored GO. If the Soldier fails any steps, show what was done wrong and how to do it correctly.

References
    Required
    None

Related
TACTICAL EMERGENCY CARE
USAMRIID

TREAT A RADIATION CASUALTY
081-833-0202

Conditions: You encounter a casualty with symptoms consistent with radiological exposure. You will need a fully stocked aid bag, intravenous (IV) administration equipment and fluids, oxygen, suction and ventilation equipment (if available), selected medications, documentation forms, and personal chemical protective equipment. You are not in a CBRNE environment.

Standards: Effectively manage a casualty exposed to ionizing radiation.

Performance Steps

1. Perform an assessment of the casualty to identify any conventional injuries. (See task 081-833-4149.)

2. Recognize the course of radiation sickness.
    a. Initial stage (before rash and fever).
        (1) Symptoms - relatively rapid onset of nausea, vomiting, and malaise.
        (2) Short duration - generally a few hours.
        (3) Incapacitation should not be severe enough to warrant evacuation.
    b. Latent phase.
        (1) Relatively symptom-free.
        (2) Duration varies with the dose.
    c. Clinical phase.
NOTE: Symptoms frequently occur in the whole-body irradiated casualties within the first few hours of post exposure.
        (1) Nausea and vomiting occur with increasing frequency as the radiation exceeds 100-200 centigrays (cGy).
            (a) Onset may be as long as 6 to 12 hours post exposure.
            (b) Vomiting within the first hours is associated with fatal doses.
        (2) Hyperthermia.
            (a) Significant rise in body temperature within the first few hours of potentially lethal radiation injury.
            (b) Fever and chills are associated with severe and life-threatening radiation dose.
        (3) Erythema.
NOTE: Developed within the first day of post exposure if casualty received a whole-body dose of more than 1000-2000 cGy. Erythema is restricted to the affected area.
        (4) Hypotension.
            (a) A noticeable decline in systemic blood pressure if received a lethal dose of whole-body radiation.
            (b) Severe hypotension after irradiation is associated with a poor prognosis.
        (5) Neurologic dysfunction.
            (a) Almost all persons who demonstrate obvious signs of damage to the central nervous system within the first hours post exposure have received a lethal dose.
            (b) Symptoms include mental confusion, seizures, and coma.

3. Treat conventional injuries first.

4. Remove the casualty's clothing.

Performance Steps

5. Wash exposed body surfaces with soap and water.

6. Start an IV.

7. Administer antibiotics as appropriate.

8. Provide supportive care for the casualty depending on the situation.

NOTE: The amount of supplies required by radiation casualties can exceed the capabilities of small clinics. Consider the chance of survival of the casualty and the amount of supplies on hand when providing supportive care for the casualties.

Evaluation Preparation: This task is best evaluated by performance of the steps. Give the Soldier a simulated casualty and a scenario in which he must manage casualties exposed to ionizing radiation.

Performance Measures	GO	NO-GO
1. Performed a casualty assessment.	——	——
2. Recognized the course of radiation sickness.	——	——
3. Treated conventional injuries first.	——	——
4. Removed the casualty's clothing.	——	——
5. Washed exposed body surfaces with soap and water.	——	——
6. Started an IV.	——	——
7. Administered antibiotics as appropriate.	——	——
8. Provided supportive care for the casualty depending on the situation.	——	——

Evaluation Guidance: Score each Soldier according to the performance measures. Unless otherwise stated in the task summary, the Soldier must pass all performance measures to be scored GO. If the Soldier fails any steps, show what was done wrong and how to do it correctly.

References

Required	Related
None	FM 4-02.285 (FM 8-285)

Subject Area 9: Triage and Evacuation

GUIDE A HELICOPTER TO A LANDING POINT
071-334-4001

Conditions: You have to guide a helicopter to a landing site. You will need a prepared landing site for a UH-1 or UH-60 helicopter (the location of which is known to the pilot), individual TOE equipment, night vision goggles, FM radio (SINCGAR), and the appropriate arm-and-hand signals to guide the helicopter to the landing site and land the helicopter on the landing site. You are not in a CBRNE environment.

Standards: Guide the helicopter to a safe landing by MEDEVAC request, identifying the landing site to the pilot and controlling the landing using the correct arm-and-hand signals.

Performance Steps
CAUTION: During training, dispose of all batteries IAW unit SOP.
CAUTION: During training, dispose of all batteries IAW unit SOP.

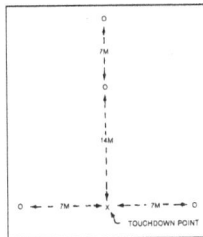

Figure 3-18. Landing site

1. As the aircraft approaches, provide the pilot with tactical and security information. Tell him of conditions that may affect his landing such as terrain, weather, landing site markings, and possible obstacles. Refer to Figure 3-18.
   a. Confirm information or answer any questions the pilot may have pertaining to the landing site.
   b. Maintain communications with the pilot during the entire operation.

2. Identify the landing site and guide the pilot in.
   a. Once the pilot is within your area, he establishes radio contact with the unit for positive identification.
   b. The pilot will be oriented to the landing site by using the clock method (12 o'clock is always the direction of flight). Tell the pilot the time position of your location. (For example: "The LZ is now at 3 o'clock to your position.")
   c. Mark or identify the landing site:
      (1) Day--The only signals required are colored smoke and a signalman. VS-17 marker panels may be used to mark the landing site, but are NOT used any closer than 50 feet to the touchdown point. In addition to identifying the landing site, the smoke will give the pilot the wind direction and speed.
      (2) Night--The landing site and touchdown point are marked by an inverted "Y" composed of four lights. Refer to Figure 3-19.

Performance Steps

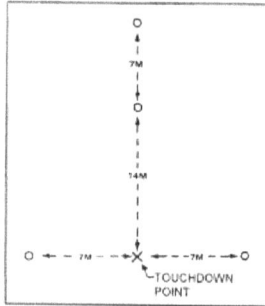

Figure 19 Night marking

3. Use arm-and-hand signals. Refer to Figures 3-20 through 3-27.

Figure 20. Arm guidance

Performance Steps

Figure 3-21. "Hover" signal

Figure 3-22. "Move ahead" signal

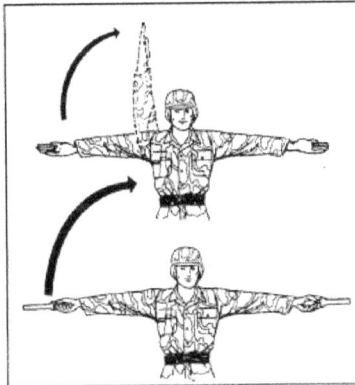

Figure 3-23. "Move to right" signal

Performance Steps

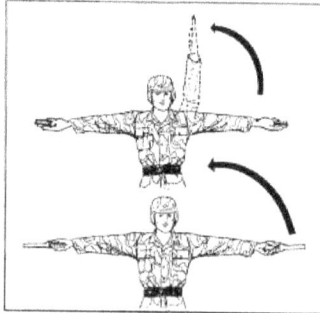

Figure 3-24. "Move to left" signal

Figure 3-25 " Move upward" signal

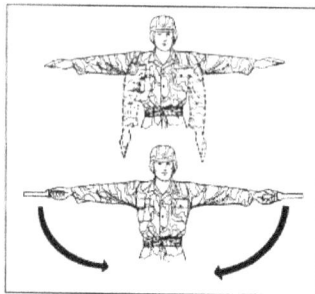

Figure 3-26. "Move downward" signal

Performance Steps

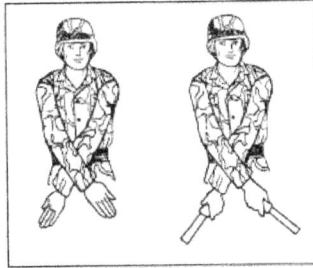

Figure 3-27. "Land" signal

3. Use arm-and-hand signals. Refer to Figures 3-28 through 3-35.
   a. The signal man's position when directing a helicopter is to the right front of the aircraft where he can be seen best by the pilot. The signal man's position for utility helicopters is 30 meters to the right front of the aircraft during day or night operations.
   a. The signalman's position when directing a helicopter is to the right front of the aircraft where he can be seen best by the pilot. The signalman's position for utility helicopters is 30 meters to the right front of the aircraft during day or night operations.
   b. Signals at night are given using lighted batons or flashlights. In the illustrations, one of the men is using a lighted wand. This is a flashlight with a plastic wand attached to the end. The flashlight is used when visibility is decreased.
   b. Signals at night are given by using lighted batons or flashlights. In the illustrations, one of the men is using a lighted wand. This is a flashlight with a plastic wand attached to the end. The flashlight is used when there is decreased visibility.
   c. The speed of the arm movement indicates the desired speed of aircraft compliance with the signal.

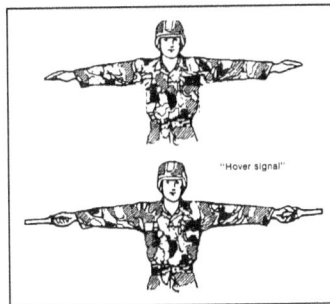

Figure 3-28. "Hover signal"

Performance Steps

Figure 3-29. Arm guidance

Figure 3-30. Move ahead" signal

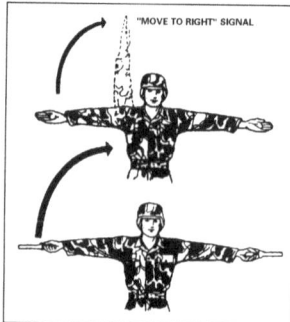

Figure 3-31. "Move to right" signal

Performance Steps

Figure 3-32. Move to left" signal

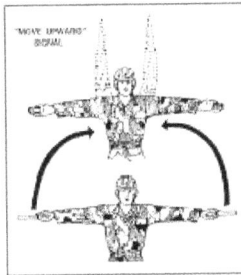

Figure 3-33. Move upward" signal

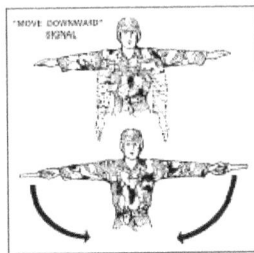

Figure 3-34. Move downward" signal

Performance Steps

Figure 3-35. "Land" signal

    c. The speed of the arm movement indicates the desired speed of aircraft compliance with the signal.

NOTE: The "hover" signal should be used to change from one arm-and-hand signal to another. For example, assume that the signalman desires to land an approaching helicopter and that the signalman has given the helicopter the "move ahead" signal. The helicopter is now positioned directly over the desired landing area. Before giving the helicopter the signal to move downward, the signalman should execute the "hover" signal. This gives the pilot time to change from the "move ahead" to the "move downward" signals.

NOTE: The "hover" signal should be used to change from one arm-and-hand signal to another. For example, assume that the signalman wants to land an approaching helicopter and has given the helicopter the "move ahead" signal. The helicopter is now positioned directly over the desired landing area. Before giving the helicopter the signal to move downward, the signalman should execute the "hover" signal. This gives the pilot time to change from the "move ahead" to the "move downward" signal.

Evaluation Preparation: SETUP: At the test site, provide all equipment and information given in the task condition statement. For test purposes, the tester may act as the pilot.

BRIEF SOLDIER: Tell the soldier to land the helicopter using arm-and-hand signals.

Performance Measures	GO	NO-GO
1. Advised the pilot of changes to the information given.	——	——
2. Identified the landing site to the pilot and guided the pilot in.	——	——
3. Controlled the landing using arm-and-hand signals.	——	——

Evaluation Guidance: Score the soldier GO if all performance measures are passed. Score the soldier NO-GO if any performance measure is failed. If the soldier scores NO-GO, show the soldier what was done wrong and how to do it correctly.

References
    Required
    None

    Related
    FM 21-60
    FM 3-21.38
    FM 3-21.8 (FM 7-8)

ESTABLISH A HELICOPTER LANDING POINT
071-334-4002

Conditions: You need to establish a helicopter landing site. You will need smoke grenades, strobe lights, flashlights or vehicle lights, marker panels, and the equipment and personnel to clear the site if required. You are not in a CBRNE environment

Standards: Identify a landing site large enough for a helicopter to land and take off. Mark and identify all obstacles that cannot be removed and designate the touchdown site.

Performance Steps

CAUTION: Comply with unit SOP and or local environmental regulations concerning the cutting of live vegetation, digging holes, and or erosion prevention.

1. Select the landing site. The factors which should be considered are:
   a. The size of the landing site.
      (1) A helicopter requires a relatively level landing area 30 meters in diameter. This does not mean that a loaded helicopter can land and take off from an area of that size. Most helicopters cannot go straight up or down when fully loaded. Therefore, a larger landing site and better approach and departure routes are required.
      (2) When obstacles are in the approach or departure routes, a 10 to 1 ratio must be used to lay out the landing site. For example, during the approach and departure, if the helicopter must fly over trees that are 15 meters high, the landing site must be at least 150 meters long (10 x 15 = 150 meters). Refer to Figure 3-36.

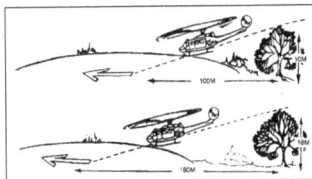

Figure 3-36. Approach or departure routes

   b. The ground slope of the landing site. When selecting the landing site, the ground slope must be no more than 15 degrees or °. Helicopters cannot safely land on a slope of more than 15 degrees.
      (1) When the ground slope is under 7 degrees, the helicopter should land up slope. Refer to Figure 3-37.
      (2) When the ground slope is 7 to 15 degrees, the helicopter must land side slope. Refer to Figure 3-38.

Performance Steps

Figure 3-37. Upslope landing

Figure 3-38. Sideslope landing

    c. Surface conditions.
        (1) The ground must be firm enough that the helicopter does not bog down during loading or unloading. If firm ground cannot be found, the pilot must be told. He can hover at the landing site during the loading or unloading.
        (2) Rotor wash on dusty, sandy, or snow-covered surfaces may cause loss of visual contact with the ground. Therefore, these areas should be avoided.
        (3) Loose debris that can be kicked up by the rotor wash must be removed from the landing site. Loose debris can cause damage to the blades or engines.
    d. Obstacles.
        (1) Landing sites should be free of tall trees, telephone lines, power lines or poles, and similar obstructions on the approach or departure ends of the landing site.
        (2) Obstructions that cannot be removed (such as large rocks, stumps, or holes) must be marked clearly within the landing site.

2. Establish security for the landing site. Landing sites should offer some security from enemy observation and direct fire. Good landing sites will allow the helicopter to land and depart without exposing it to unneeded risks. Security is normally established around the entire landing site.

3. Mark the landing site and touchdown point.
    a. When and how the landing site should be marked is based on the mission, capabilities, and situation of the unit concerned. Normally, the only mark or signals required are smoke (colored) and a signalman. VS-17 marker panels may be used to mark the landing site, but MUST NOT be used any closer than 50 feet to the touchdown point. In addition to identifying the landing site, smoke will give the pilot information on the wind direction and speed.
    b. At night, the landing site and touchdown point are marked by an inverted "Y" composed of four lights. Strobe lights, flashlights, or vehicle lights may also be used to mark the landing site. The marking system used will be fully explained to the pilot when contact is made. Refer to Figure 3-39.

Performance Steps

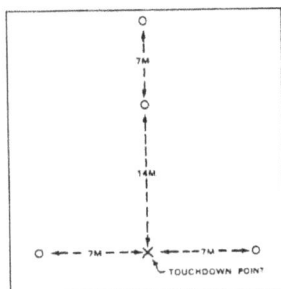

Figure 3-39. Landing site marked at night

Evaluation Preparation: SETUP: At the test site, provide all equipment, personnel, and information given in the task condition statement.

BRIEF SOLDIER: Tell the Soldier to select and prepare the helicopter landing site.

Performance Measures	GO	NO-GO
1. Selected a site large enough to permit the helicopter to land and take off.	——	——
2. Established security for the landing site.	——	——
3. Marked or identified the landing site and the touchdown point.	——	——

Evaluation Guidance: Score the Soldier GO if all performance measures are passed. Score the Soldier NO-GO if any performance measure is failed. If the Soldier scores NO-GO, show the Soldier what was done wrong and how to do it correctly.

References
 Required   **Related**
 None   FM 21-60
          FM 3-21.38
          FM 3-21.8 (FM 7-8)

REQUEST MEDICAL EVACUATION
081-831-0101

Conditions: You have a casualty requiring medical evacuation (MEDEVAC). You must request medical evacuation. You will need a patient pickup site, operational communications equipment, MEDEVAC request format, and unit signal operation instructions (SOI). You are not in a CBRNE environment.

Standards: Transmit a MEDEVAC request. As a minimum, transmit line numbers 1 through 5 during the initial contact with the evacuation unit. Transmit lines 6 through 9 while the aircraft or vehicle is en route, if not included during the initial contact.

Performance Steps

1. Collect all applicable information needed for the MEDEVAC request.
    a. Determine the grid coordinates for the pickup site. (See STP 21-1-SMCT, task 071-329-1002.)
    b. Obtain radio frequency, call sign, and suffix.
    c. Obtain the number of patients and precedence.
    d. Determine the type of special equipment required.
    e. Determine the number and type (litter or ambulatory) of patients.
    f. Determine the security of the pickup site.
    g. Determine how the pickup site will be marked.
    h. Determine patient nationality and status.
    i. Obtain pickup site chemical, biological, radiological, and nuclear (CBRN) contamination information normally obtained from the senior person or medic.
Note: CBRN line 9 information is only included when contamination exists.

2. Record the gathered MEDEVAC information using the authorized brevity codes. (See table 081-831-0101-1.)
Note: Unless the MEDEVAC information is transmitted over secure communication systems, it must be encrypted, except as noted in step 3b(1).
    a. Location of the pickup site (line 1).
    b. Radio frequency, call sign, and suffix (line 2).
    c. Numbers of patients by precedence (line 3).
    d. Special equipment required (line 4).
    e. Number of patients by type (line 5).
    f. Security of the pickup site (line 6).
    g. Method of marking the pickup site (line 7).
    h. Patient nationality and status (line 8).
    i. CBRN contamination (line 9).

3. Transmit the MEDEVAC request. (See STP 21-1-SMCT, task 113-571-1022.)
    a. Contact the unit that controls the evacuation assets.
        (1) Make proper contact with the intended receiver.
        (2) Use effective call sign and frequency assignments from the SOI.
        (3) Give the following in the clear "I HAVE A MEDEVAC REQUEST;" wait one to three seconds for a response. If no response, repeat the statement.
    b. Transmit the MEDEVAC information in the proper sequence.
        (1) State all line item numbers in clear text. The call sign and suffix (if needed) in line 2 may be transmitted in the clear.

Performance Steps

Note: Line numbers 1 through 5 must always be transmitted during the initial contact with the evacuation unit. Lines 6 through 9 may be transmitted while the aircraft or vehicle is en route.

      (2) Follow the procedure provided in the explanation column of the MEDEVAC request format to transmit other required information.

      (3) Pronounce letters and numbers according to appropriate radiotelephone procedures.

      (4) End the transmission by stating "OVER."

      (5) Keep the radio on and listen for additional instructions or contact from the evacuation unit.

Performance Steps

LINE	ITEM	EXPLANATION	WHERE/HOW OBTAINED	WHO NORMALLY PROVIDES	REASON
1	Location of pickup site	Encrypt the grid coordinates of the pickup site. When using the DRYAD Numeral Cipher, the same "SET" line will be used to encrypt the grid zone letters and the coordinates. To preclude misunderstanding, a statement is made that grid zone letters are included in the message (unless unit SOP specifies its use at all times).	From map	Unit leader(s)	Required so evacuation vehicle knows where to pick up patient. Also, so that the unit coordinating the evacuation mission can plan the route for the evacuation vehicle (if the evacuation vehicle must pick up from more than one location).
2	Radio frequency, call sign, and suffix	Encrypt the frequency of the radio at the pickup site, not a relay frequency. The call sign (and suffix if used) of person to be contacted at the pickup site may be transmitted in the clear.	From SOI	RTO	Required so that evacuation vehicle can contact requesting unit while en route (obtain additional information or change in situation or directions).
3	Number of patients by precedence	Report only applicable information and encrypt the brevity codes. A - URGENT B - URGENT-SURG C - PRIORITY D - ROUTINE E - CONVENIENCE If two or more categories must be reported in the same request, insert the word "BREAK" between each category.	From evaluation of patient(s)	Medic or senior person present	Required by unit controlling vehicles to assist in prioritizing missions.
4	Special equipment required	Encrypt the applicable brevity codes. A - None B - Hoist C - Extraction equipment D - Ventilator	From evaluation of patient/situation	Medic or senior person present	Required so that the equipment can be placed on board the evacuation vehicle prior to the start of the mission.
5	Number of patients by type	Report only applicable information and encrypt the brevity code. If requesting medical evacuation for both types, insert the word "BREAK" between the litter entry and ambulatory entry. L + # of patients - Litter A + # of patients - Ambulatory (sitting)	From evaluation of patient(s)	Medic or senior person present	Required so that the appropriate number of evacuation vehicles may be dispatched to the pickup site. They should be configured to carry the patients requiring evacuation.
6	Security of pickup site (wartime)	N - No enemy troops in area P - Possibly enemy troops in area (approach with caution) E - Enemy troops in area (approach with caution) X - Enemy troops in area (armed escort required)	From evaluation of situation	Unit leader	Required to assist the evacuation crew in assessing the situation and determining if assistance is required. More definitive guidance can be furnished the evacuation vehicle while it is en route (specific location of enemy to assist an aircraft in planning its approach).

Table 081-831-0101-1
MEDEVAC Request Format

Performance Steps

LINE	ITEM	EXPLANATION	WHERE/HOW OBTAINED	WHO NORMALLY PROVIDES	REASON
6	Number and type of wound, injury, or illness (peacetime)	Specific information regarding patient wounds by type (gunshot or shrapnel). Report serious bleeding, along with patient's blood type, if known.	From evaluation of patient(s)	Medic or senior person present	Required to assist evacuation personnel in determining treatment and special equipment needed.
7	Method of marking pickup site	Encrypt the brevity codes. A - Panels B - Pyrotechnic signal C - Smoke signal D - None E - Other	Based on situation and availability of materials	Medic or senior person present	Required to assist the evacuation crew in identifying the specific location of the pickup. Note that the color of the panels or smoke should not be transmitted until the evacuation vehicle contacts the unit (just prior to its arrival). For security, the crew should identify the color and the unit verifies it.
8	Patient nationality and status	The number of patients in each category need not be transmitted. Encrypt only the applicable brevity codes. A - US military B - US citizen C - Non-US military D - Non-US citizen E - Enemy prisoner of war (EPW)	From evaluation of patient(s)	Medic or senior person present	Required to assist in planning for destination facilities and need for guards. Unit requesting support should ensure that there is an English-speaking representative at the pickup site.
9	CBRN contamination (wartime)	Include this line only when applicable. Encrypt the applicable brevity codes. C - Chemical B - Biological R - Radiological N - Nuclear	From situation	Medic or senior person present	Required to assist in planning for the mission (determine which evacuation vehicle will accomplish the mission and when it will be accomplished).
9	Terrain description (peacetime)	Include details of terrain features in and around proposed landing site. If possible, describe relationship of site to prominent terrain feature (lake, mountain, tower).	From area survey	Personnel present	Required to allow evacuation personnel to assess route/avenue of approach into area. Of particular importance if hoist operation is required.

Table 081-831-0101-1
MEDEVAC Request Format (cont)

Evaluation Preparation:
Setup: Evaluate this task during a training exercise involving a MEDEVAC aircraft or vehicle, or simulate it by creating a scenario and providing the information as the Soldier requests it. You or an assistant will act as the radio contact at the evacuation unit during "transmission" of the request. Give a copy of the MEDEVAC request format to the Soldier.

Brief Soldier: Tell the Soldier to prepare and transmit a MEDEVAC request. State that the communication net is secure.

Performance Measures GO NO-GO

1. Collected all information needed for the MEDEVAC request line items 1 through 9. —— ——

2. Recorded the information using the authorized brevity codes. —— ——

3. Transmitted the MEDEVAC request as quickly as possible, following appropriate radiotelephone procedures. —— ——

Evaluation Guidance: Score the Soldier GO if all performance measures are passed. Score the Soldier NO GO if any performance measure is failed. If the Soldier scores NO GO, show what was done wrong and how to do it correctly.

References
    Required
    FM 8-10-6

    Related
    FM 4-02.2
    STP 21-1-SMCT

TRIAGE CASUALTIES ON A CONVENTIONAL BATTLEFIELD
081-833-0080

Conditions: You are operating on a conventional battlefield and have encountered several casualties with conventional injuries. You need to triage the casualties. You will need a DD Form 1380 (U.S. Field Medical Card). You are not in a CBRNE environment.

Standards: Complete all the steps necessary to establish priorities for medical treatment and the evacuation of casualties.

Performance Steps
CAUTION: All body fluids should be considered potentially infectious. Always observe body substance isolation (BSI) precautions by wearing gloves and eye protection as a minimal standard of protection.

1. Assess the situation.
   a. Sort the casualties and allocate treatment based on the resources available.
      (1) Assess and classify the casualties for the most efficient use of available medical personnel and supplies.
      (2) Give available treatment first to the casualties who have the best chance of survival.
      (3) A primary goal is to locate, and return to duty, troops with minor wounds; however, at no time should abandonment of a single casualty be considered.
      (4) Triage establishes the order of treatment, not whether treatment is given. It is usually the responsibility of the senior medical person present.
   b. Determine the tactical and environmental situation.
      (1) Whether casualties must be transported to a more secure area for treatment.
      (2) The number and location of the injured and the severity of the injuries.
      (3) Available assistance (self-aid, buddy-aid, combat lifesaver, and medical personnel).
      (4) Evacuation support capabilities and requirements.

2. Assess the casualties and establish priorities for treatment.
   a. Immediate--casualties whose conditions demand immediate treatment to save life, limb or eyesight. This category has the highest priority.
      (1) Airway obstruction.
      (2) Respiratory and cardiorespiratory distress from otherwise treatable injuries (for example, electrical shock, drowning or chemical exposure).
NOTE: A casualty with cardiorespiratory distress may not be classified "Immediate" on the battlefield. This casualty may be classified "Expectant", contingent upon such things as the situation, number of casualties, and available support.
      (3) Massive external bleeding.
      (4) Shock.
      (5) Burns on the face, neck, hands, feet, perineum or genitalia.
NOTE: After all life- or limb-threatening conditions have been successfully treated; give no further treatment to the casualty until all other "Immediate" casualties have been treated. Salvage of life always takes priority over salvage of limb.
   b. Delayed--casualties who have less risk of loss of life or limb if treatment is delayed.
      (1) Open wounds of the chest without respiratory distress.

Performance Steps

       (2) Open or penetrating abdominal injuries without shock.

       (3) Severe eye injuries without hope of saving eyesight.

       (4) Other open wounds.

       (5) Fractures.

       (6) Second and third degree burns (not involving the face, hands, feet, genitalia, and perineum) covering 20% or more of the total body surface area (TBSA).

    c. Minimal--"walking wounded", can be treated by self-aid or buddy-aid.

       (1) Minor lacerations and contusions.

       (2) Sprains and strains.

       (3) Minor combat stress problems.

       (4) First or second degree burns (not involving the face, hands, feet, genitalia, and perineum) covering under 20% of the TBSA.

NOTE: Minimal casualties may assist the Soldier Medic by providing buddy-aid or by monitoring other casualties.

    d. Expectant--casualties who are so critically injured that only complicated and prolonged treatment can improve life expectancy. This category is to be used only if resources are limited. If in doubt as to the severity of the injury, place the casualty in one of the other categories.

       (1) Massive head injuries with signs of impending death.

       (2) Burns, mostly third degree, covering more than 85% of the TBSA.

NOTE: Provide ongoing supportive care if the time and condition permit; keep separate from other triage categorized casualties. (See common core task 101-515-0002.)

  3. Record all treatment given on the FMC. (See task 081-831-0033.)

  4. Establish MEDEVAC priorities by precedence category.

    a. Urgent. Evacuation is required as soon as possible, but within 2 hours, to save life, limb or eyesight. Generally, casualties whose conditions cannot be controlled and have the greatest opportunity for survival are placed in this category.

       (1) Cardiorespiratory distress.

       (2) Shock not responding to IV fluid therapy.

       (3) Prolonged unconsciousness.

       (4) Head injuries with signs of increasing intracranial pressure.

       (5) Burns covering 20% to 85% of the TBSA.

    b. Urgent Surgical. Evacuation is required for casualties who must receive far forward surgical intervention to save life and stabilize for further evacuation.

       (1) Decreased circulation in the extremities.

       (2) Open chest and/or abdominal wounds with decreased blood pressure.

       (3) Penetrating wounds.

       (4) Uncontrollable hemorrhage or open fractures with severe hemorrhage.

       (5) Severe facial injuries.

    c. Priority. Evacuation is required within 4 hours or the casualty's condition could get worse and become an "Urgent" or "Urgent Surgical" category condition. Generally, this category applies to any casualty whose condition is not stabilized or who is at risk of trauma-related complications.

       (1) Closed-chest injuries, such as rib fractures without a flail segment or other injuries that interfere with respiration.

       (2) Brief periods of unconsciousness.

       (3) Soft tissue injuries and open or closed fractures.

       (4) Abdominal injuries without hypotension.

Performance Steps

      (5) Eye injuries that do not threaten eyesight.

      (6) Spinal injuries.

      (7) Burns on the hands, face, feet, genitalia, or perineum, even if under 20% of the TBSA.

  d. Routine. Evacuation is required within 24 hours for further care. Immediate evacuation is not critical. Generally, casualties who can be controlled without jeopardizing their condition or who can be managed by the evacuating facility for up to 24 hours.

      (1) Burns covering 20% to 80% of the TBSA if the casualty is receiving and responding to IV fluid therapy.

      (2) Simple fractures.

      (3) Open wounds including chest injuries without respiratory distress.

      (4) Behavioral emergencies and combat stress casualties.

      (5) Terminal cases.

  e. Convenience. Evacuation by medical vehicle is a matter of convenience rather than necessity.

      (1) Minor open wounds.

      (2) Sprains and strains.

      (3) Minor burns under 20% of TBSA.

  5. Prepare a medical evacuation request. (See STP 21-24-SMCT, task 081-831-0101.)

Evaluation Preparation:

Setup: For training and evaluation construct a scenario that involves multiple combat casualties. Moulaged Soldiers may be used as casualties. Brief each Soldier as to their wounds and what their actions should be when assessed. Tell the casualties not to assist the Soldier in any way; unless they are designated as "minimal" casualties and instructed to provide buddy-aid.

Brief the Soldier: Tell the Soldier to assess the situation and the casualties, establish priority of treatment, establish MEDEVAC priorities, and prepare the evacuation request.

Performance Measures	GO	NO-GO
1. Assessed the situation.	___	___
2. Assessed the casualties and established priorities for treatment.	___	___
3. Recorded all treatment given on the FMC.	___	___
4. Established MEDEVAC priorities by precedence category.	___	___
5. Prepared a medical evacuation request.	___	___

Evaluation Guidance: Score each Soldier according to the performance measures. Unless otherwise stated in the task summary, the Soldier must pass all performance measures to be scored GO. If the Soldier fails any steps, show what was done wrong and how to do it correctly.

References
    Required
       DD FORM 1380

    Related
       FM 4-02 (FM 8-10)
       STP 21-24-SMCT

TRIAGE CASUALTIES ON AN INTEGRATED BATTLEFIELD
081-833-0082

Conditions: You are in a chemical environment and have casualties with conventional injuries and/or signs and symptoms of chemical agent poisoning. Both you and the casualties are in MOPP level 4. You will need an aid bag and a DD Form 1380 (U.S. Field Medical Card).

Standards: Complete all the steps necessary to correctly establish priorities for the treatment and evacuation of casualties on an integrated battlefield.

Performance Steps

1. Assess the situation.
    a. Number and location of the injured.
    b. Severity of the injuries.
    c. Assistance available (self-aid or buddy-aid).
    d. Evacuation support capabilities.
    e. Type of chemical agents used, if known.

2. Assess the individual casualties.
    a. Assess for conventional injuries.
    b. Assess for signs and symptoms of chemical agent poisoning.
        (1) Determine if the casualty responds to commands.
            (a) Check the casualty's response to simple directions, such as "Hold up your right arm."
            (b) Ask the casualty to describe any symptoms.
        (2) Check for symptoms of chemical agent poisoning. (See tasks 081-833-0083 through 081-833-0086.)

3. Establish priorities for treatment.
    a. Immediate.
        (1) No signs and symptoms of chemical agent poisoning.
        (2) Presence of life-threatening conventional injuries.
    b. Chemical immediate.
        (1) Presence of signs and symptoms of severe chemical agent poisoning.
        (2) No conventional injuries.
    c. Delayed.
        (1) Presence of mild signs and symptoms of chemical agent poisoning.
        (2) Presence of conventional injuries that are not life-threatening.
    d. Minimal.
        (1) No signs and symptoms of chemical agent poisoning.
        (2) Presence of minor conventional injuries.
    e. Expectant.
        (1) Presence of severe signs and symptoms of both chemical agent poisoning and life-threatening conventional injuries.
        (2) No conventional injuries and not breathing due to chemical agent poisoning.
NOTE: Expectant casualties are so critically injured that only prolonged and complicated treatment may offer increased life expectancy.

4. Initiate treatment in the following order.
    a. Chemical agent poisoning.
    b. Conventional injuries.

Performance Steps
NOTES: 1. Employ casualties who have only minor injuries or minimal chemical agent exposure to provide buddy-aid for those with more severe injuries. 2. Sorting and treatment should be done almost simultaneously.

    5. Move the casualties to the collection point.

    6. Record all observations and treatment on a FMC.

    7. Establish evacuation priorities. (See task 081-833-0080.)

Evaluation Preparation:
Setup: You will need several Soldiers in MOPP level 4 to act as the casualties. Use a moulage kit or similar materials to simulate conventional wounds. Coach the Soldiers on signs and symptoms of nerve agent poisoning to exhibit.

Brief Soldier: Tell the Soldier to triage casualties on an integrated battlefield.

Performance Measures	GO	NO-GO
1. Assessed the situation.	____	____
2. Assessed the individual casualties.	____	____
3. Established priorities for treatment.	____	____
4. Initiated treatment.	____	____
5. Moved the casualties to the collection point.	____	____
6. Recorded all observations and treatment on a FMC.	____	____
7. Established evacuation priorities.	____	____

Evaluation Guidance: Score each Soldier according to the performance measures. Unless otherwise stated in the task summary, the Soldier must pass all performance measures to be scored GO. If the Soldier fails any steps, show what was done wrong and how to do it correctly.

References

Required	Related
DD FORM 1380	FM 4-02 (FM 8-10)

LOAD CASUALTIES ONTO GROUND EVACUATION PLATFORMS
081-833-0151

Conditions: You have completed treating and triaging multiple patients. You are in charge of loading litter and ambulatory casualties on a standard ground transport platform. You will need litters, a litter suspension kit, and an M996, M997, or M113. You are not in a CBNRE environment.

Standards: Configure the vehicle properly and load and unload casualties in the correct sequence for the transport platform.

Performance Steps

1. Determine ambulance load capacities.
    a. Truck, ambulance, 4X4, 2 litter, utility (M996).
        (1) 2 litter casualties.
        (2) 6 ambulatory casualties.
        (3) 1 litter and 3 ambulatory casualties.
    b. Truck, ambulance, 4X4, 4 litter, utility (M997).
        (1) 4 litter casualties.
        (2) 8 ambulatory casualties.
        (3) 2 litter and 4 ambulatory casualties.
    c. Carrier, personnel, full tracked, armored (M113, T113E2).
        (1) 4 litter casualties.
        (2) 10 ambulatory casualties.
        (3) 2 litter and 5 ambulatory casualties.
NOTE: Casualties are normally loaded head first. They are less likely to experience motion sickness or nausea with the head in the direction of travel. When en route care is required for an injury on one side, it may be necessary to load feet first to access from the aisle. Casualties with wounds of the chest or abdomen or those receiving IV fluids are loaded in lower berths to provide gravity flow. Casualties wearing bulky splints should be placed on lower berths.

2. Determine the loading sequence.
    a. M996.
        (1) Load casualties in the right berth first and then the left.
        (2) Load the most seriously injured casualty last.
        (3) Go to step 3.
    b. M997.
        (1) Load casualties in the upper right, lower right, upper left, and then the lower left berths.
        (2) Load the most seriously injured casualties last.
CAUTION: To install the litter suspension kit in the M113 ambulance, the spall liner must be removed. Litter casualties cannot be safely moved if the litter suspension kit is not installed.
        (3) Go to step 4.
NOTE: Unload the M113 ambulance in reverse sequence.
    c. M113, T113E2.
        (1) Load casualties in the upper right, lower right, upper left, and then the lower left berths.
        (2) Load the most seriously injured casualties last.

3. Load and unload the M996 for ambulatory or litter casualties.

Performance Steps

   a. Prepare the M996 for ambulatory casualties. Refer to Figures 3-40 and 3-41.

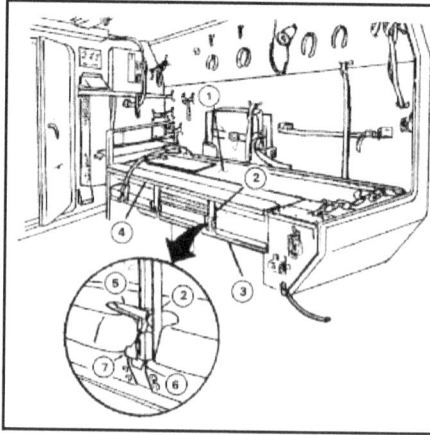

Figure 3-40. Prepare the M996

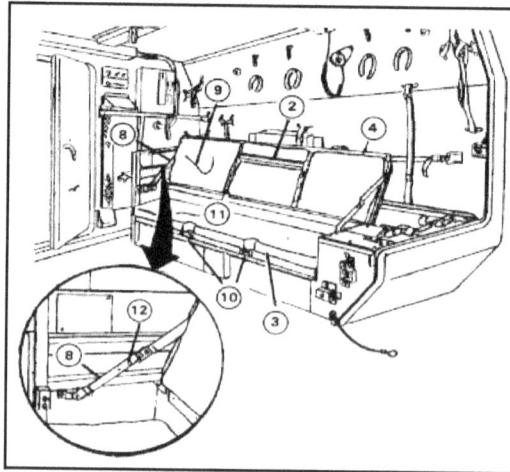

Figure 3-41. Prepare the M996

   (1) Ensure litters are in stowed position.
   (2) Pull out and up on seat latch handle (5) and remove latch (7) from catch (6).
   (3) Lift seat back (4) to open position and fold seat back support (2) into recesses
       between seat cushions (9).
   (4) Ensure that seat braces (8) are fully extended and locked in position.
  b. Prepare the M996 for litter casualties (Figures 3-x1 and 3-x2).

Performance Steps

    (1) Press lock buttons (12) on seat braces (8) and fold braces (8) toward seat back (4).

    (2) Fold seat back support (2) outward and fold seat back (4) into closed position. Ensure that guide pins (11) on seat back support engage holes (10) in seat base (3).

    (3) Install seat back (4) to seat base (3) with seat latch (7) and secure with latch handle (5). If necessary to ensure security of seat back (4), adjust seat latch (7) to proper length by turning clockwise or counterclockwise.

  c. Prepare the M996 to utilize the litter rail extension for loading and unloading of casualties.

    (1) Assemble litter rail extension for M996. Refer to Figures 3-42 and 3-43.

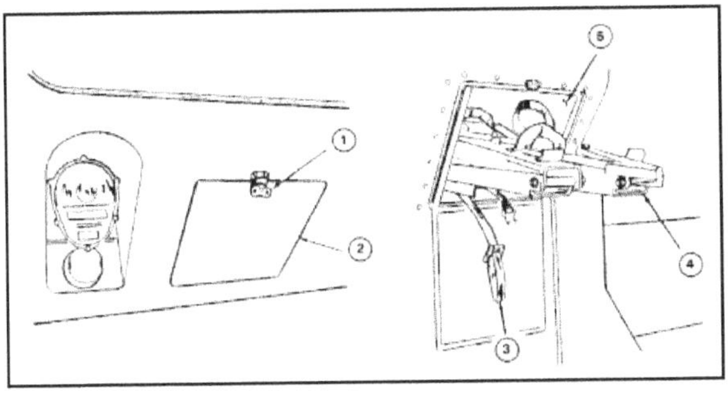

Figure 3-42. Assemble litter rail

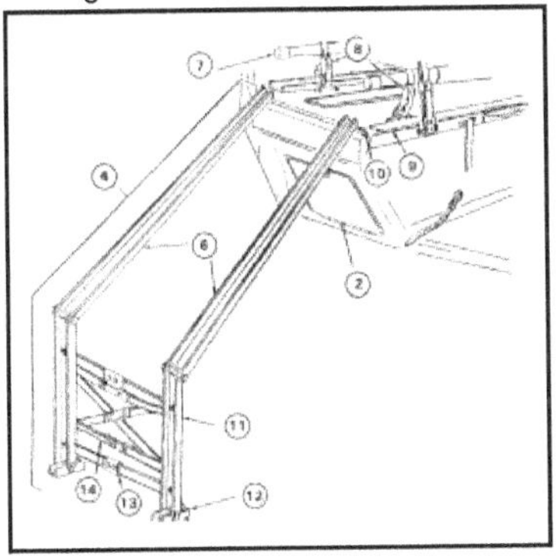

Figure 3-43. Assemble litter rail

    (a) Turn latch (1) counterclockwise and open stowage compartment door (2).

    (b) Loosen and disconnect securing strap (3) and remove folded litter rail extension (4) from stowage compartment (5).

    (c) Pull left and right rails (6) apart and let legs (11) drop down. Ensure feet (12) are flat on ground.

Performance Steps

        (d) Lock support braces (13) and adjust straps (14) as necessary.

    (2) Load litters on litter rack. Refer to Figure 3-44.

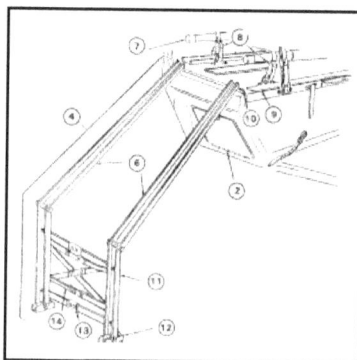

Figure 3-44. Assemble litter rail

        (a) Secure both rails (6) of litter rail extension (4) into slots (10) on litter rack (9).

        (b) Place litter (7) on litter rail extension (4).

WARNING: Ensure straps and equipment do not inhibit litter loading operations. Load litters carefully to prevent casualty injury.

        (c) Slide litter (7) onto litter rack (9).

        (d) Secure litter (7) to litter rack (9) with front and rear litter handle straps (8).

    (3) Unload litters from the litter rack (Figure 3-x4).

        (a) Release front and rear litter handle straps (8) securing litter (7) to litter rack (9).

        (b) Secure both rails (6) of litter rail extension (4) into slots (10) on lower litter rack (9).

        (c) Slide litter (7) from lower litter rack (9) onto litter rail extension (4). Lift up and remove litter (7) from litter rail extension (4).

    (4) Fold and stow litter rail extension for M996 (Figures 3-x3 and 3-x4).

        (a) Unlock support braces (13).

        (b) Fold left and right rails (6) together.

        (c) Fold left and right litter rail legs (11) and feet (12) against rails (6).

        (d) Place folded litter rail extension (4) into stowage compartment (5) and secure with strap (3).

        (e) Close door (2) and turn latch (1) clockwise to secure door (2).

  4. Load and unload the M997 for litter and ambulatory casualties.

    a. Prepare the upper litter rack. Refer to Figure 3-45.

Performance Steps

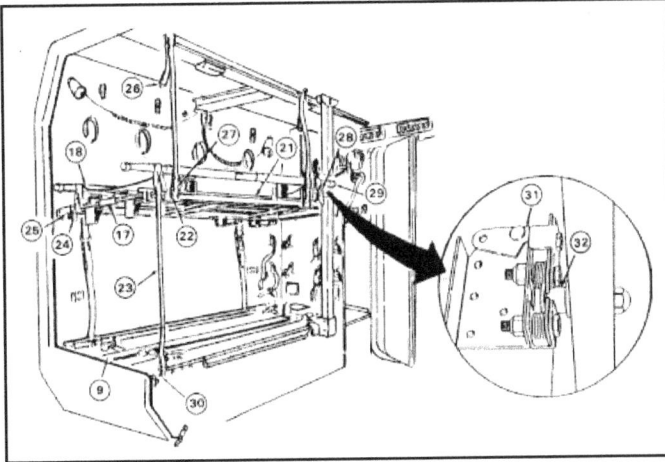

Figure 3-45. Prepare the upper litter rack

(1) Unhook tension strap (23) from footman loop (30) on lower litter rack (9).

(2) Pull out upper litter rack handle (17) and support weight of upper litter rack (21).

WARNING: The rear end of the upper litter must be supported before releasing the suspension strap hook. Injury to personnel may result if rear end of upper litter is not supported.

(3) Unhook rear suspension strap hook (27) from loop (22) on upper litter rack (21). Clip suspension strap hook (27) to eye (26).

(4) Release litter support latch stop (25), push latch (24) in, and lower upper litter rack (21) onto lower litter rack (9).

(5) Slide litter rack handle (17) into upper litter rack (21).

b.   Assemble the litter rail extension. Refer to Figures 3-46 and 3-47.

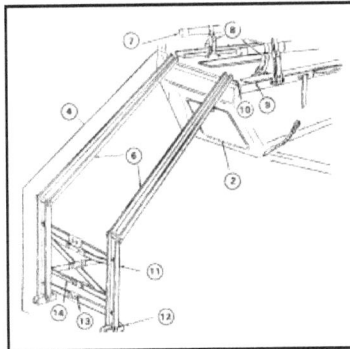

Figure 3-46. Assemble the litter rail extension

Performance Steps

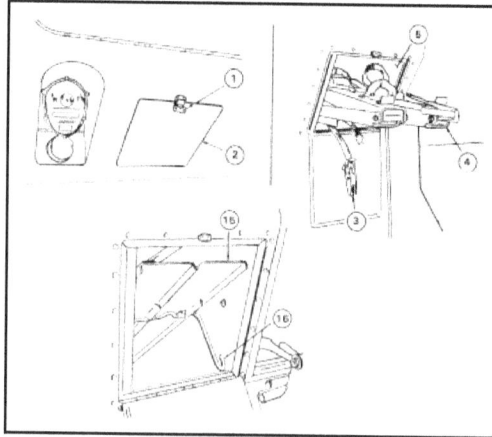

Figure 3-47. Assemble the litter rail extension

(1) Turn latch (1) counterclockwise and open stowage compartment door (2).

(2) Loosen and disconnect securing strap (3) and remove folded litter rail extension (4) from stowage compartment (5).

(3) Lift tray (15) slightly and push in tray supports (16) to lower tray (15) for access to stowed litters.

(4) Pull left and right rails (6) apart and let legs (11) drop down. Ensure feet (12) are flat on ground.

(5) Lock support braces (13) and adjust straps (14) as necessary.

b. Load litters on upper litter racks. Refer to Figures 3-48 and 3-49.

Performance Steps

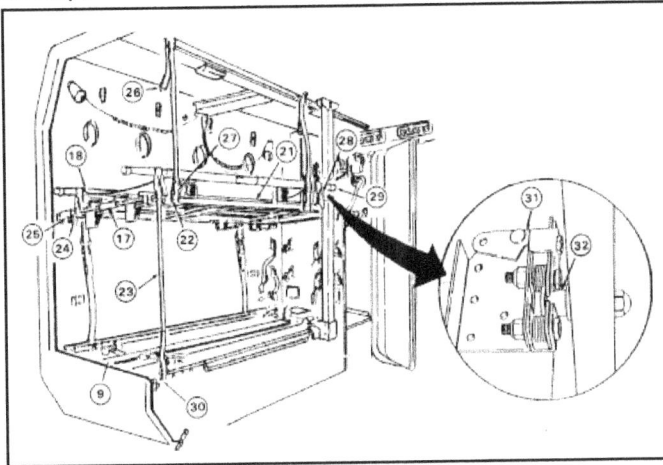

Figure 3-48. Load litters on upper litter racks

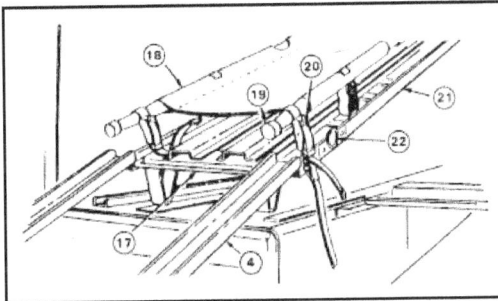

Figure 3-49. Load litters on upper litter racks

(1) Secure both rails of litter extension (4) into slots in upper litter rack (21).

(2) Place litter (18) on litter rail extension (4).

(3) Slide litter (18) up rails (4) until litter (18) is clear of litter rail extension (4).

(4) Secure rear litter handles (19) to upper litter rack (21) with rear litter handle straps (20).

(5) Remove litter rail extension (4) from upper litter rack (21).

(6) Unhook suspension strap hook (27) from eye (26).

(7) Pull out upper litter rack handle (17).

(8) Raise upper litter rack (21), push into litter support latch (24), and secure with latch stop (25).

(9) Secure front litter handles (29) to litter rack (21) with front litter handle straps (28).

(10) Hook tension strap (23) to footman loop (30) on lower litter rack (9) and adjust strap.

Performance Steps

    (11) Slide litter rack handle (17) into upper litter rack (21).

  c. Load litters on lower litter rack. Refer to Figure 3-50.

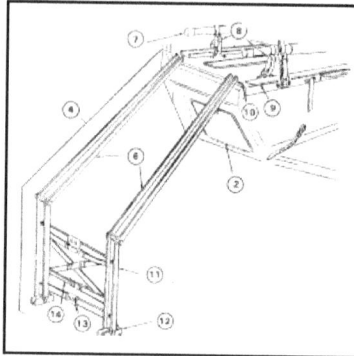

Figure 3-50. Load litters on lower litter rack

    (1) Secure both rails (6) of litter rail extension (4) into slots (10) on lower litter rack (9).

    (2) Place litter (7) on litter rail extension (4).

    (3) Slide litter (7) onto lower litter rack (9).

    (4) Secure litter (7) to lower litter rack (9) with front and rear litter handle straps (8).

  c. Unload litters from the lower litter rack. Refer to Figure 3-51.

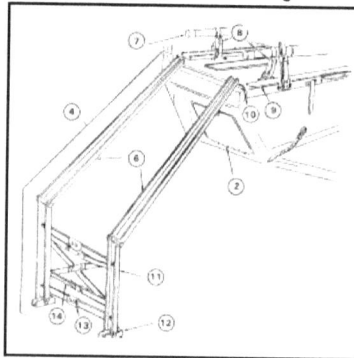

Figure 3-51. Unload litters from the lower litter rack

WARNING: When unloading more than two litter casualties, lower litter rack casualties must be unloaded first. Ensure that straps and equipment do not inhibit unloading operations. Unload litters carefully to prevent casualty injury.

    (1) Release front and rear litter handle straps (8) securing litter (7) to lower litter rack (9).

Performance Steps

(2) Secure both rails (6) of litter rail extension (4) into slots (10) on lower litter rack (9).

(3) Slide litter (7) from lower litter rack (9) onto litter rail extension (4).

(4) Lift up and remove litter (7) from litter rail extension (4).

d. Unload litters from upper litter racks. Refer to Figures 3-52 and 3-53.

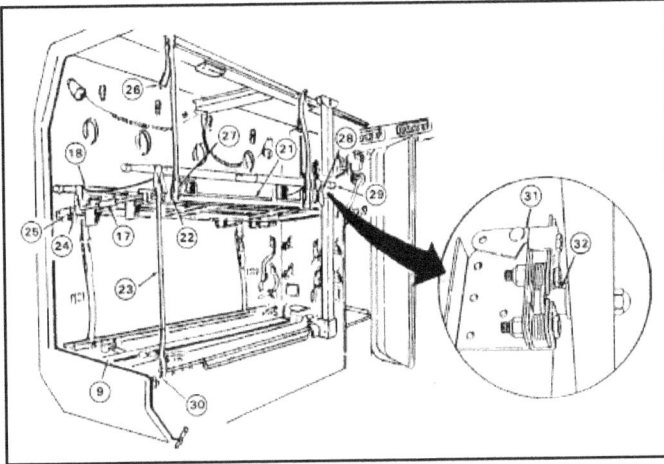

Figure 3-52. Unload litters from upper litter racks.

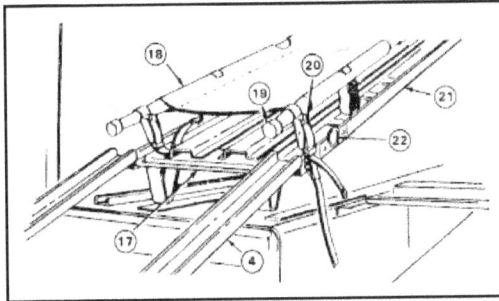

Figure 3-53. Unload litters from upper litter racks.

(1) Release litter support latch stop (28) from litter handles (29).

(2) Unhook tension strap (23) from footman loop (30) on lower litter rack (9).

(3) Pull out upper litter rack handle (17) and support weight of upper litter rack (21).

(4) Unhook rear suspension strap hook (27) from loop (22) on upper litter rack (21). Clip suspension strap hook (27) to eye (26).

(5) Release litter support latch stop (25), push latch (24) in, and lower upper litter rack (21) onto lower litter rack (9).

Performance Steps

    (6) Slide litter rack handle (17) into upper litter rack (21).

    (7) Secure rails of litter rail extension (4) into slots in upper litter rack (21).

    (8) Release rear litter handle straps (20) from litter handles (19).

    (9) Slide litter (18) down litter rail extension (4) until litter (18) is clear of upper litter rack (21).

    (10) Lift and remove litter (18) from litter rail extension (4).

    (11) Remove litter rail extension (4) from upper litter rack (21).

  e. Fold and stow litter rail extension. Refer to Figures 3-54 and 3-55.

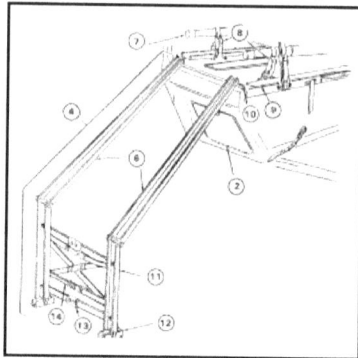

Figure 3-54. Fold and stow litter rail extension

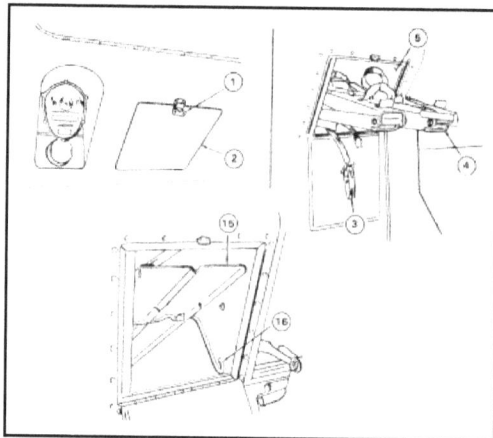

Figure 3-55. Fold and stow litter rail extension

Performance Steps

    (1) Unlock support braces (13).
    (2) Fold left and right rails (6) together.
    (3) Fold left and right litter rail legs (11) and feet (12) against rail (6).
    (4) Lift tray (15) and push tray supports (16) in, and lower tray (15).
    (5) Slide litters into stowage compartments (5) on top of lift tray (15). Pull out
        supports (16) to place lift tray (15) in raised position.
    (6) Place folded litter rail extension (4) into stowage compartment (5) and secure with
        strap (3).
    (7) Close door (2) and turn latch (1) clockwise to secure door (2).
  f. Fold upper litter rack to the backrest position. Refer to Figure 3-56.

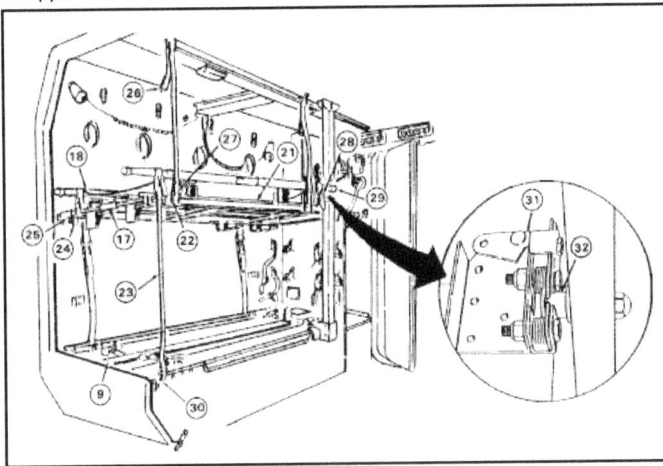

Figure 3-56. Fold upper litter rack to the backrest position

    (1) Unhook litter rack tension strap (23) from lower litter rack footman loop (30).
    (2) Unhook two upper litter rack suspension straps hooks (27) from loops (22) on
        upper litter rack (21) and reattach strap hooks (27) to eyes (26).
    (3) Release upper litter rack latch (31) and disengage rack striker (32) from latch
        (31).
    (4) Lower upper litter rack (21) onto the lower litter rack (9), forming a backrest.
  i. Cover backrest to upper litter rack. Refer to Figure 3-53.
    (1) Raise upper litter rack (21) and engage rack striker (32) into upper litter rack latch
        (31). Ensure striker (32) is locked in latch (21).
    (2) Unhook two upper litter rack suspension strap hooks (27) from eyes (26) and
        hook to loops (22) on upper litter rack (21).
    (3) Hook upper litter rack tension strap (23) to footman loop (30) on lower litter rack
        (9).
    (4) Adjust straps (23 and 27) for proper tension.

Performance Measures	GO	NO-GO
1. Determined ambulance load capacities.	——	——
2. Assessed the casualties to determine loading sequence.	——	——
3. Prepared the ambulance to receive the casualties.	——	——
4. Directed nonmedical Soldiers to--	——	——

    a. Load the casualties head first, in the proper sequence, and use the proper berths.
    b. Secure the casualties for transport.
    c. Unload the casualties in the proper sequence.

5. Did not cause further injury to the casualties.	——	——

Evaluation Guidance: Score each Soldier according to the performance measures. Unless otherwise stated in the task summary, the Soldier must pass all performance measures to be scored GO. If the Soldier fails any steps, show what was done wrong and how to do it correctly.

References
    Required   **Related**
    None   FM 8-10-6

ESTABLISH A CASUALTY COLLECTION POINT
081-833-0152

Conditions: You have a tactical battle plan for medical evacuation. You are not in a CBRNE environment.

Standards: Establish a casualty collection point (CCP).

Performance Steps

1. Select the site.
NOTE: Tactical battle plans will vary depending on the type and number of units incorporated into the plan. The plan for casualty collection points (CCP) will vary also. The medical treatment provided on the battlefield is usually accomplished by the unit's combat medics and combat lifesavers. Establishment of a CCP is essential for the rapid treatment and evacuation of the casualty.
  a. The location of the CCP will depend on the unit location, the tactical situation, and the number of casualties to be evacuated.
    (1) This location should be decided by the unit first sergeant with guidance from the company/platoon medics.
    (2) Battle drills and tactical standard operating procedures (TSOPs) should be established on how they will get the casualty from the fighting position or vehicle to the CCP.
    (3) The platoon CCP should be located at the platoon's rear.
    (4) The company CCP should be located in a covered and concealed position with the company trains.
  b. All CCPs are identified with both day and night marking systems and contain a triage area.

2. Plan movement of casualties.
  a. Casualty movement from point of wounding to the platoon CCP will be by field expedient means: individual manual carries, litter (SKED, talon, poleless) carries, or casualty evacuation vehicle.
  b. Nonstandard casualty evacuation vehicles are positioned forward with a M113A3, Stryker, M996, or M997 track or wheeled ambulance designated for litter casualties at the company CCP.
  c. The first sergeant coordinates casualty flow between the platoon CCPs and the company CCP while the senior medic conducts triage.
  d. Communication of 9 line medical evacuation requests is conducted via the company train assets or ambulances back to the battalion aid station (BAS.)
  e. The procedure/drill from the CCP to the BAS focuses largely on distance, routes, security, and operational procedures at the BAS.

Performance Steps

NOTE: "Operational procedures" refers to the set up and functionality of the BAS with regards to casualty flow, triage, treatment, and various other functions required for successful operation. In some cases the BAS will split into two treatment teams: the main aid station (MAS) and a forward aid station (FAS). Often the MAS and the FAS will conduct echelonment (bounding) during an offensive operation to maintain doctrinal distance during the fight. Whether evacuation is from the CCP to the MAS, FAS, or BAS the system remains essentially the same.

NOTE: Doctrinal distance is considered to be 1 to 4 kilometers and/or one to two terrain features behind the unit supported, emphasizing mission, enemy, terrain, troops, time available, and civilian considerations (METT-TC). Failure to maintain this is a common error. Usually the distance becomes extended due to poor planning, failure to commit medical assets forward, lack of clearly defined triggers, or communications failures.

Evaluation Preparation:

Setup: At the test site, provide all equipment, information, and personnel given in the task conditions statement.

Brief Soldier: Tell the Soldier that he must establish a casualty collection point (CCP) and may have to request medical evacuation and establish a helicopter landing point after establishing the CCP.

Performance Measures	GO	NO-GO
1. Selected a site for a CCP based on the tactical mission.	____	____
2. Planned movement of casualties.	____	____

Evaluation Guidance: Score each Soldier according to the performance measures. Unless otherwise stated in the task summary, the Soldier must pass all performance measures to be scored GO. If the Soldier fails any steps, show what was done wrong and how to do it correctly.

References

Required	Related
None	FM 4-02.4 (FM 8-10-4)
	FM 4-02.6 (FM 8-10-1)

LOAD CASUALTIES ONTO NONSTANDARD VEHICLES, 1 1/4 TON, 4X4, M998
081-833-0171

Conditions: You have completed treating and triaging multiple casualties. You are in charge of loading litter or ambulatory patients onto a non-standard transport vehicle (M998). You will need litters and a non-standard vehicle. You are not in a CBRNE environment.

Standards: Load litter casualties IAW the proper loading sequence without causing further injury.

Performance Steps

1. Determine vehicle load capacities.
    a. M998 (four-man configuration)--three litters.
    b. M998 (two-man configuration)--five litters.

2. Direct nonmedical Soldiers to load an M998 (four-man configuration). Refer to Figure 3-57.

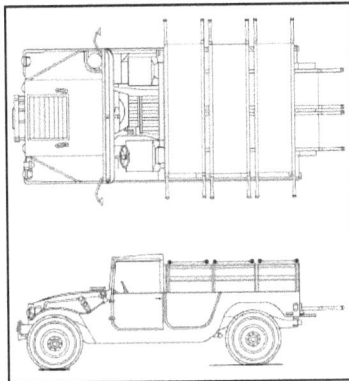

Figure 3-57. Load an M998

    a. Remove the cargo cover and metal bows. Secure them in the vehicle and lower the tailgate.
    b. Place two litters side-by-side across the back of the truck with the litter handles resting on the sides of the truck.
    c. Secure the litters to the vehicle with any available material.
    d. Place one litter lengthwise, head first, in the bed of the truck. Secure it in place.
    e. Leave the tailgate open. It is supported by the two tailgate chain hooks.

Performance Steps

3. Direct nonmedical Soldiers to load an M998 (two-man configuration). Refer to Figure 3-58.

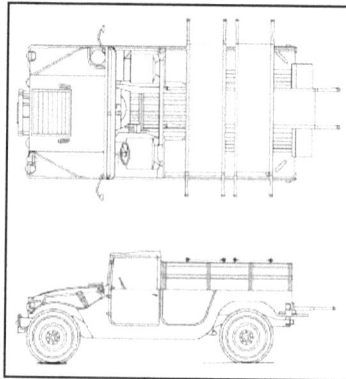

Figure 3-58. Load an M998

   a. Fold the fabric cover and metal bows forward and together as an assembly. Secure them in place. Lower the tailgate.
   b. Place three litters side-by-side across the sideboards. Secure them in place with any material available.
   c. Place two litters lengthwise, head first, in the bed of the truck. Secure them in place.
   d. Leave the tailgate open. It is supported by the two tailgate chain hooks.

4. Direct nonmedical Soldiers to unload the vehicle in reverse sequence.

Performance Measures	GO	NO-GO
1. Determined vehicle load capacity.	____	____
2. Prepared the vehicle to receive casualties.	____	____
3. Directed nonmedical Soldiers to--	____	____
a. Load the casualties in the proper sequence.		
b. Load casualties side-by-side across the back of the truck.		
c. Load casualties lengthwise on the floor, head first.		
d. Secure the casualties for transport.		
4. Directed nonmedical Soldiers to unload the vehicle in reverse sequence.	____	____
5. Did not cause further injury to the casualties.	____	____

Evaluation Guidance: Score each Soldier according to the performance measures. Unless otherwise stated in the task summary, the Soldier must pass all performance measures to be scored GO. If the Soldier fails any steps, show what was done wrong and how to do it correctly.

References
    Required   **Related**
    None   FM 8-10-6

LOAD CASUALTIES ONTO NONSTANDARD VEHICLES, 2 1/2 TON, 6X6 OR 5 TON, 6X6, CARGO TRUCK
081-833-0172

Conditions: You have completed treating and triaging multiple casualties. You are in charge of loading litter casualties onto a non-standard transport vehicle. You will need litters, and a 2 1/2 ton or 5 ton cargo truck. You are not in a CBRNE environment.

Standards: Load litter casualties IAW with the proper loading sequence without causing further injury.

Performance Steps

1. Determine vehicle load capacities.
    a. 12 litters.
    b. 16 ambulatory.

2. Direct nonmedical Soldiers to load the vehicle.
    a. Remove the canvas cover. (The cover can be rolled toward the front of the truck and secured.)
    b. Lower the seats.
    c. Place three litters crosswise on the seats as far forward as possible and three litters lengthwise in the bed of the truck as far forward as possible.
    d. Secure the litters individually to the seats.
    e. Place three additional litters crosswise on the seats and three additional litters lengthwise in the bed of the truck.
    f. Secure the litters individually to the seats.
    g. Raise and secure the tailgate as high as possible to help secure the litters in place.

3. Direct nonmedical Soldiers to unload the vehicle in reverse sequence.

Performance Measures	GO	NO-GO
1. Determined vehicle load capacity.	____	____
2. Prepared the vehicle to receive casualties.	____	____
3. Directed nonmedical Soldiers to--	____	____

    a. Load the casualties in the proper sequence crosswise on the seats as far forward as possible and three litters lengthwise in the bed of the truck as far forward as possible.

NOTE: Patients may be loaded either head to head or head to toe.

    b. Secure the casualties for transport.
    c. Load three additional litters crosswise on the seats and three additional litters lengthwise in the bed of the truck.
    d. Secure the casualties for transport.
    e. Unload casualties in the proper sequence.

4. Directed nonmedical Soldiers to unload the vehicle in reverse sequence.	____	____

Evaluation Guidance: Score each Soldier according to the performance measures. Unless otherwise stated in the task summary, the Soldier must pass all performance measures to be scored GO. If the Soldier fails any steps, show what was done wrong and how to do it correctly.

References

Required	Related
None	FM 8-10-6

LOAD CASUALTIES ONTO NONSTANDARD VEHICLES, 5 TON M-1085, M-1093, 2 1/2 TON M-1081

081-833-0173

Conditions: You have completed treating and triaging multiple casualties. You are in charge of loading litter casualties onto a nonstandard transport vehicle. You will need litters, and an M-1085, M-1093, or M-1081 vehicle. You are not in a CBRNE environment.

Standards: Load litter casualties IAW with the proper loading sequence without causing further injury.

Performance Steps

    1. Determine vehicle load capacities.
        a. Long Wheelbase, 5-Ton, M-1085.
            (1) 12 litter.
            (2) 22 ambulatory.
        b. Light Vehicle Air Drop/Air Delivery, 5 Ton, M-1093.
            (1) 8 litters.
            (2) 14 ambulatory.
        c. Light Vehicle Air Drop/Air Delivery, 2 1/2 Ton, M-1081.
            (1) 7 litters.
            (2) 12 ambulatory.

    2. Direct nonmedical Soldiers to load an M-1085. Refer to 3-59.

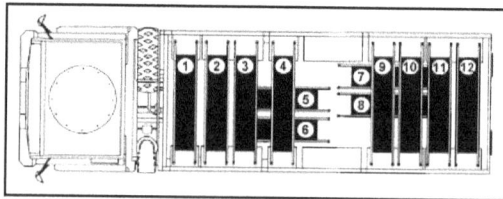

Figure 3-59. Load an M-1085

        a. Lower the seats and secure the vertical support brackets in place.
        b. Place four litters (litter numbers 1 through 4) crosswise on the seats, forward, next to the cab. Secure the litters individually to the seats.
        c. Place two litters (litter numbers 5 and 6) lengthwise on the floor, forward toward the cab, feet first, ensuring that casualty's' heads are exposed from under the upper litters. Secure the litters together and to the vertical seat supports.
        d. Place litter number 7 crosswise on the seats near the rear of the vehicle. Slide the litter as far forward as possible. Do not secure the litter at this time.
        e. Follow the same procedures in step 2d above for litter numbers 8 and 9.
        f. Place litter number 10 crosswise on the furthest seat rearward. Secure the litter to the seat.
        g. Slide litters (litter numbers 7, 8, and 9) rearward next to litter number 10. Secure the litters to the seats individually.

Performance Steps

    h. Place two litters lengthwise on the floor, head first, ensuring that the casualty's head is exposed to the center opening, between the upper litters. Secure the litters together and to the vertical seat supports.

NOTE: The combat medic or combat lifesaver rides in the center of the vehicle to monitor the casualties. If the vehicle is loaded with the maximum number of casualties, the combat medic will not be able to attend to the casualties.

  3. Direct nonmedical Soldiers to load an M-1093. Refer to Figure 3-60.

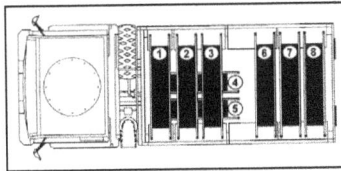

Figure 3-60. Load an M-1085

    a. Lower the seats and secure the vertical support bracket into place.

    b. Place three litters (litter numbers 1 through 3) crosswise on the seats, forward, next to the cab. Secure the litters individually to the seats.

    c. Place two litters (litter numbers 4 and 5) lengthwise on the floor, forward toward the cab, feet first. Secure the litters together and to the vertical seat support.

    d. Place litter number 6 crosswise on the seats near the rear of the vehicle. Slide the litter as far forward as possible. Do not secure the litter at this time.

    e. Place litter number 7 crosswise on the seats near the rear of the vehicle and slide it forward as in step 3d above. Secure the litter to the seats.

    f. Place litter number 8 crosswise on the seats as far rearward as possible. Secure the litter to the seats.

    g. Glide litter numbers 6 and 7 rearward next to litter number 8. Secure the litters to the seats.

    h. Raise and secure the tailgate.

NOTE: The combat medic or combat lifesaver rides in the center of the vehicle to monitor the casualties.

  4. Direct nonmedical Soldiers to load an M-1081. Refer to Figure 3-61.

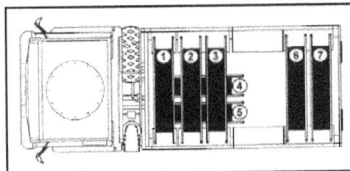

Figure 3-61. Load an M-1081

Performance Steps
    a. Lower the seats and secure the vertical support bracket into place.
    b. Place three litters (litter numbers 1 through 3) crosswise on the seats, forward, next to the cab. Secure the litters individually to the seats.
    c. Place two litters (litter numbers 4 and 5) lengthwise on the floor, forward toward the cab, feet first. Secure the litters together and to the vertical seat support.
    d. Place litter number 6 crosswise on the seats near the rear of the vehicle. Slide the litter as far forward as possible. Do not secure the litter at this time.
    e. Place litter number 7 crosswise on the seats as far rearward as possible. Secure the litter to the seats.
    f. Slide litter number 6 rearward next to litter number 7. Secure the litter to the seats.
    g. Raise and secure the tailgate.
NOTE: The combat medic or combat lifesaver rides in the center of the vehicle to monitor the casualties.

   5. Direct nonmedical Soldiers to unload vehicle in reverse sequence.

Performance Measures	GO	NO-GO
1. Determined vehicle load capacity.	——	——

2. Directed nonmedical Soldiers to--
   a. Load the casualties in the proper sequence.
   b. Load casualties side-by side across the back of the truck.
   c. Load casualties lengthwise on the floor.
   d. Secure the casualties for transport.
	——	——

3. Direct nonmedical Soldiers to load an M-1093.	——	——
4. Direct nonmedical Soldiers to load an M-1081.	——	——
5. Direct nonmedical Soldiers to unload vehicle in reverse sequence.	——	——

Evaluation Guidance: Score each Soldier according to the performance measures. Unless otherwise stated in the task summary, the Soldier must pass all performance measures to be scored GO. If the Soldier fails any steps, show what was done wrong and how to do it correctly.

References
    Required   **Related**
    None   FM 8-10-6

LOAD CASUALTIES ONTO A UH-60 SERIES HELICOPTER
081-833-0214

Conditions: You have completed treating and triaging multiple casualties. You are in charge of loading litter and ambulatory patients on a standard UH-60 helicopter. You will need a UH-60 helicopter, a medical evacuation kit, and litters. You are not in a CBRNE environment.

Standards: Configure the aircraft properly and load and unload casualties in the correct sequence for the transport platform without causing further injury.

Performance Steps

1. Follow principles of loading casualties aboard a rotary-wing aircraft.
NOTE: The UH-60A Blackhawk, as with the UH-1H/V, has a number of possible seating or cargo configurations. A major difference in preparing the UH-60A to carry litters is that a medical evacuation kit must be installed. This kit consists of a seat/converter assembly unit and a litter support unit. The seat/converter assembly unit provides for three rear-facing seats which allow the medical attendant and crew chief to monitor casualties. The litter support unit consists of a center pedestal which can be rotated 90 degrees about the vertical axis for the loading and unloading of casualties.
    a. Responsibility for loading and securing a rotary-wing aircraft.
        (1) The pilot has the overall responsibility for the proper loading and securing of litter and ambulatory casualties and related equipment on board the aircraft.
        (2) The flight medic crewmember is responsible for ensuring that the litter squad follows the prescribed methods for loading litter or ambulatory casualties and securing litters and related medical equipment.
        (3) The final decision regarding how many casualties may be safely loaded rests with the pilot.
    b. Safety measures.
        (1) Litter bearers must present as low a silhouette as possible and must keep clear of the main and tail rotors at all times.
        (2) The helicopter must not be approached until a crewmember signals to do so.
        (3) The litter bearers should approach the aircraft at a 45-degree angle from the front of the helicopter.
        (4) If the helicopter is on a slope and conditions permit, loading personnel should approach the aircraft from the downhill side.
        (5) Directions given by the crew must be followed, and litters must be carried parallel to the ground.
        (6) Smoking is not permitted within 50 feet of the aircraft.

2. Determine UH-60 rotary wing aircraft load capacities.
    a. Casualty litter capability of the support unit.
        (1) 4 to 6 litter casualties.
NOTE: The litter support unit consists of a center pedestal that can be rotated 90 degrees about the vertical axis for the loading and unloading of casualties. The litter support unit has a capacity of four to six litter casualties. The casualties can be loaded from either side of the aircraft. Only the upper litter supports in the four-litter configuration can be tilted for loading and unloading casualties. When the six-litter modification kit is installed, the center pedestal can no longer be rotated.

Performance Steps

  (2) If litter casualties are not being evacuated, a maximum of six ambulatory casualties can be seated on the litter support unit (three on each side). A seventh ambulatory casualty can be seated on a troop seat.

 3. Determine the loading sequence.

  a. The most seriously injured casualties are loaded last on the bottom pans of the litter support unit. However, if it is anticipated that a casualty's medical condition may require in-flight emergency medical care (such as cardiopulmonary resuscitation), he should be loaded onto either of the top pans to facilitate access.

  b. Casualties in traction splits should be loaded last and on a bottom pan.

  c. The UH-60A has the capability to be loaded on both sides simultaneously. Casualties should be loaded so that upon rotating the litter support, the casualty's head will be forward in the cabin. To accomplish this, casualties loaded on the left side of the aircraft should be loaded head first and casualties loaded on the right side of the aircraft should be loaded feet first (left and right sides are determined from the position of the pilot in command's seat, looking forward).

  d. When the six-litter configuration is used, the fifth and sixth litter casualties are loaded with the carousel in the fly position. The head of each casualty should face toward the front of the aircraft.

  e. Loading and securing casualties.

   (1) In loading four litter casualties with a four-man litter squad, the litters are loaded from the top to bottom. The sequence for loading litters from one side of the aircraft with the carousel turned is upper right, upper left, lower right, and then lower left. To load litters from both sides of the aircraft simultaneously, the sequence is upper then lower.

    (a) The litter support unit is rotated 90 degrees clockwise to receive the litter casualties. The flight crew lowers the top pan to accept the litter and stands by to assist. This is accomplished as the litter squad approaches the aircraft.

    (b) The litter squad moves into the semi overhead carry, lifting the litter just high enough for the litter stirrups of one end to slide onto the litter pan. The litter squad slides the litter forward. The flight crew member guides and assists the litter squad, until the litter stirrups of both ends are secured on the pan. The litter squad departs as the flight crew member raises the pan back to its upright position and secures it. The flight crew member fastens the litter straps attached to the litter support assembly.

    (c) After the first litter is loaded, the squad leaves the aircraft as a team to obtain another litter casualty. The second, third, and fourth litters are loaded in the same manner, except that the bottom pans are not tilted to receive casualties.

    (d) After having loaded four litter casualties, the litter support unit is rotated 90 degrees counterclockwise and locked in the in-flight position. The cargo doors must be closed for flight.

   (2) The loading of six litter casualties requires the repositioning of the litter support prior to loading. The loading procedure remains the same as the four-litter configuration except for the following:

    (a) The top litter support no longer tilts. This necessitates overhead loading and may require additional assistance.

Performance Steps

  (b) After four litters are loaded, the pedestal must be rotated back to the locked position. The restraint and tube assembly modification kit is then installed. The last two litters are side loaded between the restraints, with the casualty's heads toward the front of the aircraft. They are secured.

 (3) When the aircraft is to receive a mixed load of litter and ambulatory casualties, one top pan of the litter support is removed and repositioned just above the bottom pan on the same side. The aircraft can now accommodate two or three litter and four ambulatory casualties.

  (a) The litter support unit is rotated clockwise to receive the litter casualties, except for the third litter in the six-litter configuration. Upon loading and securing the litter casualties, the litter support unit is rotated counterclockwise to the in-flight position. The third litter is then loaded when the six-litter configuration is used.

  (b) Ambulatory casualties are escorted to the aircraft by ground personnel. They are assisted into their seats and secured with the seat belts attached to the litter support unit.

  (c) The cargo doors are now closed for flight.

WARNING: To prevent further injury to casualties, all end support pins of the installed litter pans must be in the locked position for flight.

 f. Unloading casualties.

 (1) The aircraft is unloaded in the reverse order of the loading procedure.

 (2) The pans are normally unloaded bottom pan first, then top, to ensure that the most seriously injured casualties are unloaded first.

Performance Measures   <u>GO</u>   <u>NO-GO</u>

1. Followed the principles of loading casualties on a rotary-wing aircraft and determined load capabilities. ——— ———

2. Assessed the casualties to determine loading sequence. ——— ———

3. Directed nonmedical Soldiers to-- ——— ———
   a. Load the casualties in the proper sequence, and use the proper berths.
   b. Secure the casualties for transport.
   c. Unload the casualties in the proper sequence.

4. Did not cause further injury to the casualties. ——— ———

Evaluation Guidance: Score each Soldier according to the performance measures. Unless otherwise stated in the task summary, the Soldier must pass all performance measures to be scored GO. If the Soldier fails any steps, show what was done wrong and how to do it correctly.

References
  Required   **Related**
  None   FM 8-10-6

LOAD CASUALTIES ONTO A STRYKER ARMORED AMBULANCE
081-833-0226

Conditions: You have completed treating and triaging multiple casualties. You are in charge of loading litter and ambulatory casualties on a STRYKER armored ambulance. You will need a STRYKER armored ambulance and litters. You are not in a CBRNE environment.

Standards: Configure the STRYKER properly, load and unload casualties in the correct sequence for the transport platform.

Performance Steps

1. Configure the casualty compartments for litter casualties.
    a. Remove litters from the stowed position.
    b. Remove the platform from the stowage mount.
        (1) Position one person at each translation beam.
        (2) Remove the quick release pins that hold the translation beams in place on the stowage mount.
        (3) Holding the translation beam with two hands, both people push out and lift up to disengage the translation beam from the stowage mounts.
        (4) Rotate platform so it is horizontal and translation beams are lined up with support mounts on sponson wall.
        (5) Insert translation beams into support mounts.
        (6) Push down on each translation beam until it is fully engaged in support mount.
        (7) Insert quick release pins to hold translation beams in place.
        (8) Unlatch forward and rear platform latches and move platform out towards sponson wall approximately 1 inch so that platform latch will not engage detent and hold platform in configure position.
        (9) Turn down platform hatch handles and continue to move platform out until both latches engage detent, locking platform in transport position.
    c. Retrieve the lift arms from stowage.
    d. Attach lift arms.
        (1) Open the litter latches.
        (2) Pull out and hold the locking pin on the underside of the lift arm.
        (3) Hold out the locking pin and position the lift arm over the lift arm mount so the litter latch pin is on the outside.
        (4) Push down on the lift arm until it is fully engaged on the lift arm mount.
        (5) Release the locking pin to lock the lift arm on the lift arm mount.
        (6) Repeat steps 1e (1) through 1e (5) to attach the lift arm to the other lift arm mount.
    e. Lower the lift arm.
        (1) Pull the E-Stop switch on the litter control box out and up to RUN.
        (2) Press and hold the litter control switch in the DOWN position until the lift arm is lowered.
        (3) Push the E-Stop switch down to DISABLE.

2. Load an upper litter casualty.
    a. Lower the ramp.
    b. If the lift arms are lowered, proceed to step 2c. If the lift arms are raised, do the following:

Performance Steps
      (1) Unlatch the forward and rear platform latches and move the platform towards the center of the vehicle.
      (2) Re-engage the forward and rear latches to hold the platform in the load position.
    c. Load the casualty.
      (1) Slide the casualty head first along the litter tracks.
      (2) Unlatch the forward and rear platform latches and slide the platform toward the wall.
      (3) Lock the forward and rear platform latches with the platform in the transport position.
    d. Lock the litter latches of the lift arms on all four litter handles.
    e. Raise the casualty to the upper position.
      (1) Pull the E-Stop switch on the litter control box out and up to RUN.
      (2) Press and hold the litter control switch in the up position until the lift arm reaches MAX HIGHT position.
      (3) Push the E-Stop switch down to DISABLE.

  3. Load a lower litter casualty.
    a. Unlatch forward and rear platform latches and move platform towards center of vehicle, re-engage forward and rear latches to hold platform in load position.
    b. Slide casualty head first along litter tracks, unlatch forward and rear platform latches and slide platform toward wall, lock forward and rear platform latches with platform in transport position.
    c. Insert litter locking pins into all four litter stirrups.

  4. Load ambulatory casualties.
    a. Seat ambulatory casualties on the bench and secure them with seat restraints.
    b. Raise the ramp making sure there are no obstructions.

Performance Measures	<u>GO</u>	<u>NO-GO</u>
1. Configured the casualty compartments for litter casualties.	——	——
2. Loaded an upper litter casualty and properly secured the litter to the lift arm.	——	——
3. Loaded a lower litter casualty and properly secured the litter to the platform.	——	——
4. Loaded and secured ambulatory casualties.	——	——
5. Did not cause further injury to the casualties.	——	——

Evaluation Guidance: Score each Soldier according to the performance measures. Unless otherwise stated in the task summary, the Soldier must pass all performance measures to be scored GO. If the Soldier fails any steps, show what was done wrong and how to do it correctly.

References
   Required
   None

                     Related
                     081-MEV1
                     TM 9-2355-311-10-1

COORDINATE CASUALTY TREATMENT AND EVACUATION
081-833-0227

Conditions: You are deployed to a unit in a forward area. There are casualties that must be treated and evacuated to receive medical aid. You will need a military vehicle (ground vehicle or rotary-wing aircraft), litters, and straps (or materials to improvise them). You are not in a CBRNE environment.

Standards: Ensure that medical and/or self-aid/buddy aid is provided to the casualties, as appropriate. Ensure that the casualties are transported to medical aid or to a pickup site using an appropriate carry or, if other Soldiers are available, by litter. Ensure that the litters are properly loaded onto a military vehicle (ground vehicle or rotary-wing aircraft) without dropping or causing further injury to the casualties.

Performance Steps

    1. Evaluate the casualties according to the tactical situation. (See task 081-833-0080.)
NOTE: Tactical combat casualty care (TC-3) can be divided into three phases.
      * Care under fire--you are under hostile fire and are very limited as to the care you can provide.
      * Tactical field care--you and the casualty are relatively safe and no longer under effective hostile fire, and you are free to provide casualty care to the best of your ability.
      * Combat casualty evacuation care--care is rendered during casualty evacuation (CASEVAC).

    2. Coordinate treatment of the casualties according to the tactical situation and available resources.

    3. Request medical evacuation (MEDEVAC). (See task 081-831-0101.)
      a. Make contact.
      b. Determine whether casualties must be moved or will be picked up at the current location. If they must be moved, continue with step 4. If they will not be moved, continue to monitor communications and go to step 6.

    4. Move a casualty, if necessary, using a four-man litter squad.
NOTE: If military vehicles and litter materials are not available, continue with step 5.

NOTE: Four-man litter squad bearers should be designated with a number from 1 to 4. The litter bearer designated as #1 is the leader of the squad.
      a. Prepare the litter.
        (1) Open a standard litter.
        (2) Lock the spreader bars at each end of the litter with your foot.
      b. Prepare the casualty.
        (1) Place the casualty onto the litter.
        (2) Secure the casualty to the litter with litter straps.
      c. Lift the litter.
        (1) Position one squad member at each litter handle with the litter squad leader at the casualty's right shoulder.
NOTE: The leader should be at the right shoulder to monitor the casualty's condition.
        (2) On the preparatory command, "Prepare to lift," the four bearers kneel beside the litter and grasp the handles.
        (3) On the command, "Lift," all bearers rise together.

Performance Steps

    (4) On the command, "Four man carry, move," all bearers walk forward in unison.

        (a) If the casualty does not have a fractured leg, carry the casualty--

              * Feet first on level ground.

              * Head first when going uphill.

        (b) If the casualty has a fractured leg, carry the casualty--

              * Head first on level ground.

              * Feet first when going uphill.

    (5) To change direction of movement, such as from feet first to head first, begin in a litter post carry position. The front and back bearers release the litter and the middle bearers rotate the litter and themselves.

5. Coordinate transportation of casualties using appropriate carries, if necessary. (See STP 21-1-SMCT, task 081-831-1046.)

6. Coordinate loading of casualties onto a military vehicle.

    a. Ground ambulance.

NOTE: Ground ambulances have medical personnel to take care of the casualties during evacuation. Follow any special instructions for loading, securing, or unloading casualties.

    (1) Secure each litter casualty to his litter.

    (2) Load the most serious casualty last.

    (3) Load the casualty head first (head in the direction of travel) rather than feet first.

    (4) Secure each litter to the vehicle.

    b. Air ambulance.

NOTE: Air ambulances have medical personnel to take care of the casualties during evacuation. Follow any special instructions for loading, securing, or unloading casualties.

    (1) Remain 50 yards from the helicopter until the litter squad is signaled to approach the aircraft.

    (2) Approach the aircraft in full view of the aircraft crew, maintaining visual confirmation that the crew is aware of the approach of the litter party. Ensure that the aircrew can continue to visually distinguish friendly from enemy personnel at all times. Maintain a low silhouette when approaching the aircraft.

    (3) Approach UH-60/UH-1 aircraft from the sides. Do not approach from the front or rear. If you must move to the opposite side of the aircraft, approach from the side to the skin of the aircraft. Then, hug the skin of the aircraft, and move around the front of the aircraft to the other side.

    (4) Load the most seriously injured casualty last.

    (5) Load the casualty who will occupy the upper birth first, then load the next litter casualty immediately under the first casualty.

NOTE: This is done to keep the casualty from accidentally falling on another casualty if his litter is dropped before it is secured.

    (6) When casualties are placed lengthwise, position them with their heads toward the direction of travel.

    (7) Secure each litter casualty to his litter.

    (8) Secure each litter to the aircraft.

    c. Ground military vehicles.

Performance Steps
NOTE: Nonmedical military vehicles may be used to evacuate casualties when no medical evacuation vehicles are available.

NOTE: FM 4-02.2 Medical Evacuation contains suggested loading plans for many common nonmedical vehicles. You should become familiar with the plans for vehicles assigned to your unit.

    (1) When loading casualties into the vehicle, load the most seriously injured casualty last.

    (2) When a casualty is placed lengthwise, load the casualty with his head pointing forward, toward the direction of travel.

    (3) Secure each litter casualty to his litter.

    (4) Secure each litter to the vehicle as it is loaded into place.

    (5) Watch the casualties closely for life-threatening conditions and provide treatment, as necessary, during CASEVAC.

NOTE: CASEVAC refers to the movement of casualties aboard nonmedical vehicles or aircraft. Care is rendered while the casualty is awaiting pickup or is being transported. A Soldier accompanying an unconscious casualty should monitor the casualty's airway, breathing, and bleeding.

Evaluation Preparation:
Setup: Evaluate this task during a training exercise involving a MEDEVAC aircraft or vehicle, or simulate it by creating a scenario, and provide the equipment needed for the evaluation.

Brief Soldiers: Tell the Soldier the scenario to include the end result desired.

Performance Measures

	GO	NO-GO
1. Evaluated the casualties according to the tactical situation.	——	——
2. Coordinated treatment of the casualties according to the tactical situation and available resources.	——	——
3. Requested medical evacuation.	——	——

4. Moved a casualty using a four-man litter squad, if necessary.
   a. Prepared the litter.
   b. Prepared the casualty.
   c. Lifted the litter.
	——	——

5. Coordinated transportation of casualties using appropriate carries, if necessary.	——	——
6. Coordinated loading of casualties onto a military vehicle.	——	——

Evaluation Guidance: Score each Soldier according to the performance measures. Unless otherwise stated in the task summary, the Soldier must pass all performance measures to be scored GO. If the Soldier fails any steps, show them what was done wrong and how to do it correctly.

References
    Required
    None

    Related
FM 4-02.2
FM 4-25.11 (FM 21-11)
STP 21-1-SMCT

Subject Area 10: Medication Administration

PREPARE AN INJECTION FOR ADMINISTRATION
081-833-0088

Conditions: You must prepare an injection for administration. You have performed a patient care hand-wash. You will need needles and syringes, medication, alcohol sponges, dry sterile gauze, and a medical officer's order. You are not in a CBRNE environment.

Standards: Select, inspect, and assemble the appropriate needle and syringe. Draw the correct medication. Follow aseptic technique throughout the procedure.

Performance Steps

1. Select an appropriate needle.
   a. Select a needle with the proper length based upon the following factors:
      (1) The type of injection to be given (subcutaneous, intramuscular, or intradermal).
      (2) The size of the patient (thin, obese).
      (3) The injection site (1 inch for deltoid, 1 1/2 inches for gluteus maximus).
   b. Select a needle with the proper gauge based upon the thickness of the medication to be injected.
   NOTE: The gauge of the needle is indicated by the numbers 104 through 27. The higher the number, the smaller the diameter (bore) of the needle. A small bore needle is indicated for thin medications. A large bore needle is indicated for thick medications.

2. Select an appropriate syringe.
   a. Check the drug manufacturer's specifications to determine whether a glass or plastic syringe should be used for the medication.
   NOTE: Some medications deteriorate in a plastic syringe. Drug manufacturer's specifications provide guidance.
   b. Ensure that the total capacity of the syringe, usually measured in cubic centimeters (cc), is appropriate for the amount of medication to be administered.
   c. Check the intervals of the calibration marks on the syringe.

3. Inspect the needle and syringe packaging for defects such as open packages, holes, and water spotting. Discard the equipment if any defect is found.

4. Unpack the syringe.
   a. If the syringe is in a flexible wrapper, peel the sides of the wrapper apart to expose the rear end of the syringe barrel.
   b. Grasp the syringe by the barrel with the free hand.
   CAUTION: The needle adapter and the shaft of the plunger are sterile. Contamination could cause infection in the patient. The outside of the syringe barrel does not have to be kept sterile.
   c. Pull the syringe from the packaging.
   d. If the syringe is packaged in a hard plastic tube container, press down and twist the cap until a distinct "pop" is heard. If the "pop" is not heard, the seal has been previously broken and the equipment must be discarded.

5. Inspect the syringe.
   a. Grasp the flared end of the syringe and pull the plunger back and forth to test for smooth, easy movement.

Performance Steps

    b. Visually check the rubber stopper (inside the syringe) to ensure that it is attached securely to the top end of the plunger, forming a seal.

    c. If the plunger is stuck or does not move smoothly, discard the syringe.

    d. Push the plunger fully into the barrel until ready to fill the syringe with medication.

6. Unpack the needle.

CAUTION: All parts of the needle are sterile. Be careful not to touch the hub. This would contaminate the needle and possibly pass an infection to the patient. Only the outside of the needle cover may be touched.

    a. If the needle is packaged in a flexible wrapper, peel the sides of the wrapper apart to expose the needle hub.

    b. If the needle is packaged in a hard plastic tube, twist the cap of the tube until a "pop" is heard. Remove the cap to expose the needle hub. If a "pop" is not heard, the seal has been previously broken, and the equipment must be discarded.

7. Join the needle and the syringe.

    a. Insert the needle adapter of the syringe into the hub of the needle.

    b. Tighten the needle by turning 1/4 of a turn to ensure that it is securely attached.

8. Inspect the needle.

    a. Hold the needle and syringe upright and remove the protective cover from the needle by pulling it straight off.

NOTE: A twisting motion may pull the needle off the hub.

    b. Visually inspect the needle for burrs, barbs, damage, and contamination. If the needle has any defects or damage, replace it with another sterile needle.

    c. Place the protective cover back on the needle utilizing the "scoop" method.

9. Place the assembled needle and syringe on the work surface.

    a. Leave the protective cover on the needle.

    b. Leave the plunger pushed fully into the barrel.

    c. Keep the assembled needle and syringe continually within range of vision.

NOTE: When you assemble a needle and syringe, you are responsible for maintaining sterility and security of the equipment.

10. Verify the drug label and check the container for defects.

    a. Compare the medication with the medical officer's orders. The medication label must be verified three times.

      (1) When obtained from the place of storage.

      (2) When withdrawing the medication.

      (3) When returning the container to storage.

    b. Examine the container.

      (1) Examine the rubber stopper for defects, such as small cores or plugs torn from the stopper.

      (2) Hold the vial to the light to check for foreign particles and changes in color and consistency. If the solution is in a dark vial, withdraw some solution to perform the checks.

      (3) Check the date a multidose vial was opened and check the expiration date of the medication.

      (4) Determine whether the medication was stored properly, such as under refrigeration.

NOTE: If there is any evidence of contamination, discard the container and obtain another.

Performance Steps

   11. Prepare and draw the medication:
      a. Draw medication from a stoppered vial which contains a prepared solution.
        (1) Remove the protective cap.
        (2) Clean the stopper and neck of the vial with an alcohol sponge.
        (3) Pick up the assembled needle and syringe and remove the protective needle cover.
        (4) Slowly draw the plunger to the prescribed cc mark of medication.
        (5) Pick up the vial and insert the needle into the rubber stopper, exerting slight downward and forward pressure. Ensure that the needle tip passes completely through the cap.

NOTE: To avoid contamination, the hub of the needle should not touch the rubber cap.

        (6) Push the plunger fully into the barrel to inject the air.
        (7) With the vial inverted (and keeping the needle tip in the solution), pull the plunger back to the desired cc mark, withdrawing the medication.
        (8) Withdraw the needle from the container.
        (9) Verify the correct dosage against the medical officer's orders by raising the syringe to eye level and ensuring that the forward edge of the plunger is exactly on the prescribed cc mark.
      b. Draw medication from a stoppered vial which contains a powdered medication which must be prepared.
        (1) Remove the protective caps from the vial containing the powdered medication and the vial containing the sterile diluent.
        (2) Clean the stoppers of both vials with alcohol sponges.
        (3) Withdraw the required diluent, using the same procedure as for a stoppered vial. (See steps 11a(3) through 11a(8).)
        (4) Hold the vial with the powdered medication horizontally, insert the needle through the stopper, and inject the diluent.

NOTE: If the vial with powdered medication contains air, the diluent may be difficult to inject. Air may have to be withdrawn to allow the diluent to be injected.

        (5) Withdraw the needle.
        (6) Gently invert the vial several times until all the powder is dissolved. Visually inspect the solution to ensure that it is well-mixed.
        (7) Change the needle (or needle and syringe) and insert it into the vial of reconstituted solution.
        (8) Withdraw the prescribed amount of medication. (See step 11a(7).)
        (9) Withdraw the needle from the container.
        (10) Verify the correct dosage. (See step 11a(9).)
      c. Draw medication from an ampule.
        (1) Lightly tap the upright ampule to force any trapped medication from the ampule neck and top.
        (2) Clean the neck of the ampule with an alcohol sponge and wrap it with the same sponge.
        (3) Grasp the ampule with both hands and snap the neck by bending it away from the break line, directing it away from yourself and others.
        (4) Inspect the ampule for minute glass particles. If any are found, discard the ampule.
        (5) Remove the protective cover from the assembled needle and syringe.
        (6) Insert the needle and withdraw the medication by holding the ampule vertically or by placing the ampule upright on a flat surface.

Performance Steps

        (7) Withdraw the prescribed medication, being careful not to touch the outside edge or bottom of the ampule with the needle.

        (8) Withdraw the needle.

        (9) Verify the correct dosage. (See step 11a(9).)

12. Check the syringe for air bubbles.

    a. Hold the syringe with the needle pointing up.

    b. Pull back on the plunger slightly to clear all the medication from the needle shaft.

    c. Tap the barrel lightly to force bubbles to the top of the barrel.

    d. Pull the plunger back slightly and push it forward until the solution is in the needle hub, clearing it of bubbles.

13. Reverify the correct dosage. (See step 11a(9).)

14. Cover the needle with the protective needle cover utilizing the "scoop" method.

15. Do not violate aseptic technique.

Evaluation Preparation:

Setup: If the performance of this task must be simulated for training and evaluation, colored solutions may be used to simulate medications. Have several sizes of needles and syringes available. Tell the Soldier what type of medication is being simulated and what the route of administration would be. Have Soldier select the appropriate needle and syringe. To test step 2, tell the Soldier of any manufacturer's specifications. Testing may be varied by using various medications to be administered by different routes.

Brief Soldier: Tell the Soldier to assemble the proper needle and syringe and draw the medication.

Performance Measures	GO	NO-GO
1. Selected the appropriate needle.	——	——
2. Selected the appropriate syringe.	——	——
3. Inspected the packaging for defects.	——	——
4. Unpacked the syringe.	——	——
5. Inspected the syringe.	——	——
6. Unpacked the needle.	——	——
7. Joined the needle and syringe.	——	——
8. Inspected the needle.	——	——
9. Placed the assembled needle and syringe on the work surface.	——	——
10. Verified the drug label and checked the container for defects.	——	——
11. Prepared and drew the medication.	——	——
12. Checked the syringe for air bubbles.	——	——

Performance Measures   GO   NO-
                                                                    GO

13. Reverified the correct dosage.   ____   ____

14. Covered the needle with the protective needle cover utilizing the "scoop"   ____   ____
    method.

15. Did not violate aseptic technique.   ____   ____

Evaluation Guidance: Score each Soldier according to the performance measures. Unless otherwise stated in the task summary, the Soldier must pass all performance measures to be scored GO. If the Soldier fails any steps, show what was done wrong and how to do it correctly.

References
   Required   **Related**
   None   BASIC NURSING 7

ADMINISTER AN INJECTION (INTRAMUSCULAR, SUBCUTANEOUS, INTRADERMAL)
081-833-0089

Conditions: You have performed a patient care hand-wash and have verified the medical officer's order. You will need syringe(s) with the prepared medication(s), antiseptic pads, alcohol sponge swabs, sterile gauze, adhesive tape, and the patient's record. You are not in a CBRNE environment.

Standards: Administer the injection IAW the medical officer's order without violating aseptic technique or causing injury to the patient.

Performance Steps

1. Verify the required injection(s) with the medical officer's order.

2. Identify the patient by asking the patient's name and checking the identification tag or band. Ask the patient if he has any allergies or has experienced a drug reaction.
WARNINGS: 1. If there is a known allergy, do not administer the injection. Consult your supervisor. 2. Determine if a female patient is pregnant because of possible side effects of certain immunizing agents on the unborn child. If there is a question, do not administer the injection without written authorization.

3. Verify that the appropriate needle, syringe, and medication are being used. (See task 081-833-0088.)
WARNING: Have an emergency tray available for the immediate treatment of serious reactions. Include a constricting band and a syringe containing a 1:1000 solution of epinephrine.

4. Do not violate aseptic technique.

5. Select and expose the injection site.
    a. Intramuscular injection.
        (1) The upper arm deltoid muscle--the outer 1/3 of the arm between the lower edge of the shoulder bone and the armpit. Approximately three finger widths below the shoulder bone is the safe area.
        (2) Buttocks--the upper-outer quadrant of either buttock.

Performance Steps

NOTE: To identify the injection site, draw an imaginary horizontal line across the buttocks from hip bone to hip bone. Then divide each buttock in half with an imaginary vertical line. Refer to Figure 3-62.

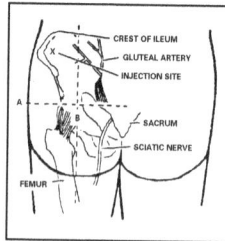

Figure 3-62. Buttock in half

WARNING: Do not give the injection in an area outside the upper-outer quadrant. The needle may do irreparable damage to the sciatic nerve or pierce the gluteal artery and cause significant bleeding.

        (3) Outer thigh--the area between a hand's width above the knee and a hand's width below the groin.
    b. Subcutaneous injection.
        (1) Upper arm.
        (2) Anterior thigh.
        (3) Abdomen.
    c. Intradermal injection.
        (1) Inner forearm.
        (2) Back of the upper arm.
        (3) On the back below the shoulder blades.
  6. Position the patient.
    a. Intramuscular.
        (1) Upper arm--standing or sitting with the area completely exposed, muscles relaxed, and the arm at the side.
        (2) Buttocks--lying face down or leaning forward and supported by a stable object with the weight shifted to the leg that will not be injected. The area is completely exposed.

NOTE: If the patient is lying in a prone position (face down), place the toes together with the heels apart. This will relax the muscles of the buttocks.

        (3) Outer thigh- -lying face up or seated with the area completely exposed.
    b. Subcutaneous.
        (1) Upper arm--see step 5a(1).
        (2) Outer thigh--lying face up or seated, with the area completely exposed.
    c. Intradermal.
        (1) Inner forearm--standing, sitting, or lying. Palm up, with the arm supported and relaxed.
        (2) Upper arm--see step 5a(1).
        (3) Back--seated, leaning forward and supported on a stable object, or lying face down.

Performance Steps

    7. Clean the injection site.
        a. Intramuscular and subcutaneous.
           (1) Open the antiseptic pad package.
           (2) Begin at the injection site and with a spiral motion, clean outward 3 inches.
        b. Intradermal.
           (1) Use ethyl alcohol or acetone germicide and a sterile sponge.
           (2) Begin at the injection site and with a spiral motion, clean outward 3 inches.
NOTE: The antiseptic pad may be held between the last two fingers for use when the needle is removed.

    8. Pull the needle cover straight off without bending or touching the needle.

    9. Prepare the skin for the injection.
        a. Intramuscular and subcutaneous. Form a fold of skin at the injection site by pinching the skin gently between the thumb and the index finger of the nondominant hand. Do not touch the injection site.
        b. Intradermal. Using the thumb of the nondominant hand, apply downward pressure directly below and outside the prepared injection site. Hold the skin taut until the needle has been inserted.
CAUTION: Do not retract or move the skin laterally.

    10. Insert the needle.
        a. Intramuscular. With the dominant hand, position the needle, bevel up, at a 90 degree angle to, and about 1/2 inch from, the skin surface. Plunge the needle firmly and quickly straight into the muscle.
        b. Subcutaneous. With the dominant hand, position the needle, bevel up, at a 45 degree angle to the skin surface. Plunge the needle firmly and quickly into the fatty tissue below the skin.
        c. Intradermal. With the dominant hand, position the needle, bevel up, at a 15 to 20 degree angle to the skin surface. Insert it just under the skin until the bevel is covered. Do not move the skin.

    11. Release the hold on the skin.

    12. Administer the medication.
        a. Intramuscular and subcutaneous.
           (1) Aspirate by pulling back slightly on the plunger of the syringe.
              (a) If blood appears, stop the procedure. Go to step 3 and begin the procedure again. Use a new needle, syringe, and medication, and select a different injection site.
              (b) If no blood appears, continue the procedure.
WARNING: Failure to aspirate could cause the medication to be injected into the blood stream.
           (2) Using a slow continuous movement, completely depress the plunger, injecting the medication.
NOTE: Rapid pressure may cause a burning pain.
           (3) Place an antiseptic pad (or sterile 2 x 2) lightly over the injection site and withdraw the needle at the same angle at which it was inserted. Gently massage the injection site with the pad, unless this is contraindicated for the medication that has been injected.
           (4) Put an adhesive bandage strip over the injection site if bleeding occurs.

Performance Steps

    b. Intradermal.

NOTE: Do not aspirate.

        (1) Push the plunger slowly forward until all medication has been injected and a wheal appears at the site of the injection.

            (a) If a wheal does not appear, go to step 3 and begin the procedure again. Use a new needle, syringe, and medication and select a different injection site.

            (b) If a wheal appears, continue the procedure.

        (2) Quickly withdraw the needle at the same angle at which it was inserted.

        (3) Without applying pressure, cover the injection site with dry sterile gauze.

        (4) Instruct the patient not to scratch, rub, or wash the injection site.

        (5) If appropriate, instruct the patient when and where to have the test read IAW local SOP.

13. Check the site for bleeding.

14. Observe the patient for anaphylactic shock symptoms IAW local SOP. (See task 081-833-0031.)

15. Dispose of the needle and syringe IAW local SOP.

16. Record the procedure on the appropriate form.

17. Do not cause further injury to the patient.

Evaluation Preparation:

Setup: If the performance of this task must be simulated for training and evaluation, have another Soldier act as the patient. If so, ensure that the prepared syringes contain no more than 0.2 cc of a safe, sterile, injectable solution. Tell the Soldier which type of injection to give. Ensure that medical coverage is available in case of reaction.

Brief Soldier: Tell the Soldier to administer the injection.

WARNING: If the Soldier violates aseptic technique or starts to do something which could injure the patient, stop the evaluation immediately.

Performance Measures	**GO**	**NO-GO**
1. Verified the required injection(s) with the medical officer's order.	——	——
2. Identified the patient and asked the patient about allergies or drug reactions.	——	——
3. Verified the appropriate needle, syringe, and medication.	——	——
4. Did not violate aseptic technique.	——	——
5. Selected and exposed the injection site.	——	——
6. Positioned the patient.	——	——
7. Cleaned the injection site.	——	——
8. Removed the needle cover.	——	——

Performance Measures	<u>GO</u>	<u>NO-GO</u>
9. Prepared the skin for injection.	——	——
10. Inserted the needle.	——	——
11. Released the skin.	——	——
12. Administered the medication.	——	——
13. Checked the site for bleeding.	——	——
14. Observed the patient for adverse reactions.	——	——
15. Disposed of the needle and syringe.	——	——
16. Recorded the procedure on the appropriate form.	——	——
17. Did not cause further injury to the patient.	——	——

Evaluation Guidance: Score each Soldier according to the performance measures. Unless otherwise stated in the task summary, the Soldier must pass all performance measures to be scored GO. If the Soldier fails any steps, show what was done wrong and how to do it correctly.

References
    Required
    None

                       Related
                       BASIC NURSING 7

ADMINISTER MORPHINE

081-833-0174

Conditions: You are caring for a conscious casualty who has sustained an injury and is suffering from severe pain. You have gained authorization from a licensed provider to administer morphine to the casualty (authorization may have already been delegated to you in certain combat operations). You will need a morphine cartridge, injector device (or autoinjector), alcohol wipes, a semi-permanent marking device, intravenous (IV) administration set, and DD Form 1380 (U.S. Field Medical Card). You are not in a CBRNE environment.

Standards: Correctly prepare and administer morphine without causing further injury to the casualty.

Performance Steps

1. Verify the five rights of medication administration.
   a. Right casualty. Verify that the casualty does not have any contraindications that preclude the use of morphine.
   b. Right medication. Check to ensure that the medication you are about to administer is correct.
   c. Right dosage.
      (1) Morphine dosage IV.
         (a) Administer an initial dose of 5 mg IV (SLOW push over 4-5 minutes). Morphine given IV should be diluted in 5 ml of sterile water for injection or normal saline prior to usage.
         (b) When morphine is given intravenously, repeat doses may be given every 10 minutes. Most adults will experience pain relief at a total dose of 10-20 mg although higher doses may be needed.
      (2) Morphine dosage intramuscular (IM). Load the prefilled cartridge into the injector device (usually at a dose of 5 or 10 mg). If the medic is not giving the full 10 mg amount to the casualty, place the unused portion in another syringe (if possible) to utilize the full amount of morphine later.
   NOTE: The U.S. military utilizes autoinjectors. The dosage is usually one autoinjector (10 mg).
   d. Right time. Check the casualty's forehead and FMC to see when the last dose was administered and how much was given.
   e. Right route. Morphine comes in multiple strengths and routes of administration (oral, IM, subcutaneous (SC), and intravenous injection). In combat situations, the intravenous route is the preferred method of administration due to a more rapid pain response over the more traditional IM methods.

2. Load the prefilled cartridge into the injector device.
   a. Insert the cartridge into the body and guard assembly.
   b. Align the finger grip assembly notches and snap into place.

3. Lock the cartridge into the injector device by turning the plunger rod until the plunger is securely in place.

4. Place the casualty in a supine position.
NOTE: Once morphine has been administered, the casualty is considered nonambulatory.

Performance Steps

5. Select the site for an intramuscular injection.
   a. Deltoid muscle.
   b. Buttocks.
   c. Outer thigh.

6. Administer the injection. (See task 081-833-0089.)

7. If using an autoinjector, complete the following steps:
   a. Remove the safety cap.
   b. Place purple end on the outer thigh and press firmly to deliver the dosage.

8. Reassess the casualty.

9. Monitor for adverse reactions.
NOTE: The most common adverse reaction is severe respiratory depression. The casualty may require assisted ventilation.
   a. If a morphine overdose is suspected, administer Narcan.
      (1) The dose of Narcan given is 0.4 mg to 2 mg SLOW IV push over 1-2 minutes. Narcan may need to be repeated 3-4 times. Some authorities recommend up to 10-20 mg of Narcan to treat suspected morphine overdose.
NOTE: An immediate positive response is usually seen when giving Narcan for morphine poisoning. The duration of action for Narcan is 1-2 hours.
      (2) Narcan's effect may wear off earlier than the morphine, permitting the casualty to lapse back into respiratory depression. Continuous monitoring of a casualty being given Narcan to counteract morphine toxicity is crucial. After every dose of Narcan given, thoroughly reassess the casualty.
      (3) Adjust administration of Narcan according to the casualty's respiratory status, not the level of consciousness.
   b. Document every dose given and the time given on the casualty's FMC.

10. Write the letter "M" and time of injection on the casualty's forehead.

11. Document morphine doses on the FMC. (See task 081-831-0033.)

Evaluation Preparation:
Setup: None.

Brief Soldier: Tell the Soldier that a casualty is in extreme pain and needs morphine administered. Tell the Soldier to demonstrate/describe the procedure. To test performance measure 8, inform the Soldier that the casualty has signs of morphine overdose.

Performance Measures	GO	NO-GO
1. Verified the five rights of medication administration.	——	——
2. Loaded the prefilled cartridge into the injector device (eliminate this step if using an autoinjector).	——	——
3. Locked the cartridge into the injector device turning the plunger rod until the plunger was securely in place (eliminate this step if using an autoinjector).	——	——

Performance Measures   <u>GO</u>   <u>NO-</u>
                                                                                   <u>GO</u>

    4. Placed the casualty in a supine position.   ____   ____

    5. Selected the site for an intramuscular injection.   ____   ____

    6. Administered the injection.   ____   ____

    7. Reassessed the casualty.   ____   ____

    8. Monitored for adverse reactions.   ____   ____

    9. Administered Narcan for suspected morphine overdose.   ____   ____

  10. Wrote the letter "M" and time of injection on the casualty's forehead.   ____   ____

  11. Documented the administration of morphine on the FMC.   ____   ____

Evaluation Guidance: Score each Soldier according to the performance measures in the evaluation guide. Unless otherwise stated in the task summary, the Soldier must pass all performance measures to be scored GO. If the Soldier fails any step, show what was done wrong and how to do it correctly.

References
    Required   **Related**
    DD FORM 1380   PHTLS

ADMINISTER MEDICATIONS
081-833-0179

Conditions: You have a patient requiring medication to be administered. A patient care handwash has been performed. You will need calibrated medicine cups, disposable medicine cups, tray, medications, medicated pads or patches, application papers, tape, non-sterile gloves, tissues, sterile gauze, sterile normal saline, dressing materials, the prescribed medications, sterile tongue blade, sterile gloves, dressing supplies, soap and water, prescribed diluent, metered dose inhaler dispenser, a spacer device, a nebulizer aerosol medication chamber, T-piece, corrugated tubing, facemask, airflow tubing, a source for compressed air or oxygen with flowmeter, DA Form 4678 (Therapeutic Documentation Care Plan (Medication)), DA Form 3949 (Controlled Substances Record), and the patient's clinical record. You are not in a CBRNE environment.

Standards: Administer medications IAW the medical officer's orders.

Performance Steps

1. Check the DA Form 4678 against the medical officer's orders.
    a. Right drug.
    b. Right dose of medication.
    c. Route of administration.
    d. Right patient.
    e. Right time to be administered.

2. Select the medication.
    a. Check the medication label three times to ensure that the correct medication is being prepared for administration.
       (1) First time--when removing the container from the storage shelf.
       (2) Second time--when preparing the medication dose.
       (3) Third time--when returning the container to the storage shelf.
    b. Check the expiration date of the medication.
    c. Handle only one medication at a time.
NOTE: If unfamiliar with a medication, look it up to determine contraindications, precautions, and side effects before preparing it for administration.

3. Administer oral medications.
    a. Calculate the amount of medication required to equal the prescribed dose.
    b. Prepare the prescribed dose of medication.
       (1) Tablets or capsules. Transfer the prescribed dose of tablets or capsules to the medicine cup.
       (2) Liquids.
           (a) Pour the prescribed dose of liquid medication into the medicine cup.
NOTE: When liquid is poured into a cylinder, it forms a meniscus. In determining the volume of liquid, a reading must be made at the bottom of the meniscus, with the level of the liquid at eye level.
           (b) Small amounts of liquid medication should be drawn up in a syringe.
       (3) Powders.
           (a) Pour the correct dose of powdered or granulated medication into the medicine cup.
           (b) Pour the required amount of water or juice into a paper cup.

Performance Steps

NOTE: When preparing medication for more than one patient, mark each container with the patient's identification.

    c. Sign for controlled drugs on DA Form 3949, IAW local SOP.

    d. Correctly identify the patient.

    e. Locate the correct medication.

    f. Give the medication to the patient at the prescribed time.

        (1) Tablets, capsules, or liquids. Observe the patient swallow the tablets, capsules, or liquids.

        (2) Sublingual medications. Instruct the patient to allow sublingual medications to dissolve in his mouth.

        (3) Powdered medication. Reconstitute powdered or granulated medications in the prepared juice or water and observe the patient drink the preparation.

  4. Administer topical medications.

    a. Prepare the prescribed dose of topical medication.

        (1) Obtain single dose packets of topical medication.

        (2) Obtain the required number of medicated patches or pads.

        (3) Apply the prescribed size ribbon of ointment to an application paper.

        (4) Obtain the jar or tube of medication identified for that individual patient's use.

        (5) Aseptically transfer the required amount of topical medication from the bulk storage container to a sterile, disposable container.

NOTE: When preparing medication for more than one patient, mark the prepared medications with the patient's identification.

    b. Correctly identify the patient and explain procedure.

    c. Prepare the skin.

        (1) Provide privacy or screen the patient, as necessary.

        (2) Expose the prescribed area of the patient's skin.

        (3) Clean the skin.

    d. Apply the medication to the patient.

        (1) Locate the correct medication.

        (2) Apply the medication to the prescribed area IAW the medical officer's orders or local SOP.

NOTE: Wear gloves when appropriate.

          (a) Secure patches, pads, and application papers with tape.

          (b) Cover topical applications with sterile dressings IAW the medical officer's orders, if required.

  5. Administer medicated eye drops and ointments.

    a. Take the medication and other supplies to the patient's bedside.

    b. Identify the patient and explain the procedure.

    c. Position the patient.

        (1) Supine in bed.

        (2) Sitting, with the head supported.

NOTE: The head must be supported for stability if the patient is seated. Support may be provided by a head rest or a high-back chair.

    d. Remove eye dressings, if present.

        (1) Perform a patient care hand-wash.

        (2) Glove.

        (3) Gently pull the dressing away from the forehead, and then pull it down and away from the eye area.

Performance Steps
    (4) Discard the contaminated dressing.
   e. Remove accumulation of secretions, if present.
    (1) Apply sterile gauze moistened with sterile normal saline to the closed eyes to soften the secretions.
    (2) Remove loosened secretions by blotting with additional moistened gauze.
   f. Prepare the medication.
    (1) Ointment tube.
      (a) Remove the cap from the tube and place the cap on a piece of sterile gauze to prevent contamination.
      (b) Squeeze a small amount of ointment onto a piece of sterile gauze to remove any crust that may have formed.
      (c) Discard this gauze.
    (2) Eye dropper.
      (a) Draw the prescribed amount of the medication into the dropper.
      (b) Do not invert the dropper after withdrawing the solution.
    (3) Squeeze vial.
      (a) Remove the cap and place it on a piece of sterile gauze.
      (b) Invert the vial.
   g. Administer the medication.
    (1) Instruct the patient to tilt the head back and look upward with the eyes open.
    (2) Steady the hand holding the medication container against the patient's forehead.
    (3) Place a finger on the skin below the lower eyelid and apply gentle, downward pressure to create a small conjunctival pocket.
    (4) Instill the correct number of drops or amount of ointment into the conjunctival pocket.
    (5) Apply ointment in a thin ribbon from the inner aspect to the outer aspect of the conjunctival pocket.
    (6) Do not instill medication directly onto the eyeball.
    (7) Instruct the patient to close the eyes gently and "roll" them to distribute the medication.
NOTE: Instruct the patient not to squeeze the eyes tightly shut.
    (8) Remove any excess solution or ointment by blotting gently with a clean tissue or gauze square.
   h. Apply fresh dressings or patches, if required.
   i. Remove gloves.

  6. Administer respiratory treatments.
   a. Prepare to administer inhalation medication.
    (1) Review the medical officer's order.
      (a) Determine the drug schedule and the number of prescribed inhalations.
      (b) Compare the medication administration record with the medical officer's order and medication label.
    (2) Assess the patient's medical history to identify contraindications/allergies to drug administration.
    (3) Identify the patient and provide privacy.
    (4) Explain the procedure to the patient.
NOTE: Always read the manufacturer's insert before administering inhalation medication.
    (5) Wash your hands and follow standard precautions.
    (6) Arrange equipment.

Performance Steps

  (7) Assess the patient's respiratory status and report unexpected findings to the medical officer.
 b. Assess the patient's knowledge and ability to handle the required equipment.
  (1) Knowledge of disease and drug therapy.
  (2) Willingness to learn.
  (3) Ability to demonstrate use of the required equipment.
   (a) Handicap such as blindness.
   (b) Ability to form an airtight seal with mouth.
   (c) Appropriate strength
 c. Administer the medication
  (1) Liquid metered dose inhaler.
   (a) Ensure the patient is in a sitting or standing position.
   (b) Insert the medication canister stem down into the longer part of the metered dose dispenser.
   (c) Hold the canister upright and shake to mix the medication and propellant before each use.
   (d) Remove the mouthpiece cover and have the patient hold the mouthpiece 2 inches from his mouth.
   (e) Have the patient take a deep breath through pursed lips and then exhale to promote greater inspiratory volume.
   (f) Instruct the patient to inhale slowly, through the mouth, as the canister is depressed. Have the patient inhale for 3 to 5 seconds.
   (g) Instruct the patient to hold his breath for 10 seconds and then exhale slowly through pursed lips.
   (h) Instruct the patient to wait 2 minutes between puffs, if more than one puff is ordered.
  (2) Dry powder inhaler.
   (a) Have the patient hyperextend the neck.
   (b) Ask the patient to place his lips around the mouth of the dispenser.
   (c) Ensure there is an airtight seal.
   (d) Have the patient depress the canister while taking a quick deep breath.
   (e) Instruct the patient to hold his breath for 10 seconds.
   (f) Have the patient exhale slowly through pursed lips.
   (g) Instruct the patient to wait 2 minutes between puffs, if more than one puff is ordered.
  (3) Inhaler with a spacing device.
NOTE: Some patients are unable to coordinate inspiration and activation of the metered dose inhaler. A spacer eliminates the need for simultaneous hand and inspiration action by trapping the medication in the chamber. Spacers also reduce the risk of oropharyngeal candidiasis associated with inhaled corticosteroids.
   (a) Shake the inhaler.
   (b) Remove the mouthpiece cover.
   (c) Insert the metered dose inhaler into the spacer device.
   (d) Have the patient place in mouth and close lips.
   (e) Instruct the patient to breathe normally through the spacer device mouthpiece.
   (f) Have the patient depress the canister one time.
   (g) Ask the patient to breathe in slowly for 5 seconds.
   (h) Have the patient hold his breath for 5 to 10 seconds and then slowly exhale.

Performance Steps
        (i) Wait the appropriate interval and repeat the procedure for the prescribed number of puffs.
    (4) Small volume nebulizer.
        (a) Assemble equipment according to manufacturer's instructions.
        (b) Add the prescribed medication and diluent to the nebulizer.
        (c) Keep the nebulizer vertical while connecting it to a T-piece side arm.
        (d) Attach corrugated tubing to one end of the T-piece.
        (e) Attach the mouthpiece to the other end of the T-piece.
        (f) Attach the airflow tubing to the nebulizer.
        (g) Attach the other end of the airflow tubing to the air or oxygen source.
        (h) Have the patient hold the mouthpiece between the lips using gentle pressure.
        (i) Turn the nebulizer on and set the driving air or oxygen to 8 L/min.
        (j) Ask the patient to take a slow deep breath, pause, and exhale passively.
        (k) Observe to determine if a mist forms. If a mist does not form, the nebulizer is not operating correctly.
        (l) Monitor for tachycardia during medication administration.
        (m) Tap the nebulizer cup periodically to prevent obstruction.
        (n) Turn off the compressor or oxygen when administration is complete.
        o) Encourage the patient to rinse his mouth after treatment is complete, especially if steroids were used.
        (p) Reset oxygen to the prescribed rate if ordered.
  d. Evaluate the patient's response to treatment and report unexpected outcomes to the charge nurse or medical officer.
    (1) Respiratory rate.
    (2) Lung sounds.

  7. Clean and store all equipment.
CAUTION: Do not leave any medication at the patient's bedside without a specific medical officer order to do so.

  8. Record the administration of all medications on the appropriate medical forms.
NOTE: Administration of all scheduled and nonscheduled (PRN) medication must be documented.
    a. Initial the DA Form 4678.
    b. Make a nursing note entry describing the location of the application topical medication and the condition of the skin at the time of application.
    c. Annotate the nursing notes when administering controlled drugs, nonscheduled (PRN) medications, and other medications as required by local policy.
      (1) Name of the medication.
      (2) Time the medication was administered.
      (3) Reason for the medication.

  9. Record the omission of a medication on the appropriate medical forms whenever a scheduled medication is not administered.
NOTE: If a patient refuses a medication, offer it again in 5 minutes. If refused a second time, record the omission on DA Form 4678 and document the reason for the omission in the nursing notes. Notify your supervisor about the patient refusal.
    a. Annotate DA Form 4678 by placing a circle in the initial block.
    b. Annotate the nursing notes.
      (1) Name of the medication.
      (2) Time it should have been administered.

Performance Steps
      (3) Reason it was not administered.
      (4) Follow-up action taken.

Performance Measures	<u>GO</u>	<u>NO-GO</u>

1. Checked DA Form 4678 against the medical officer's order.
   a. Right drug.
   b. Right dose of medication.
   c. Route of administration.
   d. Right patient.
   e. Right time to be administered.
	____	____

2. Selected the medication.
   a. Checked the medication label three times to ensure that the correct medication was being prepared for administration.
   b. Checked the expiration date of the medication.
   c. Handled only one medication at a time.
	____	____

3. Administered oral medications.	____	____
4. Administered topical medications.		
5. Administered medicated eye drops and ointments	____	____
6. Administered respiratory treatments.	____	____
7. Cleaned and stored all equipment.		
8. Recorded the administration of all medications on the appropriate medical forms.		
9. Recorded the omission of a medication on the appropriate medical forms whenever a scheduled medication was not administered.		

Evaluation Guidance: Score each Soldier according to the performance measures in the evaluation guide. Unless otherwise stated in the task summary, the Soldier must pass all performance measures to be scored GO. If the Soldier fails any step, show what was done wrong and how to do it correctly

References
    Required   **Related**
    DA FORM 3949   BASIC NURSING 7
    DA FORM 4678

Subject Area 11: Force Protection Protection

DISINFECT WATER FOR DRINKING
081-831-0037

Conditions: You are a member of a field sanitation team. You have just filled a Lyster bag or Water Buffalo from a source that is not safe for drinking. You will need calcium hypochlorite, mess kit spoon, a canteen cup, and a field chlorination kit. You are not in a CBRNE environment.

Standards: Disinfect water to a chlorine residual of 5 parts per million (ppm) or as ordered by the commander.

Performance Steps

1. Mix the stock disinfecting solution.
   a. Add the prescribed dosage of calcium hypochlorite to 1/2 canteen cup of water.
      (1) 3 ampules per 36 gallons of water.
      (2) 22 ampules or 3 plastic meals ready to eat (MRE) spoonfuls (from a bulk container) in 400 gallons of water.
   b. Stir the stock solution.

2. Add the stock solution to the water container.
   a. Pour the stock solution into the water container.
   b. Mix the solution vigorously with a clean implement.
   c. Cover the container.

3. Flush the faucets.

4. Test the chlorine residual after 10 minutes.
   a. Follow the manufacturer's instructions on the color comparator in the chlorination kit to test the chlorine residual.
   b. Add more calcium hypochlorite as needed to maintain 5 ppm/mg/l.
   c. Retest the chlorine residual after 20 minutes.

5. Retest the water two or three times daily.

Evaluation Preparation:
Setup: Test this task only when there is a need to disinfect water for drinking. Do not simulate this task for training or evaluation.

Brief Soldier: Tell the Soldier to disinfect the water. Tell the Soldier that he is not in a CBRNE environment. After the Soldier completes step 5, ask him how often the water should be retested.

Performance Measures	GO	NO-GO
1. Mixed the stock disinfecting solution.	___	___
2. Added the stock solution to the water container.	___	___
3. Flushed the faucets.	___	___

Performance Measures	GO	NO-GO
4. Tested the chlorine residual after 10 minutes.	——	——
5. Retested the water two or three times daily.	——	——

Evaluation Guidance: Score each Soldier according to the performance measures. Unless otherwise stated in the task summary, the Soldier must pass all performance measures to be scored GO. If the Soldier fails any steps, show what was done wrong and how to do it correctly.

References

Required	Related
None	FM 4-25.12

TREAT A CASUALTY FOR A HEAT INJURY
081-831-0038

Conditions: A casualty is suffering from a heat injury. You must treat the casualty for a heat injury. All other more serious injuries have been assessed and treated. You will need water, a thermometer, intravenous (IV) administration set, ringer's lactate or sodium chloride, stethoscope, sphygmomanometer, and a DD Form 1380 (U.S. Field Medical Card). You are not in a CBRNE environment.

Standards: Provide the correct treatment for the heat injury without causing further injury to the casualty.

Performance Steps
CAUTION: All body fluids should be considered potentially infectious. Always observe body substance isolation (BSI) precautions by wearing gloves and eye protection as a minimal standard of protection.

1. Identify the type of heat injury based upon the following characteristic signs and symptoms:
    a. Heat cramps--muscle cramps of the arms, legs, and/or abdomen.
    b. Heat exhaustion.
        (1) Often--
            (a) Profuse sweating and pale (or gray), moist, cool skin.
            (b) Headache.
            (c) Weakness.
            (d) Dizziness.
            (e) Loss of appetite or nausea.
            (f) Normal or slightly elevated body temperature; or as high as 104 °F (rarely).
        (2) Sometimes--
            (a) Heat cramps.
            (b) Nausea (with or without vomiting).
            (c) Urge to defecate.
            (d) Chills.
            (e) Rapid breathing.
            (f) Tingling sensation of the hands and feet.
            (g) Confusion.
NOTE: A key to distinguishing heat stroke from other heat disorders is the elevation of body temperature and altered mental status. Any casualty warm to the touch with an altered mental status should be suspected of having heat stroke and treated aggressively.
    c. Heat stroke.
        (1) Rapid onset with the core body temperature rising to above 106 °F within 10 to 15 minutes.
        (2) Hot, dry skin.
        (3) Headache.
NOTE: Early in the progression of heat stroke, the skin may be moist or wet.
        (4) Dizziness.
        (5) Headache.
        (6) Nausea.
        (7) Confusion.
        (8) Weakness.
        (9) Loss of consciousness.

Performance Steps
      (10) Possible seizures.
      (11) Pulse and respirations are weak and rapid.

  2. Provide the proper treatment for the heat injury.
     a. Heat cramps.
      (1) Move the casualty to a cool shaded area, if possible.
      (2) Loosen the casualty's clothing unless he is in a chemical environment.
      (3) Rest the cramping muscles.
      (4) Oral rehydration with water or electrolyte solution.
      (5) Evacuate the casualty if the cramps are not relieved after treatment.
NOTE: Do not give salt tablets.
     b. Heat exhaustion.
      (1) Conscious casualty.
        (a) Move the casualty to a shaded area, if possible.
        (b) Loosen and/or remove the casualty's clothing and boots unless he is in a chemical environment.
        (c) Pour water on the casualty and fan him, if possible.
        (d) Oral rehydration unless nauseated. If nauseated, initiate IV hydration.
        (e) Elevate the casualty's legs.
        (f) Provide oxygen to the casualty, if not already done as part of the initial assessment.
      (2) An unconscious casualty or one who is nauseated, unable to retain fluids, or whose symptoms have not improved after 20 minutes.
        (a) Cool the casualty as in step 2b(1).
        (b) Initiate an IV infusion of ringer's lactate or sodium chloride.
        (c) Evacuate the casualty.
        (d) Transport the casualty on his side if they are nauseated.
     c. Heat stroke.
CAUTION: Heat stroke is a medical emergency. If the casualty is not cooled rapidly, the body cells, especially the brain cells, are literally cooked; irreversible damage is done to the central nervous system. The casualty must be evacuated to the nearest medical treatment facility immediately.
      (1) Conscious casualty.
        (a) Cool the casualty with any means available, even before taking the clothes off.
        (b) Remove the casualty's outer garments and/or protective clothing.
        (c) Lay the casualty down and elevate his legs.
        (d) Immerse the casualty in cold water, if available.
        (e) Ice packs in groin, axillae and around the neck, if available.
        (f) Provide supplemental oxygen, if available.
        (g) Initiate an IV infusion of Ringer's lactate or sodium chloride.
        (h) Evacuate the casualty.
      (2) Unconscious casualty or one who is vomiting or unable to retain oral fluids.
        (a) Cool the casualty as in steps 2c(1a-f) but give nothing by mouth.
        (b) Initiate an IV.
        (c) Evacuate the casualty.

  3. Record the treatment given on a FMC. (See task 081-831-0033.)

Evaluation Preparation:
Setup: For training and evaluation, describe to the Soldier the signs and symptoms of heat cramps, heat exhaustion, or heat stroke and ask the Soldier what type of heat injury is indicated.

Brief Soldier: Ask the Soldier what should be done to treat the heat injury.

Performance Measures	GO	NO-GO
1. Identified the type of heat injury.	——	——
2. Provided the proper treatment for the heat injury.	——	——
3. Recorded the treatment given on a FMC.	——	——

Evaluation Guidance: Score each Soldier according to the performance measures. Unless otherwise stated in the task summary, the Soldier must pass all performance measures to be scored GO. If the Soldier fails any steps, show what was done wrong and how to do it correctly.

References
Required	Related
DD FORM 1380	PHTLS

TREAT A CASUALTY FOR A COLD INJURY
081-831-0039

Conditions: You have a casualty who is having symptoms of cold weather injuries. You must treat the casualty for a cold injury. All other more serious injuries have been assessed and treated. You will need dry clothing or similar material, sterile dressings, a thermometer, and a DD Form 1380 (U.S. Field Medical Card). You are not in a CBRNE environment.

Standards: Provide correct treatment based upon the signs and symptoms of the injury.

Performance Steps

1. Recognize the signs and symptoms of cold injuries.
   a. Chilblains are caused by repeated prolonged exposure of bare skin to low temperatures from 60 °F down to 32 °F.
     (1) Acutely red, swollen, hot, tender, and/or itching skin.
     (2) Surface lesions with shedding of dead tissue, or bleeding lesions.
   b. Frostbite is caused by exposure of the skin to cold temperatures that are usually below 32 °F depending on the windchill factor, length of exposure, and adequacy of protection.
NOTE: The onset is signaled by a sudden blanching of the skin of the nose, ears, cheeks, fingers, or toes followed by a momentary tingling sensation. Frostbite is indicated when the face, hands, or feet stop hurting.
     (1) First Degree.
       (a) Epidermal injury; limited to skin that has brief contact with cold air or metal.
       (b) No blister or tissue loss; healing occurs in 7-10 days.
     (2) Second Degree.
       (a) Involves epidermis and superficial dermis.
       (b) Redness of the skin in light-skinned individuals and grayish coloring of the skin in dark-skinned individuals, followed by a flaky sloughing of the skin.
       (c) Blister formation 24 to 36 hours after exposure followed by sheet-like sloughing of the superficial skin.
       (d) No permanent loss of tissue; healing occurs in 3-4 weeks.
     (3) Third Degree.
       (a) Involves the epidermis and dermis layers.
       (b) Frozen skin stiff with restricted mobility.
       (c) After tissue thaws, skin swells along with blood-filled blister.
       (d) Skin loss occurs slowly; healing is delayed.
     (4) Fourth degree.
       (a) Frozen tissue involves full thickness skin with muscle and bone involvement.
       (b) Necrotic tissue develops along with sloughing of tissue and auto amputation of nonviable tissue.
   c. Generalized hypothermia is caused by prolonged exposure to low temperatures, especially with wind and wet conditions, and it may be caused by immersion in cold water.
CAUTION: With generalized hypothermia, the entire body has cooled with the core temperature below 95 °F.
     (1) Moderate hypothermia.
NOTE: This condition should be suspected in any chronically ill person who is found in an environment of less than 50 °F.

Performance Steps

        (a) Conscious, but usually apathetic or lethargic.

        (b) Shivering, with pale, cold skin, slurred speech, poor muscle coordination, faint pulse.

    (2) Severe hypothermia.

        (a) Unconscious or stuporous.

        (b) Ice cold skin.

        (c) Inaudible heart beat or irregular heart rhythm.

        (d) Unobtainable blood pressure.

        (e) Unreactive pupils.

        (f) Very slow respirations.

  d. Immersion syndrome (immersion foot, trench foot and hand) is caused by fairly long (hours to days) exposure of the feet or hands to wet conditions at temperatures from about 50 °F down to 32 °F.

    (1) First phase (anesthetic).

        (a) There is pain sensation, but the affected area feels cold.

        (b) The pulse is weak at the affected area.

    (2) Second phase (reactive hyperemic)--limbs feel hot and/or burning and have shooting pains.

    (3) Third phase (vasospastic).

        (a) Affected area is pale.

        (b) Cyanosis.

        (c) Pulse strength decreases.

    (4) Check for blisters, swelling, redness, heat, hemorrhage, or gangrene.

  e. Snow blindness.

    (1) Scratchy feeling in the eyes as if from sand or dirt.

    (2) Watery eyes.

    (3) Pain, possibly as late as 3 to 5 hours later.

    (4) Reluctant or unable to open eyes.

2. Treat the cold injury.

  a. Chilblains.

    (1) Apply local re-warming within minutes.

    (2) Protect lesions (if present) with dry sterile dressings.

CAUTION: Do not treat with ointments.

  b. Frostbite.

    (1) Apply local re-warming using body heat.

CAUTION: Avoid thawing the affected area if it is possible that the injury may refreeze before reaching the treatment center.

    (2) Loosen or remove constricting clothing and remove jewelry.

    (3) Increase insulation and exercise the entire body as well as the affected body part(s).

CAUTION: Do not massage the skin or rub anything on the frozen parts.

    (4) Move the casualty to a sheltered area, if possible.

    (5) Protect the affected area from further cold or trauma.

    (6) Evacuate the casualty.

NOTE: For frostbite of a lower extremity, evacuate the casualty by litter, if possible.

CAUTION: Do not allow the casualty to use tobacco or alcohol.

  c. Generalized hypothermia.

    (1) Moderate.

Performance Steps

        (a) Remove the casualty from the cold environment.

        (b) Replace wet clothing with dry clothing.

        (c) Cover the casualty with insulating material or blankets.

        (d) If available, apply heating pads to the casualty's armpits, groin, and abdomen.

NOTE: If far from a medical treatment facility and the situation and facilities permit, immerse the casualty in a tub of 104-108°F water. Avoid rewarming with intense sources of heat (campfire).

        (e) If available, slowly give sugar and sweet warm fluids.

CAUTION: Do not give the casualty alcohol or caffeine drinks.

        (f) Wrap the casualty from head to toe.

        (g) Evacuate the casualty lying down.

    (2) Severe.

CAUTION: Handle the casualty very gently.

        (a) Cut away wet clothing and replace it with dry clothing.

        (b) Maintain the airway. (See task 081-831-0018.)

           1) Administer oxygen.

           2) Assist with ventilation if the casualty's respiration rate is less than five per minute.

NOTE: Do not use artificial airways or suctioning devices.

CAUTION: Do not hyperventilate the casualty. Keep the rate of artificial ventilation at approximately 8 to 10 per minute.

        (c) Monitor the casualty's pulse. (See task 081-831-0011.) If none is detected, apply automated external defibrillator (AED), if available. (See task 081-833-3027.) Begin cardiopulmonary resuscitation (CPR). (See tasks 081-831-0046 and 081-831-0048.)

        (d) Evacuate the casualty positioned on his back with the head in a 10 degree head-down tilt.

NOTE: The treatment of moderate hypothermia is aimed at preventing further heat loss and re-warming the casualty as rapidly as possible. Re-warming a casualty with severe hypothermia is critical to saving his life, but the kind of care re-warming requires is nearly impossible to carry out in the field. Evacuate the casualty promptly to a medical treatment facility. Use stabilizing measures en route.

    d. Immersion syndrome.

      (1) Dry the affected part immediately and gradually re-warm it in warm air.

CAUTION: Never massage the skin. After re-warming the affected part, it may become swollen, red, and hot. Blisters usually form due to circulation return.

      (2) Protect the affected part from trauma and secondary infection.

      (3) Elevate the affected part.

      (4) Evacuate the casualty as soon as possible.

    e. Snow blindness. Cover the eyes with a dark cloth and evacuate the casualty to a medical treatment facility.

Evaluation Preparation:

Setup: For training and evaluation have another Soldier act as the casualty. Select one of the types of cold injuries on which to evaluate the Soldier. Coach the simulated casualty on how to answer questions about symptoms. Physical signs and symptoms that the casualty cannot readily simulate, for example blisters, must be described to the Soldier.

Brief Soldier: Tell the Soldier to determine what cold injury the casualty has. After the cold injury has been identified, ask the Soldier to describe the proper treatment.

Performance Measures	GO	NO-GO
1. Identified the type of cold injury.	——	——
2. Provided treatment for the injury.	——	——

NOTE: Although not evaluated, the Soldier would record the treatment given on a FMC and evacuate the casualty as necessary.

Evaluation Guidance: Score each Soldier according to the performance measures. Unless otherwise stated in the task summary, the Soldier must pass all performance measures to be scored GO. If the Soldier fails any steps, show what was done wrong and how to do it correctly.

References

Required	Related
DD FORM 1380	PHTLS

IMPLEMENT SUICIDE PREVENTION MEASURES
081-831-9018

Conditions: You have a Soldier whose conduct and work has changed from outstanding to poor. He has also been heard making threats of self-harm. You must interview/counsel the individual to determine the problem and where to refer the individual for professional help to avoid the possibility of suicide. The interview/counseling session should be done as quickly as possible, if not immediately, and in a location that will provide privacy. You are not in a CBRNE environment.

Standards: Apply the seven-step procedure to identify whether a Soldier needs to be referred for professional help or not.

Performance Steps

1. Gather information about the individual, if not already known and time permits. (See task 158-100-1260.)

2. Apply these seven steps to assist in the prevention of suicide.
   a. Take suicide threats seriously.
   b. Answer cries for help.
   c. Confront the problem.
   d. Be direct.
   e. Listen and be a friend.
   f. Tell the individual you care.
   g. Refer to professional assistance. DO NOT leave the person alone.

Evaluation Preparation:
Setup: Have a fellow Soldier role play a simulated possible suicidal individual. Provide the role player with an example to follow. A private area should be made available (simulated).

Brief Soldier: Tell the Soldier to evaluate the simulated possible suicidal individual for appropriate action to be taken.

Performance Measures	GO	NO-GO
1. Determined if the individual was exhibiting the warning signs of suicide.	——	——
2. Applied the seven step procedure to determine if Soldier needed to be referred.	——	——

Evaluation Guidance: Score each Soldier according to the performance measures. Unless otherwise stated in the task summary, the Soldier must pass all performance measures to be scored GO. If the Soldier fails any steps, show what was done wrong and how to do it correctly.

References
Required	Related
None	AR 600-63
	DA PAM 600-24

References
    Required

    Related
DA PAM 600-70

TREAT A CASUALTY FOR INSECT BITES OR STINGS
081-833-0072

Conditions: You have a casualty who needs to be treated for insect bites. All other more serious injuries have been assessed and treated. You will need antiseptic cleanser, tweezers, sphygmomanometer, stethoscope, thermometer, ice packs, and a DD Form 1380 (U.S. Field Medical Card). You are not in a CBRNE environment.

Standards: Treat the casualty without causing further injury.

Performance Steps

1. Remove the casualty's clothing, shoes, or jewelry to expose the sting or bite area.
NOTE: Remove rings, watches, and other constricting items that are in the area of the bite or sting to prevent circulatory impairment in the event swelling of the extremity occurs.

2. Ask the casualty to identify, if possible, what bit or stung him.

3. Check the casualty for the signs and symptoms of insect bites and stings.
    a. Black widow spider.
NOTE: There are five species of widow spiders. Most are a glossy black with a red or orange hourglass shape on the underside of the abdomen. The brown widow may be either gray or light brown with a red or orange hourglass marking. The red widow has brilliant red spots or a yellow marking on its back.
        (1) An immediate pin-prick sensation from the bite.
        (2) A dull, numbing pain at the bite site.
        (3) Severely painful muscular or abdominal spasms.
            (a) Begin in 10 to 40 minutes.
            (b) Peak in 1 to 3 hours.
            (c) Persist for 12 to 48 hours.
        (4) Rigid, board-like abdomen.
        (5) Tightness in the chest and painful breathing.
        (6) Dizziness.
        (7) Nausea.
        (8) Vomiting.
        (9) Sweating.
        (10) Skin rash.
    b. Brown recluse spider.
NOTE: The brown recluse spider is medium sized, yellowish to medium dark brown, and covered with fine short hairs. It has a distinct groove between its chest and abdominal body parts, and a violin shaped mark on its back.
        (1) Mild to severe pain within hours.
        (2) The area becomes red, swollen, and tender.
        (3) The area develops a pale, mottled, cyanotic center.
        (4) A small blister may form.
        (5) A large scab of dead skin, fat, and debris forms (over several days).
    c. Scorpion.
NOTE: There are two general types of scorpions. The Arizona (black) scorpion is the only deadly type in the United States.
        (1) Harmless species.
            (a) Severe pain and burning sensation at the sting site.

Performance Steps
        (b) Local swelling and discoloration.
        (c) The symptoms last for 24 to 72 hours.
    (2) Deadly species.
        (a) "Pins and needles" sensation at the sting site.
        (b) No swelling at the sting site.
        (c) Excessive salivation.
        (d) Severe muscle contractions.
        (e) Hypertension
        (f) Convulsions.
        (g) Circulatory collapse.
        (h) Cardiac failure.
  d. Bee, wasp, hornet, and yellow jacket.
NOTE: A wasp or yellow jacket (slender body with elongated abdomen) retains its stinger and can sting repeatedly. A honey bee (rounded abdomen) usually leaves its stinger in the casualty.
    (1) Mild reaction.
        (a) Pain at the sting site.
        (b) A wheal, redness, and swelling.
        (c) Itching.
        (d) Anxiety.
    (2) Severe reaction.
        (a) Generalized itching and burning.
        (b) Urticaria (hives).
        (c) Chest tightness and cough.
        (d) Swelling around the lips and tongue.
        (e) Bronchospasm and wheezing.
        (f) Dyspnea.
        (g) Abdominal cramps.
        (h) Anxiety.
        (i) Respiratory failure.
        (j) Anaphylactic shock.
  e. Fire ant.
NOTE: Fire ants inject a very irritating toxin into the skin. They bite repeatedly and in a very short period of time.
    (1) Burning sensation.
    (2) Wheal within minutes.
    (3) Clear, fluid-filled bubble or blister within minutes.
    (4) Cloudy, fluid-filled bubble within 2 to 4 hours.
    (5) Bubble on red base within 8 to 10 hours.
    (6) Ulceration (with scarring after healing).
    (7) Anaphylactic shock.
  f. Tick.
NOTE: Hard ticks can transmit Rocky Mountain Spotted Fever and Lyme's disease, and may even cause anemia if the infestation is severe enough.
    (1) Itching and redness at the site.
    (2) Headache.
    (3) Moderate to high fever, which may last 2 to 3 weeks.
    (4) Pain in the joints or legs.

Performance Steps

       (5) Swollen lymph nodes in the bitten area.

       (6) Paralysis and other central nervous system disorders are possible after several days.

NOTE: Generally, a tick must remain attached to the body for 4 to 6 hours in order to transmit infections. Early detection and proper removal may prevent transmission.

    g. Unknown, nonspecific insects.

       (1) Pain and swelling at the site.

       (2) Breathing difficulty.

       (3) Shock.

  4. Treat the bite or sting.

    a. Black widow spider, brown recluse spider, and scorpion.

       (1) Keep the casualty quiet and calm.

       (2) Remove jewelry.

       (3) Cleanse the bite site using antiseptic.

       (4) Apply ice or an ice pack to the site.

       (5) Treat the casualty for anaphylactic shock, if necessary.

    b. Bee, wasp, hornet, and yellow jacket.

       (1) Scrape the stinger from the site, if still in place.

CAUTION: Do not squeeze the stinger or attempt to pull it out. More venom will be injected into the casualty.

       (2) Cleanse the site with soap and water.

       (3) Apply ice or an ice pack to the site.

       (4) Treat the casualty for anaphylactic shock, if necessary.

    c. Fire ant.

       (1) Cleanse the bite site using antiseptic.

       (2) Apply ice, an ice pack, or a cold compress to the site.

       (3) Treat the casualty for anaphylactic shock, if necessary.

    d. Tick.

       (1) Remove all parts of the tick. Leave nothing embedded in the skin.

          (a) Using tweezers, grasp the tick as close to the skin as possible. Using steady pressure, pull the tick straight out.

          (b) If tweezers are not available, use an absorbent material (gauze, toweling) to protect your skin. Grasp the tick as close to the skin as possible and pull straight out using steady pressure.

       (2) If the tick breaks, thoroughly clean your hands with antiseptic.

       (3) Cleanse the bite site using antiseptic.

NOTE: Ticks harbor pathogenic bacteria in their bodies. Adequate removal and cleansing is essential to prevent infection.

    e. Unknown, nonspecific insect.

       (1) Cleanse the site using antiseptic.

       (2) Apply ice, an ice pack, or a cold compress to the site.

       (3) Monitor the vital signs.

       (4) Treat the casualty for anaphylactic shock, if necessary.

  5. Record the treatment on a FMC.

  6. Evacuate the casualty, if necessary.

NOTE: It is necessary to evacuate any casualty who shows signs of respiratory distress, shock, anaphylaxis, or who does not respond to initial treatment.

Evaluation Preparation:
Setup: For training and evaluation, have another Soldier act as the casualty. Indicate the area of the bite or sting. To test step 3, coach the casualty on how to answer the Soldier's questions regarding signs and symptoms such as pain. Tell the Soldier what signs and symptoms, such as respiratory distress or shock, the casualty is exhibiting.

Brief Soldier: Tell the Soldier to treat the casualty for an insect bite or sting.

Performance Measures	GO	NO-GO
1. Exposed the bite or sting site.	——	——
2. Asked the casualty what bit or stung him.	——	——
3. Checked for the signs and symptoms of the insect bite or sting.	——	——
4. Treated the bite or sting.	——	——
5. Recorded the treatment on a FMC.	——	——
6. Evacuated the casualty, if necessary.	——	——

Evaluation Guidance: Score each Soldier according to the performance measures. Unless otherwise stated in the task summary, the Soldier must pass all performance measures to be scored GO. If the Soldier fails any steps, show what was done wrong and how to do it correctly.

References
Required
DD FORM 1380

Related
EMERG CARE AND TRANS 9

TREAT A CASUALTY FOR SNAKEBITE
081-833-0073

Conditions: You have a casualty with a snakebite. All other more serious injuries have been assessed and treated. You will need antiseptic cleaning solution, iodine, water, soap, and a DD Form 1380 (U.S. Field Medical Card). You are not in a CBRNE environment.

Standards: Determine the type of snakebite and provide treatment without causing further injury to the casualty.

Performance Steps

1. Expose the injury site.

2. Determine the type of snakebite.
CAUTION: If the bite cannot be positively identified as nonpoisonous, the bite should be treated as a poisonous bite. Do not delay treatment during this step.
    a. Nonpoisonous.
        (1) Four to six rows of teeth.
        (2) No fangs.
    b. Poisonous.
        (1) Two rows of teeth.
        (2) Two fangs which create puncture wounds.
NOTES: 1. Coral snakes are neurotoxic and leave only one or more tiny scratch marks in the area of the bite. 2. If the snake can be killed without risk of another bite, it should be brought to the MTF for identification.

3. Check the casualty for signs and symptoms of a poisonous bite.
NOTE: The casualty may exhibit any or all of the symptoms. Symptoms may develop in 1 to 8 hours.
    a. Pain and progressive swelling at the bite site.
    b. Drowsiness.
    c. General skin discoloration.
    d. Blurred vision.
    e. Difficulty hearing.
    f. Fever, chills, or sweating.
    g. Nausea and vomiting.
    h. Shock.
    i. Difficulty breathing.
    j. Paralysis.
    k. Seizures.
    l. Coma.
CAUTION: Antivenom is indicated in casualties who, within 30 to 60 minutes following the bite, show progressive swelling involving the injured area, complain of paresthesia of the mouth, scalp, fingertips, or toes, or who have any signs or symptoms of poisoning.

4. Initiate treatment.
CAUTION: Do not give the casualty any sedatives, alcohol, food, or tobacco.
    a. Nonpoisonous bite.
        (1) Clean and disinfect the wound.
            (a) Use soap and water or antiseptic solution.
            (b) Apply iodine (betadine) if the casualty is not allergic to it.

Performance Steps

      (2) If the casualty has a current tetanus toxoid series, refer the casualty to the medical officer.

      (3) If the casualty does not have a current tetanus toxoid series or does not know, refer the casualty to a medical treatment facility for an immunization.

  b. Poisonous bites.

      (1) Clean the wound with soap and water or antiseptic solution.

      (2) Immobilize the casualty.

         (a) Have the casualty lie down.

         (b) Tell the casualty not to move.

         (c) Monitor the casualty's vital signs.

         (d) Mark the skin with a pen over the area that is swollen, proximal to the swelling, to note whether the swelling is spreading.

         (e) If there are any signs of shock, place the casualty on supplemental oxygen.

         (f) Keep the casualty calm and reassured.

NOTE: Keeping the casualty calm and still will delay venom absorption.

         (g) If the bite is on an extremity, do not elevate the limb but rest it in a position of function at heart level.

         (h) Explain to the casualty what will be done.

NOTE: Remove jewelry.

      (3) Monitor the casualty for development of breathing problems.

      (4) Monitor distal pulse of the bitten extremity.

WARNING: Antivenom, if available, may be administered only by specifically authorized personnel.

  5. Record the procedure on a FMC.

  6. Evacuate the casualty.

Evaluation Preparation:

Setup: For training and evaluation, have another Soldier act as the casualty. Simulate a snakebite on the casualty's arm or leg or describe its appearance to the Soldier. Coach the casualty on how to answer the Soldier's questions regarding signs and symptoms such as pain. To test step 2, ask the Soldier what type of bite the casualty has. To test step 3, have the Soldier tell you the symptoms of a poisonous snakebite. You may vary the testing by telling the Soldier that the casualty cannot be evacuated for more than 1 hour, or that the casualty is having difficulty breathing.

Brief Soldier: Tell the Soldier to treat a casualty for a snakebite.

Performance Measures	GO	NO-GO
1. Exposed the injury site.	——	——
2. Determined the type of snakebite.	——	——
3. Checked the casualty for signs and symptoms of a poisonous bite.	——	——
4. Initiated treatment.	——	——
5. Recorded the procedure on a FMC.	——	——
6. Evacuated the casualty.	——	——

Evaluation Guidance: Score each Soldier according to the performance measures. Unless otherwise stated in the task summary, the Soldier must pass all performance measures to be scored GO. If the Soldier fails any steps, show what was done wrong and how to do it correctly.

References

Required	Related
DD FORM 1380	EMERG CARE AND TRANS 9

APPLY RESTRAINING DEVICES TO PATIENTS
081-833-0076

Conditions: You have a patient that needs to be restrained. You will need a bed, wrist and ankle restraining devices, abdominal (ABD) pads, padding materials, litters, flexible gauze (Kerlix/Kling), rifle slings, web belts, elastic bandages, bandoleers, cravats, and sheets. You are not in a CBRNE environment.

Standards: Apply restraining devices to a patient without causing injury to the patient or yourself.

Performance Steps
NOTE: In a field environment, the need for restraints may be your own decision, especially in the absence of senior medical personnel.

1. Apply wrist and ankle restraints.
NOTE: If you apply ankle restraints, also apply wrist restraints.

WARNINGS: 1. Do not attempt to apply restraining devices by yourself. Get adequate help. 2. A patient who is depressed or has an altered level of consciousness should be positioned on the stomach with the head turned to the side. 3. Position restraints to avoid causing further injury to a wound or interfering with IV lines, catheters, and tubes.
    a. Adjustable limb holders (cuff and strap).
        (1) Pad the limb with ABD pads or similar material.
        (2) Position the restraint cuff over the padded limb.
        (3) Thread the strap through the loop on the cuff. Pull the straps snugly enough to restrict free movement of the limb.
NOTE: If two fingers can be comfortably inserted under the cuff, the restraint is snug enough. The patient, however, must not be able to wiggle his hand out of the cuff.
        (4) Wrap the strap around the bed frame.
        (5) Lock the buckle and position it facing the outside of the bed frame for quick access.
        (6) Repeat steps 1a(1) through 1a(6) for each limb.
NOTE: The keys to the locked restraints must be readily available.
    b. Improvised restraints.
        (1) Pad the limb with any soft cloth such as towels, gauze, cravats, clean handkerchiefs, or clothing.
        (2) Secure the restraining material (gauze or roller bandage) to the limb with a clove hitch.
        (3) Pull the knot to fit the limb snugly.
        (4) Using a bow knot, tie both free ends to the bed frame in a location inaccessible to the patient.
        (5) Repeat steps 1b(2) through 1b(5) for each limb.

2. Apply mitt restraints.
    a. Place the patient's hand in a naturally flexed position.
    b. Place a soft rolled dressing or similar material in the patient's hand and close the hand.
    c. Wrap the entire hand snugly with a flexible gauze bandage (Kerlix, Kling).
    d. Secure the bandage with tape, not clips.

Performance Steps
CAUTION: Remove and replace mitts at least every 8 hours. Perform range-of-motion exercises.

   3. Apply sheet restraints.
NOTE: This procedure requires the assistance of another person.
      a. Litter or stretcher.
         (1) Unfold a sheet. Hold it at opposite corners and fold it lengthwise.
         (2) Twirl the sheet into a tight roll.
         (3) Place the patient on his stomach on a litter. Turn the head to the side.
WARNING: Check the patient frequently because he may suffocate while in the prone position.
         (4) Place the middle of the rolled sheet diagonally across the patient's upper back and one shoulder.
         (5) Bring both ends of the sheet under the litter, cross the ends, and bring the ends up over the other shoulder and upper back. Tie snugly in the middle of the upper back.
         (6) Secure one wrist to the litter, parallel to the thigh, using a wrist restraint.
         (7) Secure the other wrist above the head by attaching it to the nearest litter handle using a wrist restraint.
CAUTION: Use litter or stretcher restraints only as a TEMPORARY restraint for a patient who is combative or uncontrollable.
      b. Bed.
         (1) Fold a sheet in half lengthwise.
         (2) Tuck approximately 2 feet of one end of the sheet under one side of the mattress at the patient's chest level.
         (3) Bring the other end of the sheet over the patient's chest, keeping the sheet over the arms. Tuck the free end of the sheet snugly under the other side of the mattress.
         (4) If further restriction is necessary, apply sheets in the same manner at the level of the patient's abdomen, legs, knees, and ankles.
NOTE: Use this method of restraint only for limiting movement. It is not a secure method of restraining a violent patient.

   4. Apply field expedient restraints.
NOTE: Field expedient restraints should not be used for long periods of time and should be replaced with regular restraining devices as soon as possible.
      a. Mixed equipment. Restraints may be improvised from such items as rifle slings, web belts, bandoleers, or cravats.
         (1) Lay the patient on the ground.
         (2) Restrain the patient's arms and legs tight enough to restrict movement but not so tight as to restrict circulation.
      b. Evacuate the patient as soon as possible.

   5. Check the patient at least once every half hour for signs of distress and security of restraints.
WARNINGS: The use of restraints has the following hazards: 1. Tissue damage under the restraints. 2. Development of pressure areas. 3. Nerve damage. 4. Injury or death in case of fire or other emergencies. 5. Inability to effectively resuscitate a patient. 6. Possibility of shoulder dislocations in combative patients or those with seizure activity.

   6. Change the patient's position at least once every 2 hours, day and night. Exercise the limbs through normal range-of-motion activities.

Performance Steps

7. Evacuate the patient, if necessary.

Performance Measures	GO	NO-GO
1. Applied wrist and ankle restraints, as applicable.	——	——
2. Applied mitt restraints, as applicable.	——	——
3. Applied sheet restraints, as applicable.	——	——
4. Applied field expedient restraints, as applicable.	——	——
5. Checked the patient.	——	——
6. Changed the patient's position.	——	——
7. Evacuated the patient, if necessary.	——	——
8. Did not cause further injury to the patient.	——	——

Evaluation Guidance: Score each Soldier according to the performance measures. Unless otherwise stated in the task summary, the Soldier must pass all performance measures to be scored GO. If the Soldier fails any steps, show what was done wrong and how to do it correctly.

References

Required
None

Related
BTLS FOR PARAMEDICS

Skill Level 2

Subject Area 12: Advanced Procedures (SL 2)

PLACE A PATIENT ON A CARDIAC MONITOR
081-833-0167

Conditions: You have a conscious patient requiring continuous cardiac monitoring. You will need cardiac monitor, leads, electrodes, and SF 600 (Medical Record - Chronological Record of Medical Care). You are not in a CBRNE environment.

Standards: Correctly connect the patient to the monitor.

Performance Steps

1. Identify the patient.
    a. Have the patient state his name.
    b. Check the patient's arm band, if one is available.
    c. Explain the procedure to the patient.

2. Prepare the equipment.
    a. Turn the power switch to the on position. Make sure the cord is connected to a power source.
    b. Attach the lead cable to the monitor and observe for flat line on the monitor.
    c. Check for the presence of recording paper and replace it as needed.
    d. Turn the lead selection knob to the Lead II position, if so equipped.

3. Attach the patient to the monitor.
    a. Place an electrode on the right anterior superior chest just inferior to the clavicle (right arm lead).
    b. Place a second electrode on the left anterior superior chest just inferior to the clavicle (left arm lead).
    c. Place a third electrode on the left lateral aspect of the abdomen (this electrode may also be placed on the left upper leg).
    d. Attach the cable from the monitor to the corresponding electrodes.
    e. Observe the monitor and note the type of pattern.
    f. Set the monitoring and alarm parameters IAW with local SOP.

4. Document the procedure on a SF 600.

Performance Measures	GO	NO-GO
1. Identified the patient.	____	____
2. Prepared the equipment.	____	____
3. Attached the patient to the monitor.	____	____
4. Documented the procedure on a SF 600.	____	____

Evaluation Guidance: Score each Soldier according to the performance measures in the evaluation guide. Unless otherwise stated in the task summary, the Soldier must pass all performance measures to be scored GO. If the Soldier fails any step, show what was done wrong and how to do it correctly.

References
    Required   **Related**
    SF 600   BASIC NURSING 7

MAINTAIN IMMUNIZATION PROGRAM
081-833-0189

Conditions: Your job is to maintain the unit's immunization program. You will need immunization cards and/or shot records, SF 601 (Immunization Record), patient records, vaccines, alcohol swabs, needles, and syringes. You are not in a CBRNE environment.

Standards: Establish and maintain an immunization program so that all Soldiers receive the proper immunizations at the proper time in the proper dosage and all immunizations are properly recorded.

Performance Steps

1. Gather all required medical supplies, immunizations, patient records, and blank forms.

2. Ensure all necessary equipment is on hand to manage an anaphylactic reaction when administering injections.

3. Establish an area sufficient for administering the immunizations and recording the paperwork.

4. Issue each Soldier an immunization card or shot record.
NOTE: If Soldiers already have an immunization record or shot record, check which immunizations they do not require.

5. Administer appropriate immunizations according to schedule.
NOTE: The immunization schedules listed below are for reference only. Immunization schedules change frequently. Review a current vaccination guideline prior to the administration of vaccines.
    a. Adenovirus.
        (1) Based on likelihood of transmission, adenovirus types 4 and 7 vaccines are administered orally on a one-time basis to recruits.
    b. Anthrax.
        (1) Administered to people 18 to 65 years of age with potential for exposure to large amounts of B. anthracis bacteria on the job, such as lab workers.
        (2) Administered to deploying military personnel who may be at risk of anthrax exposure.
        (3) If a dose is not given at the scheduled time, the series does not have to be started over. Resume the series as soon as practical.
        (4) Dosage schedule.
            (a) Full immunity requires six doses. The first three doses are given at two-week intervals. Three additional doses are given, each one 6 months after the previous dose. Annual boosters are required.
            (b) Doses of the vaccine should not be administered on a compressed or accelerated schedule. (For example, shorter intervals between doses or more doses than required).
        (5) Adverse events.
            (a) Localized injection site reactions-redness, pain, itching, lump at injection site.
            (b) Muscle or joint aches, headaches, fatigue, chills, nausea.
            (c) Serious adverse reactions are rare.
        (6) Contraindications.
            (a) Previous allergic reaction.

Performance Steps

        (b) Anyone who has a history of cutaneous (skin) anthrax.

        (c) As a precaution, pregnant women should not be routinely vaccinated with anthrax although this is not an absolute contraindication. Refer to medical officer for determination.

  c. Cholera.

    (1) Cholera vaccine is not administered routinely.

    (2) Only administered to military personnel, upon travel or deployment to countries requiring cholera vaccination as a condition for entry.

    (3) Adverse events.

        (a) Pain at injection site, mild systemic complaints, and temperature > 100.4.

        (b) Local reaction may be accompanied by fever, malaise, and headache.

        (c) Serious reactions, including neurologic reactions, after cholera vaccinations are extremely rare.

  d. Hepatitis A.

    (1) Hepatitis A vaccine is required for all deployments.

    (2) Dosage schedule.

    (3) 1st dose should be given at least 4 weeks prior to deployment.

    (4) 2nd dose given 6-12 months after initial dose.

    (5) Adverse events - rare.

  e. Hepatitis B.

    (1) Given to health care workers and Soldiers.

    (2) Dosage schedule.

        (a) Series of 3 injections.

        (b) 2nd injection given 30 days after 1st shot.

        (c) 3rd injection given 6 months after 1st shot.

    (3) Adverse events - pain at injection site, mild systemic complaints, and temperature > 100.4.

  f. Influenza.

    (1) All active duty and reserve military personnel entering active duty for periods in excess of 30 days are immunized against influenza soon after entry on active duty.

    (2) The vaccine is provided to all military personnel and others considered being at high risk for influenza infection.

    (3) The vaccine is provided annually beginning in October.

    (4) Adverse events - local reactions, fever/malaise, severe allergic reactions (egg allergies), and neurological reactions (rare).

  g. Japanese B Encephalitis (JE).

    (1) Required for the military during deployments and travel to endemic areas in Eastern Asia and the Western Pacific Islands.

    (2) Adverse events - fever, headache, myalgia, malaise (common), general urticaria, angioedema, respiratory distress, and anaphylaxis (rare).

  h. Measles, Mumps, and Rubella (MMR).

    (1) Measles and rubella are administered to all recruits regardless of prior history.

    (2) Mumps or MMR vaccine is administered to persons considered to be mumps susceptible. Written documentation of medical officer diagnosed mumps or a documented history of prior receipt of live virus mumps vaccine or MMR vaccine is adequate evidence of immunity.

    (3) All military and civilian personnel engaged in the delivery of health care and having patient contact are appropriately immunized against measles, mumps, and rubella.

Performance Steps

        (4) Adverse events - low grade fever, parotitis (inflammation of the parotid gland located near the ear), rash, pruritus (mild), deafness (rare).

    i. Meningococcus.

        (1) Meningococcal vaccine is administered on a one-time basis to recruits.

        (2) Given as soon as practical after in-processing.

        (3) Adverse events - rare.

    j. Plague.

        (1) There are no requirements for routine immunization. Plague vaccine is administered to Soldiers who are likely to be assigned to areas where the risk of endemic transmission or other exposure is high.

        (2) The addition of antibiotic prophylaxis is recommended for such situations.

        (3) The primary series consists of three doses of vaccine. The first is followed by the second dose 4 weeks later. The third dose is administered six months after the first dose.

        (4) Adverse events - general malaise, headache, fever, mild lymphadenopathy, and/or erythema, and induration at the injection site.

    k. Polio.

        (1) A single dose of Oral Polio Vaccine (OPV) is administered to all enlisted recruits. Officer candidates, ROTC cadets, and other Reserve Components on initial active duty for training receive a single dose of OPV unless prior booster immunization as an adult is documented.

        (2) Booster doses of OPV are not routinely administered.

        (3) Adverse events - paralytic poliomyelitis, more likely in immunodeficient persons, no procedure available for identifying persons at risk of paralytic disease, rarely vaccine induced.

        (4) The live OPV is currently being phased out and being replaced with an injectable Inactivated Polio Vaccine (IPV).

    l. Rabies.

        (1) Pre exposure Series. Rabies vaccine is administered to personnel with a high risk of exposure (animal handlers; certain laboratory, field, and security personnel; and personnel frequently exposed to potentially rabid animals in a non occupational or recreational setting).

        (2) Post exposure Series. Rabies vaccine and rabies immune globulin (RIG) administration will be coordinated with appropriate medical authorities following current ACIP recommendations.

        (3) Adverse events - Anaphylaxis (rare) smallpox, person can become infected with the vaccinia virus and may experience soreness, fatigue, fever, and body aches.

        (4) This vaccine is currently administered only under the authority of the Immunization Program for Biological Warfare Defense.

    m. Tetanus-Diphtheria.

        (1) A primary series of tetanus-diphtheria (Td) toxoid is initiated for all recruits lacking a reliable history of prior immunization. Individuals with previous history of Td immunization receive a booster dose upon entry to active duty and every 5-10 years thereafter.

        (2) Adverse events - local reactions (erythema, induration), nodule at injection site, fever and systemic symptoms uncommon.

    n. Typhoid.

        (1) Typhoid vaccine is administered to alert forces and personnel deploying to endemic areas. Either oral or intramuscular (IM) vaccine is used.

        (2) IM - 1 shot, booster every 2 years.

Performance Steps

NOTE: Was previously a two shot series, second injection given 30 days after the first. Booster every three years.

        (3) Oral - 4 Tablets, given day 0,3,5,7. Booster every 5 years.

        (4) Adverse events - local reactions may be accompanied by fever, malaise, and headache (common), nausea, abdominal cramps, vomiting, skin rash and urticaria.

   o. Yellow Fever.

        (1) Yellow fever immunization is required for all alert forces, active duty personnel or reserve components traveling to yellow fever endemic areas.

        (2) Administered subcutaneously. Booster in 10 years.

        (3) Adverse events - mild headache, myalgia, low grade fever, other minor symptoms.

        (4) Immediate hypersensitivity reactions: rash, urticaria, and asthma. Uncommon and occur periodically among people with a history of egg allergies.

6. Record all vaccinations in the patient's shot record.

7. Ensure immunized personnel wait for a minimum of 15 minutes under observation before departing the area.

Evaluation Preparation: This task is best evaluated by verbalization of the steps. Give the Soldier a scenario in which he must establish an immunization program.

Performance Measures	GO	NO-GO
1. Gathered all required medical supplies, immunizations, patient records, and blank forms.	——	——
2. Ensured that all necessary equipment was on hand to manage an anaphylactic reaction when administering injections.	——	——
3. Established an area sufficient for administering the medications and recording the paperwork.	——	——
4. Issued each Soldier an immunization card or shot record or reviewed existing records.	——	——
5. Administered appropriate immunizations according to schedule.	——	——
6. Recorded all vaccinations in the patient's shot record.	——	——
7. Ensured that immunized personnel were under observation for 20 minutes before they departed the area.	——	——

Evaluation Guidance: Score the Soldier go if all steps are passed. Score the Soldier no-go if any steps are failed. If the Soldier fails any step, show what was done wrong and how to do it correctly.

References
    Required   **Related**
    SF 601   AR 40-562

REMOVE A TOENAIL
081-833-0196

Conditions: You have a patient in need of a toenail removal. You will need appropriate antimicrobial solution, 3-5ml syringe with two long (1-1.5") needles (25 and 21 gauge), 1% lidocaine local anesthetic without epinephrine, periosteal elevator or sterile, straight bladed scissors (Metzenbaum), two sterile hemostats (straight), sterile cotton-tipped applicators, sterile rubber band, 4x4 sterile gauze sponges, dressing materials, tape, topical antibiotic ointments, and SF 600 (Medical Record - Chronological Record of Medical Care). You are not in a CBRNE environment.

Standards: Perform partial or complete toenail removal IAW established protocols.

Performance Steps

1. Solicit a patient history and have patient sign written consent for procedure.

2. Gather equipment.

3. Prepare the patient.
   a. Place the patient supine with knees flexed and feet flat.
   b. Cleanse the digit with an antimicrobial scrub.
   c. Apply sterile drapes to completely surround the wound and to cover all unprepriated areas adjacent to the site
   d. Administer local anesthetic by ring block technique
      (1) Digital cutaneous nerves run along the medial and lateral aspects of each digit and can be blocked at any level above the distal phalanx.
      (2) Use the 25 gauge needle to raise a skin wheal by administering approximately 0.25ml of the anesthetic directly over the lateral and medial cutaneous nerve.
      (3) Change to 21 gauge needle and advance the needle perpendicular to the nerve until bone is reached; inject approximately 1ml of the anesthetic.
      (4) Slide the needle up and down on the dorsal and volar aspects of the digit; injecting approximately 0.5ml of the anesthetic in each side
      (5) It takes 5 to 10 minutes for complete anesthesia to develop.

4. Partial toenail removal.
   a. Perform a patient care hand-wash and put on gloves.
   b. Once anesthesia has been achieved, use a straight hemostat to firmly secure a wide rubber band around the base of the digit to serve as a tourniquet.
   c. Insert a single blade of the other straight hemostat between the nail bed and the nail to loosen and lift the nail. Split the nail with scissors or nail splitter in a longitudinal direction (distal to proximal) to include the base of the nail that rests beneath the cuticle.
   d. With the second straight hemostat, grasp the portion of the loosened nail and remove it using a steady pulling motion with a simultaneous upward twist of the hand toward the affected side completely removing the section of the nail.
   e. Debride the nail groove.
   f. Remove the tourniquet and assess for hemostasis
   g. Apply a topical antibiotic ointment to the nail bed and cover the digit with a sterile gauze sponge dressing or tubular gauze and tape in place
   h. Remove gloves.

Performance Steps

5. Complete toenail removal.
   a. Perform a patient care hand-wash and put on gloves.
   b. Once anesthesia has been achieved, use a straight hemostat to firmly secure a wide rubber band around the base of the digit to serve as a tourniquet.
   c. Insert a single blade of the other straight hemostat (or the periosteal elevator) between the nail bed and the toenail to loosen and lift the nail; advance the instrument with a continued upward pressure against the nail and away from the nail bed to minimize injury and bleeding

CAUTION: It is important to completely free the proximal nail at its base (under the edge of the cuticle) to allow removal and to expose the germinal tissue of the nail bed.
   d. With the second straight hemostat, grasp the loosened nail and remove it using a steady pulling motion with a simultaneous upward twist of the hand toward the affected side completely removing the nail.
   e. Debride the nail grooves as needed.
   f. Remove the tourniquet and assess for hemostasis.
   g. Apply a topical antibiotic ointment to the nail bed and cover the digit with a sterile gauze sponge dressing or tubular gauze and tape in place
   h. Remove gloves.

6. Provide patient follow-up instructions.
   a. Rest the foot (toe) during the initial 24 hours after the procedure.
   b. Elevate the extremity when possible.
   c. Return in 24 hours for dressing change, at which time you should re-apply the topical antibiotic ointment, apply a less bulky dressing and encourage ambulation and a return to normal activity within the next 48 hours.
   d. After 48-hours, soak in warm water for 20 minutes twice daily.
   e. Tell patient to expect some clear to yellow fluid drainage (exudate) from the toe that may continue for three weeks.
   f. Emphasize proper toenail hygiene and schedule a follow-up visit for 30 days to assess healing.

7. Document the procedure.

Evaluation Preparation: This task is best evaluated by verbalization and demonstration of the steps. Give the Soldier a scenario in which he must perform partial or complete toenail removal.

Performance Measures	GO	NO-GO
1. Solicited a patient history and obtained consent.	——	——
2. Gathered equipment.	——	——
3. Prepared the patient.	——	——
4. Removed partial toenail.	——	——
5. Removed complete toenail.	——	——
6. Provided patient follow-up instructions.	——	——
7. Documented the procedure.	——	——

Evaluation Guidance: Score each Soldier according to the performance measures in the evaluation guide. Unless otherwise stated in the task summary, the Soldier must pass all performance measures to be scored GO. If the Soldier fails any step, show what was done wrong and how to do it correctly

References

Required	**Related**
SF 600	ISBN 072166055X
	ISBN 0-07-065351-9

REMOVE A URINARY CATHETER
081-833-0197

Conditions: You have a patient that needs a urinary catheter removed. The patient has been draped. You will need a catheter, tape, 10 ml syringe, paper towels, gloves, and SF 600 (Medical Record - Chronological Record of Medical Care). You are not in a CBRNE environment.

Standards: Remove a urinary catheter without violating aseptic technique or causing further injury to the patient.

Performance Steps

1. Perform a patient care hand-wash and put on gloves.

2. Clamp the catheter.

3. Remove tape that attaches the catheter to the patient's leg.

4. Insert an empty 10 ml syringe into the balloon port of the catheter.
NOTE: Do not cut the catheter. The balloon may not deflate completely when cut.

5. Withdraw fluid from the balloon (usually 5 to 10 ml in balloon).

6. Pull gently on the catheter to ensure that the balloon is deflated before attempting to remove it.
NOTE: Damage to the urethra can occur if the balloon is not completely deflated.

7. Hold a paper towel under the catheter with your nondominant hand.

8. If resistance is not met, pinch the catheter with your fingers to clamp it and slowly withdraw the catheter.

9. Disconnect the catheter bag from the bed frame.

10. Dispose of the catheter and used equipment IAW local SOP for infectious waste and clean the area.

11. Remove gloves.

12. Record the procedure.

Performance Measures	GO	NO-GO
1. Performed a patient care hand-wash and put on gloves.	——	——
2. Clamped the catheter.	——	——
3. Removed tape that attaches the catheter to the patient's leg.	——	——
4. Inserted an empty 10 ml syringe into the balloon port of the catheter.	——	——
5. Withdrew fluid from the balloon.	——	——
6. Pulled gently on the catheter to ensure that the balloon is deflated before attempting to remove it.	——	——

Performance Measures	GO	NO-GO
7. Held a paper towel under the catheter with the nondominant hand.	——	——
8. Determined no resistance, pinched the catheter with the fingers to clamp it, and slowly withdrew the catheter.	——	——
9. Disconnected the catheter bag from the bed frame.	——	——
10. Disposed of the catheter and used equipment IAW local SOP for infectious waste and cleaned the area.	——	——
11. Removed gloves.	——	——
12. Recorded the procedure.	——	——
13. Did not cause further injury to the patient.	——	——

Evaluation Guidance: Score each Soldier according to the performance measures. Unless otherwise stated in the task summary, the Soldier must pass all performance measures to be scored GO. If the Soldier fails any steps, show what was done wrong and how to do it correctly.

References

Required	Related
SF 600	BASIC NURSING 7

TREAT PARONYCHIA
081-833-0207

Conditions: You have a patient with a fingernail injury. You will need a scalpel handle, scalpel blades of various sizes, antibiotic ointment, sterile dressing, gloves, and SF 600 (Medical Record - Chronological Record of Medical Care). You are not in a CBRNE environment.

Standards: Provide drainage for a patient's paronychia without causing further injury to the patient.

Performance Steps

1. Recognize signs and symptoms of a paronychia.
   a. Swelling and tenderness of the soft tissue along the base or side of a fingernail.
   b. Pain, often around a hangnail.
   c. Infection begins as a cellulitis and may form an abscess.

2. Obtain a history of the patient's complaint.

3. Gather the materials for the procedure.

4. Perform hand-washing procedures.

5. Put on gloves.

6. Explain the procedure to the patient.

7. Drain the paronychia.
   a. Often, an incision is unnecessary. Instead, insert the tip of a #11 blade approximately 5 mm under the surface of the nail, uplifting the cuticle, thus providing an escape for the collected suppurative material.
NOTE: This procedure alone provides for adequate drainage in most paronychias.
   b. Allow the pus to drain.
   c. Irrigate the cavity with normal saline.
   d. Apply antibiotic ointment and absorbent dry, sterile dressing.
   e. Instruct the patient to perform frequent soaks in warm tap water at home and continue to keep the wound covered as long as drainage persists.
NOTE: Culture and antibiotic therapy are usually unnecessary unless the patient shows signs of systemic infection.

8. Remove gloves.

9. Instruct the patient to follow up with you or another care provider in 24-48 hours. Instruct the patient to follow up earlier if there are signs of systemic infection.

10. Record all treatment given.

11. Seek the advice and assistance of a higher medical authority whenever possible.

Evaluation Preparation: This task is best evaluated by verbalization of the steps. Give the Soldier a simulated patient and a scenario in which he must drain a paronychia.

Performance Measures   <u>GO</u>   <u>NO-</u>
                                                                                            <u>GO</u>

1. Recognized signs and symptoms of a paronychia.   ____   ____

2. Obtained a history of the patient's complaint.   ____   ____

3. Gathered the materials for the procedure.   ____   ____

4. Performed hand washing procedure.   ____   ____

5. Put on gloves.   ____   ____

6. Explained the procedure to the patient.   ____   ____

7. Drained the paronychia.   ____   ____

8. Removed gloves.   ____   ____

9. Instructed the patient to follow up with you or another care provider in 24- ____   ____
   48 hours. Instructed the patient to follow up earlier if there are signs of
   systemic infection.

10. Recorded all treatment given.   ____   ____

11. Sought the advice and assistance of a higher medical authority whenever   ____   ____
    possible.

Evaluation Guidance: Score each Soldier according to the performance measures. Unless otherwise stated in the task summary, the Soldier must pass all performance measures to be scored GO. If the Soldier fails any steps, show what was done wrong and how to do it correctly.

References
    Required   **Related**
    SF 600   HABIF, T. B.

INSERT A URINARY CATHETER
081-833-3017

Conditions: You have a patient that needs a urinary catheter. The patient has been draped and all equipment has been prepared. You will need cotton balls or swabs, catheter, lubricant, catheterization kit, collection basin, sterile gloves, a sterile specimen container, and SF 600 (Medical Record - Chronological Record of Medical Care). You are not in a CBRNE environment.

Standards: Insert a urinary catheter without violating aseptic technique or causing further injury to the patient.

Performance Steps

   1. Put on gloves.

   2. Clean the urinary meatus with the prepared cotton balls or swabs.
NOTE: Cotton balls should be held with forceps.
      a. Females.
        (1) Gently spread the labia open with the nondominant hand.
NOTE: This hand is now considered contaminated.
          (a) Place the thumb and forefinger between the labia minora.
          (b) Separate the labia and pull up slightly.
        (2) With the dominant hand, clean the labia with cotton balls or swabs, moving from the clitoris toward the anus.
        (3) Use a cotton ball or swab to clean down the center, directly over the urinary meatus.
        (4) Keep the labia spread throughout the remainder of the procedure.
      b. Males.
        (1) Support the penis with the nondominant hand.
NOTE: This hand is now considered contaminated.
        (2) With the dominant hand, clean the penis with a cotton ball or swab, moving in a circular motion from the urinary meatus toward the base of the penis.
        (3) Repeat the procedure, using a second and third cotton ball or swab.

   3. Lubricate the catheter.
      a. Pick up the catheter with the dominant hand about 4 inches from the tip.
      b. Keep the distal end of the catheter coiled in the palm of the hand.
      c. Apply lubricant to the catheter tip.

   4. Instruct the patient to relax and breathe through the mouth.

   5. Insert the catheter.
      a. Female.
        (1) Gently insert the catheter into the urethra about 2 to 3 inches until resistance is met.
        (2) Continue to advance the catheter until urine begins to flow (about 2 to 3 inches further).
        (3) Release the labia and hold the catheter securely with the nondominant hand.
        (4) Place the distal end of the catheter in the collection basin.

Performance Steps
NOTE: If the vagina is inadvertently catheterized, do not remove the catheter. Assemble new equipment and repeat the procedure. Leaving the first catheter in place temporarily will prevent catheterizing the vagina a second time.

    b. Male.
        (1) Draw the penis upward and forward to a 60 to 90 degree angle to the legs.
        (2) Gently insert the catheter into the urethra, advancing it about 7 to 8 inches or until resistance is felt.
        (3) Continue to advance the catheter until urine begins to flow (about 2 to 3 inches further).
        (4) Lower the penis and hold the catheter securely with the nondominant hand, resting the hand on the patient's pubis for support.
        (5) Place the distal end of the catheter in the collection basin.
NOTE: With some commercially prepared catheterization kits, the catheter is pre-connected to the drainage tubing of the collecting bag.

   6. Obtain a urine specimen, if ordered.
      a. Place the sterile specimen container from the kit into the collection basin.
      b. Pinch the catheter with the nondominant hand to stop urine flow.
      c. With the dominant hand, pick up the distal end of the catheter and hold it over the specimen container.
      d. Release the pinch and allow sufficient urine to drain into the specimen container (about 30 cc).
      e. Re-pinch the catheter, place the distal end into the collection basin, and release the pinch, allowing the urine to flow.
      f. Place the lid on the specimen container and set it aside.
NOTE: If using a commercial kit with the catheter and drainage set pre-connected, do not disconnect the catheter to obtain a specimen. Obtain the specimen from the drainage bag at the end of the procedure. The first specimen taken from a new sterile drainage set is considered sterile.

   7. Inflate the balloon if an indwelling catheter has been inserted.
      a. Inflate the balloon with the water in the pre-filled syringe.
NOTE: If the balloon is difficult to inflate, advance the catheter another 1/2 to 1 inch to ensure that the catheter tip is fully within the bladder.
      b. Tug gently on the catheter to ensure that the balloon is fully inflated and seated in the bladder.
      c. Remove the syringe from the catheter using a twisting motion.

   8. Attach the distal end of the catheter to the drainage tubing of the collection set, if not pre-connected by the manufacturer.

   9. Remove the drapes.

 10. Tape the catheter in place.
      a. Female--to the inner thigh.
      b. Male--to the abdomen or inner thigh.
NOTE: The penis may be positioned up or down (facing the patient's head or feet), depending upon the patient's diagnosis, the medical officer's order, and/or the patient's comfort preference.

 11. Secure the drainage bag to the side of the bed on the bottom of the bed frame.
CAUTION: Do not secure the drainage bag to the bed side rails or loop the drainage tubing over or through the side rails.

Performance Steps

12. Remove gloves.

13. Reposition the patient.

14. Dispose of the used equipment and clean the area.
NOTE: Destroy the syringe and dispose of it IAW local SOP for infectious waste.

15. Record the procedure.

Performance Measures	GO	NO-GO
1. Put on gloves.	——	——
2. Cleaned the urinary meatus with the prepared cotton balls or swabs.	——	——
3. Lubricated the catheter.	——	——
4. Instructed the patient to relax and breathe through the mouth.	——	——
5. Inserted the catheter.	——	——
6. Obtained a urine specimen, if ordered.	——	——
7. Inflated the balloon if an indwelling catheter has been inserted.	——	——
8. Attached the distal end of the catheter to the drainage tubing of the collection set, if not pre-connected by the manufacturer.	——	——
9. Removed the drape.	——	——
10. Taped the catheter in place.	——	——
11. Secured the drainage bag to the side of the bed on the bottom of the bed frame.	——	——
12. Remove gloves.	——	——
13. Repositioned the patient.	——	——
14. Disposed of the used equipment and cleaned the area.	——	——
15. Recorded the procedure.	——	——
16. Did not violate aseptic technique, or cause further injury to the patient	——	——

Evaluation Guidance: Score each Soldier according to the performance measures in the evaluation guide. Unless otherwise stated in the task summary, the Soldier must pass all performance measures to be scored GO. If the Soldier fails any step, show what was done wrong and how to do it correctly.

References

Required	**Related**
SF 600	BASIC NURSING 7

MAINTAIN AN INDWELLING URINARY CATHETER
081-835-3010

Conditions: You have a patient with a urinary catheter in place. You will need a basin of water, soap, a hand towel, a wash cloth, antibacterial ointment, protective pads, 4 X 4 gauze, clamps, a drainage set, gloves, sterile needle and syringe, sterile specimen container, thermometer, DD Form 792 (Nursing Service – Twenty-Four Hour Patient Intake and Output Worksheet), and the patient's clinical record. You are not in a CBRNE environment.

Standards: Perform catheter care and maintain the indwelling catheter without contaminating the equipment or causing further injury to the patient.

Performance Steps

1. Provide privacy for the patient.

2. Place the patient in a comfortable position for catheter care.
    a. If the patient is awake and alert, place him in the semi-Fowler's position.
    b. If the patient is unconscious, place him in the supine position.
NOTE: Catheter care should be performed as a part of the normal morning and evening patient care, and as necessary.

3. Perform a patient care hand-wash (See task 081-831-0007.)

4. Put on gloves.

5. Inspect the catheter, drainage tubing, all connections, and the drainage bag for cracks, leaks, kinks, or obstruction of drainage.

6. Observe the urinary meatus and the surrounding area for erythema, and leakage of urine.

7. Clean the urinary meatus and the surrounding area with a cleaning solution (IAW local SOP), rinse thoroughly, and blot dry.
NOTE: Apply antibacterial ointment to the urinary meatus only if ordered by the medical officer or IAW local SOP.

8. Ensure that the catheter is secured to the patient without causing pressure within the bladder.
    a. Tape the catheter to the skin of the inner thigh for a female patient.
    b. Tape the catheter to the skin of the lower abdomen or inner thigh for a male patient.
NOTE: The penis may be positioned up or down (facing the patient's head or feet), depending upon the patient's diagnosis, the medical officer's order/the patient's comfort determine preference.

9. Maintain patency of the drainage tubing.
    a. Keep the tubing free from kinks and twists.
    b. Keep the tubing free of pressure caused by bed rails, mattress, or the patient's body.
    c. Keep the tubing above the level of the drainage bag to ensure free gravity drainage.

10. Maintain the correct position of the drainage bag at all times.
    a. Hang the drainage bag from the bed frame, not the bed rails.
    b. Do not allow the drainage bag to rest on the floor.

Performance Steps

CAUTION: The drainage bag must be kept below the level of the patient's bladder to prevent urinary reflux. If the bag must be raised to bladder level for any reason, it must be clamped first.

11. Assess the patient for indications of urinary tract infections: chills, fever, back or flank pain, hematuria, and cloudy or foul smelling urine.
NOTE: If a urinary tract infection is suspected, collect a urine specimen for culture.

12. Collect a sterile urine specimen without contaminating or disconnecting the closed system.
   a. Clamp off the drainage tubing just below the aspiration port.
   b. Wait until a sufficient quantity of urine has pooled above the aspiration port (about 15 minutes).
NOTE: Post a sign at the patient's bed indicating the urinary drainage system is temporarily clamped off.
   c. Swab the aspiration port with alcohol.
   d. Withdraw the desired amount of urine using a sterile needle and syringe.
   e. Remove the clamp from the drainage tubing.
   f. Transfer the urine in the syringe to a sterile specimen container.

13. Irrigate the catheter with the prescribed sterile solution IAW the medical officer's order.

14. Empty the drainage bag.
   a. Empty the drainage bag without disconnecting or contaminating the closed system.
   b. Measure and discard, or save, the urine as indicated by the medical officer's orders.
NOTE: The drainage bag must be emptied before it overfills to prevent reflux of urine into the drainage tubing.

15. Replace the urinary bag and tubing IAW local SOP.

16. Replace the urinary catheter IAW local SOP.

17. Remove gloves.

18. Perform a patient care hand-wash.

19. Maintain an accurate I/O record. (See task 081-833-0006.)

20. Document the procedure and all significant nursing observations on the appropriate forms IAW local SOP.

Performance Measures	GO	NO-GO
1. Provided privacy for the patient.	——	——
2. Placed the patient in a comfortable position for catheter care.	——	——
3. Performed a patient care hand-wash.	——	——
4. Put on gloves.	——	——
5. Inspected the catheter, drainage tubing, all connections, and the drainage bag for cracks, leaks, kinks, or obstruction of drainage.	——	——
6. Observed the urinary meatus and the surrounding area for erythema, and leakage of urine.	——	——

Performance Measures	GO	NO-GO
7. Cleaned the urinary meatus and the surrounding area IAW local SOP, rinsed thoroughly, and blotted dry.	——	——
8. Ensured that the catheter was secured to the patient without causing pressure within the bladder.	——	——
9. Maintained patency of the drainage tubing.	——	——
10. Maintained the correct position of the drainage bag at all times.	——	——
11. Assessed the patient for indications of urinary tract infections: chills, fever, back or flank pain, hematuria, and cloudy or foul smelling urine.	——	——
12. Collected a sterile urine specimen without contaminating or disconnecting the closed system.	——	——
13. Irrigated the catheter with the prescribed sterile solution IAW the medical officer's order.	——	——
14. Emptied the drainage bag.	——	——
15. Replaced the urinary bag and tubing IAW local SOP.	——	——
16. Replaced the urinary catheter IAW local SOP.	——	——
17. Removed gloves.	——	——
18. Performed a patient care hand-wash.	——	——
19. Maintained an accurate I/O record.	——	——
20. Documented the procedure and all significant nursing observations on the appropriate forms IAW local SOP.	——	——

Evaluation Guidance: Score each Soldier according to the performance measures in the evaluation guide. Unless otherwise stated in the task summary, the Soldier must pass all performance measures to be scored GO. If the Soldier fails any step, show what was done wrong and how to do it correctly.

References

Required	**Related**
DD FORM 792	BASIC NURSING 7

Skill Level 3

Subject Area 13: Advanced Procedures (SL 3)

PROCESS ITEMS FOR STERILIZATION
081-825-0001

Conditions: This task is performed in CMS when contaminated medical items are returned. You will need two basins, detergent, sponges, a hand brush, clean towels, distilled water, personal protective equipment (PPE) - disposable gloves and safety glasses or face shield, washer-sterilizer, a mechanical dryer, and an ultrasonic cleaner. You are not in a CBRNE environment.

Standards: Clean items by the most effective means available and appropriate to the characteristics of the item, without damage to the items or injury to personnel.

Performance Steps

1. Put on personal protective equipment (PPE) - pair of disposable gloves and face/eye shield.

2. Sort soiled items into common characteristics group.
   a. Heat sensitive--submergible.
   b. Heat resistant--submergible.
   c. Heat sensitive--not submergible.

3. Sort the common characteristic group into major categories of construction.
   a. Sharps, including scissors and cutting-edge instruments.
   b. Endoscopic, including cystoscopes and bronchoscopes.
   c. Delicate, including plastic instruments.
   d. Regular, including retractors and clamps.
   e. Special, including electric bone saws and electric dermatomes.

4. Manually clean heat sensitive submergible instruments.
   a. Prepare the basins.
      (1) Fill the first basin with a warm detergent solution.
      (2) Fill the second basin with warm water 27° to 44° C (80° to 110° F).
   b. Disassemble the instruments, as necessary, according to manufacturer's instructions.
   c. Wash the instruments.
      (1) Submerge in the detergent solution.
      (2) Use a fairly stiff brush on hinged, ratchet and box lock, and serrated edge instruments.

NOTE: Keep the brush submerged during the scrubbing process.

      (3) Support delicate non-sharp instruments with the fingers and clean them with a small soft brush.
      (4) Clean delicate instruments which are sharp or pointed with a cotton tip applicator.
      (5) Disassemble and clean the inner and outer surfaces of endoscopic equipment with a soft brush.
      (6) Use a pipe cleaner to clean the small diameter, tubular inner areas of the endoscopic equipment.
   d. Rinse the instruments.
      (1) Rinse non-delicate instruments in a basin of warm water.

Performance Steps

        (2) Rinse delicate and endoscopic instruments by:
           (a) Transferring the instruments into a basin of warm water.
           (b) Emptying the basin and refilling it with fresh warm water.
           (c) Repeat steps 4d(2)(a) and 4d(2)(b) until all the detergent has been removed.
     e. Air dry the heat sensitive instruments in a warm area on an instrument tray and blow dry cannulated and rubber tubing.
     f. Place the instruments in an instrument tray.
         (1) Open all ratchet and box lock instruments.
         (2) Place large bulky instruments in the bottom of the tray.
         (3) Place sharp and delicate instruments on top.
NOTE: Do not let the sharp points or edges contact other surfaces, points, or edges.

   5. Mechanically clean non-heat sensitive submergible instruments.
CAUTION: Items which are not heat resistant and submergible are not to be cleaned by this method.
NOTE: The use of a washer-sterilizer eliminates hand-washing of instruments that are not damaged by heat. Such items can be cleaned by using the manual method.
     a. Rinse gross soil from each instrument.
     b. Place the sharp pointed, sharp edged, and delicate instruments in the perforated washer-sterilizer pan.
     c. Load the pan in the washer-sterilizer.
     d. Add detergent.
     e. Close and secure the door.
     f. Turn on the washer-sterilizer.
     g. Open the door at the end of the sterilizing cycle.
     h. Allow the instruments to dry thoroughly while in the washer-sterilizer.
NOTE: If an ultrasonic cleaning unit is available, remove the instruments immediately from the washer-sterilizer and put them in the ultrasonic cleaner.
     i. Take the instruments to the prep area.

   6. Ultrasonically clean submergible instruments.
NOTE: Heat sensitive submergible and heat resistant submergible items may be cleaned in the ultrasonic cleaner.
     a. Open all jointed instruments.
     b. Disassemble instruments, as necessary, IAW manufacturer's instructions.
     c. Load the instruments loosely in the tray.
         (1) Do not stack the instruments over 3 inches high in the tray.
         (2) Do not mix stainless steel instruments with aluminum, brass, or copper items.
     d. Place small delicate instruments so they will not move.
     e. Fill the wash chamber with water and add detergent.
     f. Close the lid.
     g. Turn on the ultrasonic cleaner.
     h. Dry the instruments thoroughly with a clean towel when the cycle is completed.

   7. Manually clean non-submergible instruments.
     a. Wipe the instruments with a damp cloth that has been soaked with a detergent-disinfectant.
     b. Wipe the instruments again with a damp cloth soaked in clean water.
NOTE: If the instrument has an optic lens, wipe the lens with a cotton tip applicator dipped in alcohol.

Performance Steps

  8. Lubricate hinged and box lock instruments.

  9. Wrap the instrument sets. (See task 081-825-0002.)

Performance Measures	GO	NO-GO
1. Put on PPE - a pair of disposable gloves and face shield.	——	——
2. Sorted the soiled items.	——	——
3. Sorted the common characteristic groups.	——	——
4. Manually cleaned heat sensitive submergible instruments.	——	——
5. Mechanically cleaned nonheat sensitive submergible instruments.	——	——
6. Ultrasonically cleaned submergible instruments.	——	——
7. Manually cleaned nonsubmergible instruments.	——	——
8. Lubricated hinged, ratchet, and box lock instruments.	——	——
9. Wrapped the instrument sets.		

Evaluation Guidance: Score each soldier according to the performance measures in the evaluation guide. Unless otherwise stated in the task summary, the soldier must pass all performance measures to be scored GO. If the soldier fails any step, show what was done wrong and how to do it correctly.

References

Required	**Related**
None	FM 8-38

INTUBATE A PATIENT
081-830-3016

Conditions: You have an unconscious, non-breathing casualty with no gag reflex. An X-ray technician with portable X-ray machine is available. You will need the patient's chart with SF 600 or equivalent, local SOP, an assistant (assistant will have mask, gown, eye protection and non-sterile gloves on), batteries (appropriate for laryngoscope size), laryngoscope with blades (straight and curved, sizes 1-4), extra light bulb for laryngoscope blade, endotracheal tubes (7-8.5 cm), manual resuscitator, non-sterile gloves, bite block, commercial ET tube holder if available, suction equipment (either wall mounted or portable), suction kit, pulse oximeter, hemodynamic monitor, ventilator, 10 cc syringe, 1/2 inch adhesive tape, scissors, tincture benzoin, cuff pressure manometer, stethoscope, stylet, mask, gown, eye protection or face shield, disinfectant soap, two trash containers, and one red trash bag.

Standards: Insert an endotracheal tube in the airway and verify tube placement by checking for bilateral breath sounds and chest X-ray.

Performance Steps
CAUTION: Wear gloves to protect yourself against the transmission of contaminants whenever handling body fluids. Other personal protective equipment should be worn IAW local SOP.

1. Perform a patient care hand-wash with disinfectant soap.

2. Put on non-sterile gloves.

3. Inspect and prepare equipment for intubation (laryngoscope, ET tube, stylet).
    a. Attach the laryngoscope blade to the laryngoscope.
        (1) Hook the blade to the connector on the top of the laryngoscope.
        (2) Lift the blade at a 90 degree angle to lock the blade in place.
NOTE: At this time the light on the blade should be on. If not, check the light bulb and/or batteries, replace if necessary, and try again.
        (3) Unlock and collapse laryngoscope blade and set aside until ready for use.
    b. Select the appropriate size of ET tube for the patient (average adult male - 8.0-8.5 cm; average adult female - 7.0-7.5 cm).
    a. Fill the 10 cc syringe with air and attach the syringe to the ET tube cuff valve (pilot balloon), inflate the cuff, and inspect for leaks. Refer to Figure 3-63.

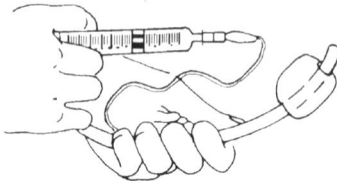

Figure 3-63. ET tube cuff valve

NOTE: If you detect a leak, discard ET tube in trash container and get a new one.
    d. Deflate cuff by pulling back on the plunger until all the air is out.
    e. Insert stylet into ET tube.
        (1) The stylet should be inserted into the ET tube so the tip of the stylet is recessed ½ inch from the tip of the ET tube.

Performance Steps
> (2) Bend the other end of the stylet at a 90 degree angle to prevent it from going further into the ET tube.

4. Preoxygenate the patient with the manual resuscitator.
NOTE: Maintain patient oxygen saturation at 100%. This can be monitored through a pulse oximeter or hemodynamic monitor.

5. Position the patient's head by hyperextending the neck.
NOTE: This will allow for visualization of the vocal cords.

6. Open the patient's mouth and hold it open by pushing down on the jaw.

7. Insert the laryngoscope blade into mouth and visualize vocal cords.
   a. Stand or kneel at the top of the patient's head.
   b. Hold the laryngoscope with your left hand.
   c. Open and lock the blade at a 90 degree angle.
   d. Place the blade into the right side of the patient's mouth.
   e. Move the laryngoscope to the center of the mouth by sliding the laryngoscope to the left side of the mouth; this will in turn move the patient's tongue out of the way. Refer to Figure 3-64.

PLACEMENT OF LARYNGOSCOPE

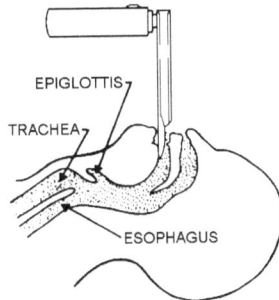

Figure 3-64. Placement of laryngoscope

Performance Steps

    f. Retract the epiglottis and observe the vocal cords. Refer to Figure 3-66.

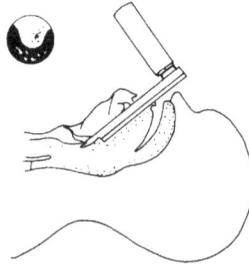

Figure 3-66. Retract the epiglottis

(1)When using a Macintosh blade (curved), apply anterior pressure to the vallecula with the tip of the laryngoscope blade. This will fold back the epiglottis and expose the vocal cords. Refer to Figure 3-67.

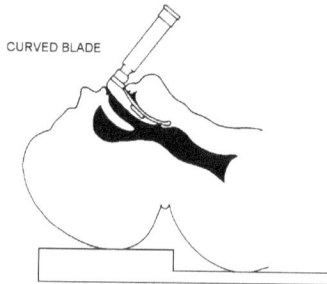

Figure 3-67 Macintosh blade

Performance Steps

    (2) When using a Miller blade (straight), hook the blade tip under the epiglottis and pull up to fold back the epiglottis and expose the vocal cords. Refer to Figure 3-68.

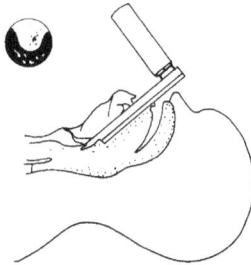

Figure 3-68. Miller blade

8. Insert the ET tube into the trachea. Refer to Figure 3-69.

Figure 3-69. Insert the ET tube

    a. Grasp the ET tube with your right hand.

Performance Steps

    a. Carefully guide the tip of the tube between the Insert the ET tube until the cuff is just below the level of the vocal cords. Refer to Figure 3-70.

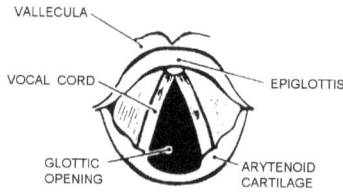

Figure 3-70. Vocal cords level

9. Remove the laryngoscope from the airway.

10. Remove the stylet from the ET tube.

11. Inflate the cuff of the ET tube by injecting the required amount of air (5 to 10 cc) to create seal by pressing the plunger of the syringe. Refer to Figure 3-71.

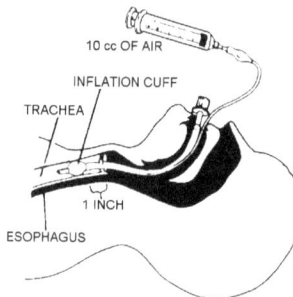

Figure 3-671 Inflate the cuff of the ET tube

12. Check placement of the ET tube.
    a. Place manual resuscitator over the end of the ET tube and ventilate patient.
    b. Instruct assistant to auscultate the patient's lung fields and epigastric area while you manually ventilate the patient.
        (1) If patient has strong bilateral breath sounds and no sounds of air movement are heard over the epigastric area, proceed to step 13.
        (2) If sound is heard over one lung field only, you must partially deflate the cuff, withdraw the ET tube slightly, reinflate the cuff, and listen again.
        (3) If a rushing sound is heard over the epigastric area, deflate the cuff, withdraw the ET tube completely, reoxygenate the patient, and repeat procedure.
NOTE: For tube placement verification, you must have a chest X-ray done.

13. Check cuff pressure.

Performance Steps

    a. Use a pressure manometer. Connect pressure manometer to pilot balloon to ensure that cuff pressure is less than 25 cm H2O. Adjust to achieve desired pressure and remove manometer from pilot balloon. Refer to Figure 3-72.

Figure 3-72. Pressure manometer

    b. Use minimal leak technique.
        (1) Attach 10 cc syringe to pilot balloon and partially deflate the cuff by pulling back on the plunger of the syringe.
        (2) Add air into cuff until a slight leak is heard around the cuff during peak inspiration.
        (3) Remove the syringe by holding the cuff valve in one hand and simultaneously twist and pull the syringe with your other hand.

NOTE: Before performing minimal leak technique, suction the patient thoroughly (see task 081-830-0097).

14. Insert a bite block between the back teeth to prevent biting of the ET tube.

15. Secure the ET tube with 1/2 inch adhesive tape.
NOTE: Commercial ET tube holder may be used to secure ET tube if you have one available.

16. Manually ventilate patient.
NOTE: At this time, if patient is to remain intubated for a prolonged period of time, you may place patient on a mechanical ventilator.

17. Using the cardiac monitor, assess the patient's blood pressure and oxygen saturation. Ensure correct tube placement is maintained by auscultating the lungs for bilateral breath sounds. Chart findings in patient's chart (SF 600 or equivalent).
NOTE: The tip of the tube should be 2 to 3 centimeters above the carina. Proper tube placement is confirmed by taking an X-ray of the patient's chest.

18. Discard trash and gloves in trash can. Perform a patient care hand-wash using disinfectant soap.

19. Record procedure in patient's chart (SF 600 or equivalent).
NOTE: Continue to monitor patient every 15 minutes initially. If placed on a mechanical ventilator, you include patient monitoring during your Q2 hour ventilator checks.

Performance Measures $\underline{\text{GO}}$ $\underline{\text{NO-}}$
$\underline{\text{GO}}$

1. Performed patient care handwash. ____ ____

2. Put on non-sterile gloves. ____ ____

3. Inspected and prepared equipment for intubation. ____ ____

4. Pre-oxygenated the patient with the manual resuscitator. ____ ____

5. Positioned the patient's head by hyperextending the neck. ____ ____

6. Opened the patient's mouth and held it open. ____ ____

7. Inserted the laryngoscope blade into mouth and visualized vocal cords. ____ ____

8. Inserted the ET tube into the trachea. ____ ____

9. Removed the laryngoscope from the airway. ____ ____

10. Removed the stylet from the ET tube. ____ ____

11. Inflated the cuff of the ET tube. ____ ____

12. Checked placement of the ET tube. ____ ____

13. Checked cuff pressure. ____ ____

14. Inserted a bite block between the back teeth. ____ ____

15. Secured the ET tube with 1/2 inch adhesive tape. ____ ____

16. Manually ventilated patient. ____ ____

17. Assessed the patient's blood pressure and oxygen saturation. ____ ____

18. Discarded trash and gloves in trash can. Performed a patient care hand- ____ ____
wash using disinfectant soap.

19. Recorded procedure in patient's chart (SF 600). ____ ____

Evaluation Guidance: Score each soldier according to the performance measures. Unless otherwise stated in the task summary, the soldier must pass all performance measures to be scored GO. If the soldier fails any steps, show what was done wrong and how to do it correctly.

References
  Required
  None

  Related
  FUNDAMENTALS RESP CARE
  ISBN 0803606621

PUT ON STERILE GLOVES
081-831-0008

Conditions: You need to put on sterile gloves. You will need hand-washing facilities, sterile gloves, and a flat, clean, dry surface. You are not in a CBRNE environment.

Standards: Put on sterile gloves without contaminating self or the gloves.

Performance Steps

1. Select and inspect the package.
    a. Select the proper size of glove.
    b. Inspect the package for possible contamination.
        (1) Water spots.
        (2) Moisture.
        (3) Tears.
        (4) Any other evidence that the package is not sterile.

2. Perform a patient care hand-wash.

3. Open the sterile package.
    a. Place the package on a flat, clean, dry surface in the area where the gloves are to be worn.
    b. Peel the outer wrapper open to completely expose the inner package.

4. Position the inner package.
    a. Remove the inner package touching only the folded side of the wrapper.
    b. Position the package so that the cuff end is nearest you.

5. Unfold the inner package.
    a. Grasp the lower corner of the package.
    b. Open the package to a fully flat position without touching the gloves.

6. Expose both gloves.
    a. Grasp the lower corners on the folder.
    b. Pull gently to the side without touching the gloves.

7. Put on the first glove.
    a. Grasp the cuff at the folded edge and remove it from the wrapper.
    b. Step away from the table or tray.
    c. Keeping your hands above the waist, insert the fingers of the other hand into the glove.
    d. Pull the glove on touching only the exposed inner surface of the glove.
NOTE: If there is difficulty in getting your fingers fully fitted into the glove fingers; make the adjustment after both gloves are on.

8. Put on the second glove.
    a. Insert the fingertips of the gloved hand under the edge of the folded over cuff.
NOTE: You may keep the gloved thumb up and away from the cuff area or may insert it under the edge of the folded over cuff with the fingertips.
    b. Keeping your hands above the waist, insert the fingers of the ungloved hand into the glove.
    c. Pull the glove on.
    d. Do not contaminate either glove.

Performance Steps

9. Adjust the gloves to fit properly.
   a. Grasp and pick up the glove surfaces on the individual fingers to adjust them.
   b. Pick up the palm surfaces and work your fingers and hands into the gloves.
   c. Interlock the gloved fingers and work the gloved hands until the gloves are firmly on the fingers.

NOTE: If either glove tears while putting them on or adjusting the gloves, remove both gloves and repeat the procedure.

Evaluation Preparation:

Setup: If performance of this task must be simulated for training and evaluation, the same gloves may be used repeatedly as long as they are properly re-wrapped after each use. You may give the Soldier a torn or moist glove package to test step 1.

NOTE: If the Soldier does not know his glove size, have several different sizes available to try on to determine the correct size.

Brief Soldier: Tell the Soldier to put on the sterile gloves. Tell the Soldier that they are not in a CBRNE environment.

Performance Measures

	GO	NO-GO
1. Selected and inspected the package.	——	——
2. Performed a patient care hand-wash.	——	——
3. Opened the sterile package.	——	——
4. Positioned the inner package.	——	——
5. Unfolded the inner package.	——	——
6. Exposed both gloves.	——	——
7. Put on the first glove without contaminating.	——	——
8. Put on the second glove without contaminating.	——	——
9. Adjusted the gloves to fit properly.	——	——

Evaluation Guidance: Score each Soldier according to the performance measures. Unless otherwise stated in the task summary, the Soldier must pass all performance measures to be scored GO. If the Soldier fails any steps, show what was done wrong and how to do it correctly.

References

Required	Related
None	BASIC NURSING 7

ESTABLISH A STERILE FIELD
081-833-0007

Conditions: You need to establish a sterile field and introduce items into the sterile field. You will need sterile packs, sterile drapes and towels, a small solution basin, sterile liquids, sterile needles and syringes, sterile gloves, and a flat, clean, dry surface. You are not in a CBRNE environment.

Standards: Establish a sterile field and introduce items into the sterile field without violating aseptic technique.

Performance Steps

1. Obtain sterile equipment and supplies IAW local SOP.

2. Select a flat, clean, dry surface.
NOTE: Choose a surface away from drafts, if possible.

3. Create a sterile field with a double-wrapped sterile package.
   a. Lift the top flap of the sterile pack away from the body without crossing your hand or arm over the sterile field.
   b. Lift the remaining flaps, one at a time, away from the center without crossing your hand or arm over the sterile field.

4. Introduce sterile items onto the sterile field.
NOTE: The outer one inch border of the sterile field is considered contaminated. Items that fall into that area are considered contaminated and should not be used. If an item rolls from the one inch border onto the sterile field, the sterile field is considered contaminated and the procedure must be stopped immediately and the procedure must be repeated using a new sterile pack.
   a. Commercially prepacked items (syringes, sutures, needles, etc.).
      (1) Keeping your hands on the outside of the sterile wrapper, grasp the opening edge of the package.
      (2) Carefully fold (roll) each end of the wrapper back toward your wrists.
      (3) Without contaminating the contents, drop them onto the sterile field.
NOTE: If the wrapper has been punctured, torn, or has water marks, the item is no longer sterile.
   b. Centralized materiel service (CMS) items (wrapped in double muslin wrappers).
      (1) Remove the outer wrapper.
      (2) Grasp the edge of the item being unwrapped, keeping your hand on the outside of the inner wrapper.
      (3) Fold each edge of the wrapper slowly back over your wrist of the hand holding the item.
      (4) Drop the item onto the sterile field.

5. Open sterile liquids.
NOTES: 1. Liquids prepared in CMS are considered sterile if a vacuum release sound is heard when the bottle is opened. If there is no sound, the bottle is considered unsterile, and a new bottle must be obtained before continuing the procedure. 2. Some commercially prepared bottles of sterile solution may not make a vacuum release sound.
   a. Remove the outer protective bottle seal, if necessary, and remove the cap.
   b. Hold the cap in one hand, or place the cap so the top rests on the table.

Performance Steps
NOTE: The bottle rim and inside of the cap are considered sterile.

CAUTION: Discard the sterile solution under any of the following conditions: 1. Anyone touches the bottle rim. 2. The lip of the bottle touches non-sterile items. 3. Someone touches the inside of the cap or the part of the cap that touches the container is placed on the table.

    6. Pour sterile liquids.
        a. Hold the bottle with the label against your palm.
        b. Pour a small amount of the liquid from the bottle into a waste receptacle.
        c. Hold the bottle about 6 inches above the container into which the liquid is to be poured.
        d. Slowly pour a steady stream to avoid splashing, thus preventing contamination.
        e. Replace the cap without contaminating the bottle.
        f. Write the date and time the bottle was opened and your initials on the label. Return the bottle to the storage area or discard it IAW local SOP.
NOTE: If the sterile field is contaminated at any time, the procedure must be stopped immediately. Repeat all steps using new sterile equipment.

Evaluation Preparation:
Setup: Place all necessary materials and equipment including sterile packs, sterile drapes and towels, a small solution basin, sterile liquids, sterile needles and syringes, and sterile gloves on a table. Place another table adjacent to the first table for the sterile field. Have a waste receptacle in place to receive the sterile liquid poured.

Brief Soldier: Tell the Soldier to establish a sterile field (they have already performed a patient care hand-wash) and introduce items and liquids into the sterile field without violating aseptic technique. Tell the Soldier that they are not in a CBRNE environment.

Performance Measures	GO	NO-GO
1. Obtained sterile equipment and supplies IAW local SOP.	____	____
2. Selected a flat, clean, dry surface.	____	____
3. Created a sterile field with a double-wrapped sterile package.	____	____
4. Introduced sterile items onto the sterile field.	____	____
5. Opened sterile liquids.	____	____
6. Poured sterile liquids.	____	____
7. Did not violate aseptic technique.	____	____

Evaluation Guidance: Score each Soldier according to the performance measures. Unless otherwise stated in the task summary, the Soldier must pass all performance measures to be scored GO. If the Soldier fails any steps, show what was done wrong and how to do it correctly.

References
    Required
    None
                Related
                BASIC NURSING 7

CHANGE A STERILE DRESSING
081-833-0010

Conditions: You need to change a sterile dressing. You will need a protective pad, scissors, forceps, sterile gloves, sterile gauze, basin, sponges, face mask, swabs, towels, tape, dressings, sterile cleaning solution, adhesive solvent, and hand-washing facilities. You are not in a CBRNE environment.

Standards: Change a sterile dressing without violating aseptic technique.

Performance Steps

1. Identify the patient.

2. Gather the equipment.

3. Prepare the patient.
   a. Explain the procedure to the patient.
   b. Expose the wound by moving the patient's clothing and folding the bed linens away from the wound area, if necessary.
   c. Position the patient to provide maximum wound exposure.
   d. Place a protective pad under the patient.

4. Prepare the work area.
   a. Clear the bedside stand or table.
   b. Cut the required tape strips and attach them where they are accessible.

5. Put on a mask and exam gloves.

6. Remove the outer dressing.
   a. Loosen the ends of the tape by peeling toward the wound while supporting the skin around the wound.
   WARNING: Do not peel the tape away from the wound.
   b. Grasp the edge of the dressing and gently remove it from the wound.
   c. Note any drainage, color, and odor associated with the dressing.
   d. If the dressing is grossly saturated, discard the dressing and the gloves in a contaminated waste container otherwise, dispose of in regular trash.

7. Perform a patient care hand-wash.

8. Establish a sterile field. (See task 081-833-0007.)
   a. Open and place all sterile equipment and supplies on the sterile field.
   b. Pour the sterile cleaning solution into a basin.

9. Put on a mask and sterile gloves.

10. Remove the inner dressings.
    a. Using forceps, remove the dressings one at a time.
    b. Note any drainage, color, and odor associated with the dressings.
    c. Discard the dressings in a contaminated waste container.
    d. Drop the forceps on the glove wrap.

Performance Steps

11. Check the wound for the following conditions.
    a. Redness, swelling, foul odor, and/or bleeding.
    b. Drainage that contains blood, serum, or pus (usually yellow but may be blood-tinged, greenish, or brown).

CAUTION: Notify the supervisor if any of the above conditions are present.
    c. If drainage is present, seek permission from the medical officer to irrigate the wound. (See task 081-833-0012.)

12. Clean the wound with sterile gauze soaked with a sterile cleaning solution.
    a. Linear wound.
       (1) First stroke. Clean the area directly over the wound with one wipe and discard the gauze.
       (2) Second stroke. Clean the skin area on one side next to the wound with one wipe and discard the gauze.
       (3) Third stroke. Clean the skin area on the other side next to the wound with one wipe and discard the gauze.
       (4) Continue the procedure alternating sides of the wound, working away from the wound until the area is cleaned.
    b. Circular wound. .
       (1) First stroke. Start at the center of the wound, wipe the wounded area with an outward spiral motion, and then discard the gauze.
       (2) Second stroke. Clean the skin area next to the wound using an outward spiral motion, approximately one and one half revolutions, and then discard the gauze.
       (3) Using successive outward, spiral strokes of approximately one and one half revolutions, clean the entire area around the wound.

13. Change gloves.

14. Remove adhesive from around the wound, if necessary.
    a. Using a solvent-soaked cotton tipped applicator or gauze pad, rub gently over the adhesive residue.
    b. Observe the skin for signs of irritation.

15. Apply a sterile dressing.
    a. Lay the first dressing over the wound so that it extends over the edge.
    b. Overlap the first dressing with a second dressing.
    c. Overlap the second dressing with a third dressing.
    d. Cover all of the dressings with a large outer dressing.

NOTE: If the wound has a drain inserted, cut the dressing halfway through and position it around the drain.

16. Remove sterile gloves and face mask.

17. Secure the dressing with tape.

NOTE: The tape should not form a constricting band around the wound.

    a. Apply tape to the edge of the dressing with half of the tape on the dressing and the other half on the skin.
    b. Write the date and time the dressing was changed on a piece of tape, initial it, and secure the tape to the dressing.

18. Dispose of contaminated materials in a contaminated waste container.

Performance Steps

19. Perform a patient care hand-wash.

20. Record the procedure on the appropriate form.
    a. Enter the date and time of the dressing change.
    b. Enter a description of the wound's appearance.
       (1) Type and amount of drainage, if any.
       (2) Characteristics of the wound before and after cleaning.

Evaluation Preparation:
Setup: If the performance of this task must be simulated for training or evaluation, have another Soldier act as the patient. A moulage kit or similar materials may be used to simulate an injury. Apply a dressing to the patient.

NOTE: For testing purposes, the dressing may be reused.

Brief Soldier: Tell the Soldier to change the patient's sterile dressing. Tell the Soldier that his is not in a CBRNE environment.

Performance Measures	GO	NO-GO
1. Identified the patient.	——	——
2. Gathered the equipment.	——	——
3. Prepared the patient.	——	——
4. Prepared the work area.	——	——
5. Put on a mask and exam gloves.	——	——
6. Removed the outer dressing.	——	——
7. Performed a patient care hand-wash.	——	——
8. Established a sterile field.	——	——
9. Put on a mask and sterile gloves.	——	——
10. Removed the inner dressings.	——	——
11. Checked the wound.	——	——
12. Cleaned the wound with sterile gauze soaked with a sterile cleaning solution.	——	——
13. Changed gloves.	——	——
14. Removed adhesive from around the wound, if necessary.	——	——
15. Applied a sterile dressing.	——	——
16. Removed sterile gloves and face mask.	——	——
17. Secured the dressing with tape.	——	——

Performance Measures	GO	NO-GO
18. Disposed of contaminated materials in the appropriate waste container.	——	——
19. Performed a patient care hand-wash.	——	——
20. Recorded the procedure on the appropriate form.	——	——
21. Did not violate aseptic technique.	——	——

Evaluation Guidance: Score each Soldier according to the performance measures. Unless otherwise stated in the task summary, the Soldier must pass all performance measures to be scored GO. If the Soldier fails any steps, show what was done wrong and how to do it correctly.

References
    Required
    None

    Related
    BASIC NURSING 7

PERFORM A WOUND IRRIGATION
081-833-0012

Conditions: You need to perform a wound irrigation. You will need protective pads, irrigating syringe, examination gloves, sterile gloves, mask, prescribed irrigating solution, sterile dressing, catch basin, sterile gauze sponges, a sterile solution basin, and SF 600 (Medical Record - Chronological Record of Medical Care). You are not in a CBRNE environment.

Standards: Perform a wound irrigation without violating aseptic technique or causing further injury to the patient.

Performance Steps
CAUTION: All body fluids should be considered as potentially infectious so always observe body substance isolation (BSI) precautions by wearing gloves and eye protection as a minimal standard of protection.

1. Identify the patient.

2. Explain the procedure to the patient.

3. Provide privacy, if possible, and position the patient to provide maximum wound exposure.

4. Place a protective pad directly under the wound area.

5. Prepare the irrigation equipment.
   a. Establish a sterile field using the wrapper of the sterile solution basin.
   b. Open and place all other sterile equipment and supplies on the sterile field.
   c. Verify the prescribed irrigating solution and pour it into the sterile basin.

6. Put on a mask, eye protection, and examination gloves.

7. Remove the soiled outer dressing.

8. Remove the examination gloves.

9. Place a catch basin on the protective pad, against the body, to collect the used solution.

10. Put on sterile gloves.

11. Use sterile forceps to remove the inner dressings.

12. Irrigate the wound.
   a. Fill the irrigating syringe with solution from the sterile basin.
   b. Hold the tip of the syringe as close to the wound as possible without touching it. Depress the bulb or plunger, directing the flow of solution to all parts of the wound in a slow, steady stream.
   c. Repeat steps 12a and 12b until the wound is clear of debris and/or drainage.
   d. Observe the drainage for blood or characteristics such as unusual color, odor, or consistency.
NOTE: If signs of infection are observed, notify the medical officer immediately.

CAUTION: Use extra care when irrigating a wound in which an abscess has formed. Check all internal surfaces of the wound to inspect for "sinus tract" (resembles tunnels in which purulence or "pus" may be collected). This may require using the gloved hand or a sterile object to gently pull back the flesh. Be careful not to tear healing tissue.

Performance Steps

13. Dry the wound and apply a sterile dressing.
    a. Pat the wound dry with sterile gauze sponges.
       (1) Start at the center of the wound.
       (2) Move outward toward the wound edges.
    b. Apply a sterile dressing to the wound. (See task 081-833-0010.)
    c. Remove the catch basin and protective pad, if they are still in place.

14. Remove the mask, eye protection, and gloves.

15. Reposition the patient for comfort, if necessary.

16. Clean and store the equipment IAW local SOP.

17. Perform a patient care hand-wash.

18. Record the procedure on the appropriate medical form.

Evaluation Preparation:
Setup: If the performance of this task must be simulated for training or evaluation, have another Soldier act as the patient. Designate a wound site or use a moulage kit or similar material to simulate an injury. Prepare a medical officer's order specifying the type and amount of solution to be used.

Brief Soldier: Give the Soldier the medical officer's order and tell the soldier to irrigate the wound. Tell the Soldier that they are not in a CBRNE environment.

Performance Measures	**GO**	**NO-GO**
1. Identified the patient.	____	____
2. Explained the procedure to the patient.	____	____
3. Provided privacy, if possible, and positioned the patient to provide maximum wound exposure.	____	____
4. Placed a protective pad directly under the wound area.	____	____
5. Prepared the irrigation equipment.	____	____
6. Put on a mask, eye protection, and examination gloves.	____	____
7. Removed the soiled outer dressing.	____	____
8. Removed the examination gloves.	____	____
9. Placed a catch basin on the protective pad, against the body, to collect the used solution.	____	____
10. Put on sterile gloves.	____	____
11. Used sterile forceps to remove the inner dressings.	____	____
12. Irrigated the wound.	____	____
13. Dried the wound and applied a sterile dressing.	____	____

Performance Measures	GO	NO-GO
14. Removed the mask, eye protection, and gloves.	——	——
15. Repositioned the patient for comfort, if necessary.	——	——
16. Cleaned and stored the equipment IAW local SOP.	——	——
17. Performed a patient care hand-wash.	——	——
18. Recorded the procedure on the appropriate medical form.	——	——
19. Did not violate aseptic technique, or cause further injury to the patient.	——	——

Evaluation Guidance: Score each Soldier according to the performance measures. Unless otherwise stated in the task summary, the Soldier must pass all performance measures to be scored GO. If the Soldier fails any steps, show what was done wrong and how to do it correctly.

References
   Required
   SF 600

   Related
   BASIC NURSING 7

SET UP A CASUALTY DECONTAMINATION STATION
081-833-0093

Conditions: You are assigned to a division level medical facility (battle aid station (BAS) or division clearing station (DCS)). Chemical agents are being used against the units supported by your medical treatment facility. The commander has ordered that a decontamination station and protective shelter be established. Your current location is in a non-contaminated area, upwind from the chemical hazard.

Standards: Set up a fully operational decontamination station in a non-contaminated area upwind from the chemical hazard. Establish the decontamination area on the downwind side of the protective shelter or other clean treatment area and clearly mark a hot line. Construct a shuffle pit as the only point of access to the clean areas. Install chemical agent alarms.

Performance Steps

1. Select sites for the location of the operation.
   a. Primary and alternate sites must be selected in advance of operations.
NOTE: Alternate sites must be selected in conjunction with selection of the primary site. If the prevailing winds change direction, use of the primary site may no longer be possible.
   b. Site selection factors.
     (1) The direction of the prevailing winds.
     (2) The downwind chemical hazard.
     (3) The availability of protective shelters or buildings to house clean treatment facilities.
     (4) The terrain.
     (5) Availability of cover and concealment.
NOTE: The protective shelter may possess visual, audible, and infrared signatures. Therefore, concealment may be compromised.
     (6) The general tactical situation.
     (7) The availability of evacuation routes (contaminated and clean).
     (8) The location of the supported unit's vehicle decontamination point, personnel decontamination point, and MOPP exchange point.
NOTE: It is sometimes best to co-locate with these unit decontamination sites. The arrangement of the operational areas must be kept flexible and adaptable to both the medical and tactical situations.

Performance Steps

2. Set up the decontamination area. Refer to Figure 3-73.

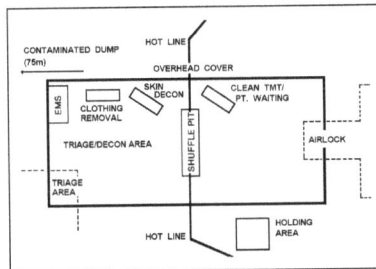

Figure 3-73. Decontamination area

    a. Triage area.
    b. Emergency treatment area.
NOTE: Sometimes, triage and emergency treatment are conducted in the same area.
    c. Clothing removal area.
    d. Skin decontamination area.
    e. Overhead cover.
       (1) Erect an overhead cover, at least 20 x 50 feet, to cover the decontamination area and the clean waiting and treatment area. If the protective shelter is used, the overhead cover should overlap the air lock entrance.
       (2) If plastic sheeting is not available, alternate materials such as trailer covers, ponchos, or tarpaulins may be used.

3. Set up the clean side of the decontamination station on the upwind side of the contaminated areas.
NOTE: Erect a windsock for easy determination of wind direction.
    a. Clean waiting area.
    b. Clean treatment area.

4. Set up the shuffle pit as the only point of access between the decontamination area and the clean waiting and treatment area.
    a. Turn over the soil in an area that is 1 to 2 inches deep, and of sufficient length and width to accommodate a litter stand.
NOTE: The shuffle pit should be wide enough that the litter bearers are not able to straddle the pit.
    b. Mix super tropical bleach (STB) with the soil in a ratio of two parts STB to three parts soil.

5. Set up the protective shelter on the upwind side of the clean waiting and treatment area.
    a. Set up the protective shelter with the air lock adjoining the clean side of the decontamination station.
    b. When a protective shelter is not available for use, set up a protected medical treatment facility 30 to 50 meters upwind from the shuffle pit.

Performance Steps

    6. Set up the evacuation holding area.
        a. Set up an overhead cover of plastic sheeting at least 20 x 25 feet.
        b. Make sure the cover overlaps part of the clean treatment area and part of the protective shelter.
        c. When the protective shelter is used, set up the cover on the side opposite the generator.

    7. Mark the hot line.
        a. Use wire, engineer's tape, or other similar material to mark the entire perimeter of the hot line.
        b. Ensure that the hot line is clearly marked.

    8. Establish ambulance points on both the "clean" and "dirty" evacuation routes.
        a. Establish a "dirty" ambulance point downwind from the triage area in the decontamination station.
        b. Establish a "clean" ambulance point upwind from the evacuation holding area on the clean side of the decontamination station.

    9. Set up a contaminated (dirty) dump.
        a. Establish the contaminated dump 75 to 100 meters downwind from the decontamination station.
        b. Clearly mark the dump with NATO chemical warning markers.

    10. Place chemical agent alarms upwind from the clean treatment area.

    11. Camouflage areas IAW tactical directives.

Performance Measures	GO	NO-GO
1. Selected primary and alternate sites.	——	——
2. Set up the decontamination area.	——	——
3. Set up the clean treatment/waiting area.	——	——
4. Set up the shuffle pit.	——	——
5. Set up the protective shelter.	——	——
6. Set up the evacuation holding area.	——	——
7. Marked the hot line.	——	——
8. Established ambulance points.	——	——
9. Set up a contaminated (dirty) dump.	——	——
10. Placed chemical agent alarms.	——	——
11. Camouflaged areas.	——	——

Evaluation Guidance: Score each Soldier according to the performance measures. Unless otherwise stated in the task summary, the Soldier must pass all performance measures to be scored GO. If the Soldier fails any steps, show what was done wrong and how to do it correctly.

References
    Required   **Related**
    None   TF-TVT 3-6

INSERT A CHEST TUBE
081-833-0168

Conditions: You have a casualty suffering from a hemothorax or pneumothorax who requires the insertion of a chest tube. You will need a chest tube (16-36 French), sterile gloves, one-way valve, scalpel handle and blades (#10 and #15), Kelly forceps, large hemostat, betadine solution, suture material (size 0 nylon), lidocaine 1% for injection, needle, syringe, intravenous (IV) administration set, and DD Form 1380 (U.S. Field Medical Card). You are not in a CBRNE environment.

Standards: Insert a chest tube and correct the hemothorax or pneumothorax without causing further injury to the casualty.

Performance Steps

1. Assess the casualty.
    a. If necessary, open the airway. (See task 081-831-0018.)
    b. Ensure adequate respiration and assist as necessary.
    c. Provide supplemental oxygen, if available.
    d. Connect the casualty to a pulse oximeter, if available.
    e. Initiate an IV. (See task 081-833-0033.)

2. Prepare the casualty.
    a. Place the casualty in the supine position.
    b. Raise the arm on the affected side above the casualty's head.
    c. Select the insertion site at the anterior axillary line over the 4th or 5th intercostal space.
    d. Clean the site with betadine solution.
    e. Put on sterile gloves.
    f. Drape the area.
    g. Liberally infiltrate the area with the 1% lidocaine solution.

3. Insert the tube.
    a. Make a 2 to 3 cm transverse incision over the selected site and extend it down to the intercostal muscles.
NOTE: The skin incision should be 1 to 2 cm below the interspace through which the tube will be placed.
    b. Insert the Kelly forceps through the intercostal muscles in the next intercostal space.
    c. Puncture the parietal pleura with the tip of the forceps and slightly enlarge the hole by opening the clamp 1.5 to 2 cm.
CAUTION: Avoid puncturing the lung. Always use the superior margin of the rib to avoid the intercostal nerves and vessels.
    d. Immediately insert a gloved finger into the pleural cavity to clear any adhesions, clots, etc.
    e. Grasp the tip of the chest tube with Kelly forceps. Insert the tip of the tube in the incision as you withdraw your finger.
    f. Advance the tube until the last side hole is 2.5 to 5 cm inside the chest wall.
    g. Connect the end of the tube to a one-way drainage valve (e.g., Heimlich valve).
    h. Secure the tube using the suture materials.
    i. Apply an occlusive dressing to the site.
    j. Radiograph the chest to confirm placement, if available.

Performance Steps

    4. Reassess the casualty.
        a. Check for bilateral breath sounds.
        b. Monitor and record vital signs every 15 minutes.

    5. Document the procedure on a FMC.

Performance Measures	GO	NO-GO
1. Assessed the casualty.	——	——
2. Prepared the casualty.	——	——
3. Inserted the tube.	——	——
4. Reassessed the casualty.	——	——
5. Documented the procedure on a FMC.	——	——

Evaluation Guidance: Score each Soldier according to the performance measures in the evaluation guide. Unless otherwise stated in the task summary, the Soldier must pass all performance measures to be scored GO. If the Soldier fails any step, show what was done wrong and how to do it correctly.

References
    Required   **Related**
    DD FORM 1380   BTLS FOR PARAMEDICS

ESTABLISH AN AMBULANCE EXCHANGE POINT
081-833-0184

Conditions: You need to establish an ambulance exchange point (AXP). You will need combat service support and operation overlays. You are not in a CBRNE environment.

Standards: Establish an ambulance exchange point to facilitate evacuation of casualties.

Performance Steps

1. Select the site for an ambulance exchange point (AXP) based on the tactical mission.
    a. The location of the AXP will depend on the location and number of units being supported.
    b. This location should provide the required support to reduce ambulance turnaround time to supported units.
    c. When supporting tracked vehicles the AXP should be located as close as possible to the supported unit to reduce the time and distance requirements for the tracked vehicles.
    d. The AXP may be an established point in an ambulance shuttle or it may be designated independently.

2. Establish the AXP.
    a. AXPs may be staffed or unstaffed.
        (1) Points that are not staffed may serve as rendezvous points for the rapid transfer of a patient from one transportation mode to another.
        (2) In most cases AXPs will not be staffed.
        (3) The ambulance platoon leader/sergeant coordinates/establishes the AXPs as required by the medical evacuation mission.
    b. The medical evacuation plan should include an overlay depicting (at a minimum) the location of supported units, casualty collection points, Echelon I facilities, and AXPs.
        (1) The platoon leader should also obtain the combat service support and operations overlays for the tactical operation. These overlays provide valuable information on:
            (a) Mine fields.
            (b) Obstacles and barriers.
            (c) Artillery target reference points.
            (d) Air corridors.
        (2) Supported units.
            (a) An AXP may serve two to three battalions/squadrons (brigade support medical company/medical troop) or a specific number of non divisional Echelon I facilities (area support medical company).
            (b) In these cases the AXP should be centrally located to reduce ambulance turnaround and enhance the timely execution of the medical evacuation mission.

Evaluation Preparation:
Evaluation Preparation: Setup: At the test site, provide all equipment, information, and personnel given in the task conditions statement.

Brief Soldier: Tell the Soldier that he is to select and prepare an APX.

Performance Measures	GO	NO-GO
1. Selected a site for an AXP based on the tactical mission.	——	——
2. Established the AXP.	——	——

Evaluation Guidance: Score each Soldier according to the performance measures in the evaluation guide. Unless otherwise stated in the task summary, the Soldier must pass all performance measures to be scored GO. If the Soldier fails any step, show what was done wrong and how to do it correctly.

References
Required
None

Related
FM 8-10
FM 8-10-6

PERFORM ABSCESS INCISION AND DRAINAGE
081-833-0192

Conditions: You have a patient with an abscess that needs incision and drainage (I&D). You will need povidone-iodine 1% betadine) antimicrobial solution, 10ml syringe, 25 gauge needle, ¼" or 1" iodoform or plain sterile gauze packing, 4x4 inch sterile gauze sponge, adhesive tape, #3 scalpel handle, #11 blade, hemostat (curved), tissue forceps, surgical scissors, cotton-tipped sterile applicators, gloves, a culture swab, and SF 600 (Medical Record - Chronological Record of Medical Care). You are not in a CBRNE environment.

Standards: Perform an I&D using sterile technique.

Performance Steps

1. Solicit a patient history.

2. Gather equipment.

3. Perform a patient care hand-wash.

4. Put on gloves.

5. Prepare the site.
   a. Position the patient with the abscess easily accessible.
   b. Prepare the abscess and surrounding area with the antimicrobial solution.
   c. Apply sterile drapes to completely surround the abscess and to cover all unprepared areas adjacent to the abscess.
   d. Perform a field block by anesthetizing the perimeter surrounding the abscess with lidocaine 1% with epinephrine.
NOTE: Do not inject the anesthetic directly into the abscess because it will not be effective in the acidic medium.
   e. Wait 5-10 minutes before proceeding to ensure complete anesthesia.

6. Perform the incision, drainage, and packing of abscess.
   a. Using the #11 blade, incise the abscess deeply from one side of the fluctuant area, to the opposite side of the fluctuant area. This is necessary to ensure complete evacuation of the purulent drainage.
   b. Allow the material to drain, using the 4x4 gauze sponges to soak up any purulence and blood. Use the sterile cotton-tipped applicator to swab the inside of the abscess cavity for culture.
   c. Use the hemostat to gently explore the abscess cavity and to break up any sacs or adhesions within the abscess.
   d. Clean the cavity with 4-6 sterile cotton-tipped applicators soaked with 3% hydrogen peroxide. You may also irrigate the cavity with 0.9% normal saline solution under moderate pressure.
   e. Loosely pack the abscess with ¼" to 1" iodoform or plain sterile gauze packing allowing a small portion of the packing to protrude outside the cavity and dress the incision site with sterile gauze.

7. Remove gloves.

8. Document the procedure.
   a. Description of the abscess prior to performing the procedure.

Performance Steps

    b. Description of the procedure (anesthesia, length of incision, approximate amount of material exuded).

    c. Type of packing and dressing material.

    d. How the patient tolerated the procedure.

    e. Any follow-up care or instructions given to the patient.

        (1) Stress the importance of keeping the area dry for the first 24 hours. Have the patient leave the packing in place and the initial dressing should not be disturbed until the following day. Elevate the affected extremity when possible.

        (2) On the second day, the patient should remove the outside dressing, leaving the packing undisturbed and apply warm water compresses or submerge the site in a warm water bath for 20-30 minutes. This will hasten resolution of the inflammatory process and promote rapid healing.

        (3) The patient should return as soon as possible if he develops a fever, if the area of erythema increases in size or if flu-like symptoms develop.

        (4) The patient may require medications for pain (rarely) and possibly antibiotics (based on subsequent gram-stain analysis of the culture).

9. Re-evaluate the patient in 24-48 hours after the I&D procedure.

    a. Perform a patient care hand-wash.

    b. Put on gloves.

    c. Remove the external dressing.

    d. Gently remove the packing material from the abscess site.

    e. Cleanse the abscess site with a sterile cotton-tipped applicator soaked in hydrogen peroxide 3%.

NOTE: Anesthesia is rarely necessary.

    f. Do not repack the cavity, especially if it is clean and pain and tenderness are significantly diminished.

    g. Re-apply a sterile dressing to the open abscess site.

    h. Remove gloves.

    i. Document your findings and any follow-on care plan.

Evaluation Preparation: This task is best evaluated by verbalization and demonstration of the steps. Give the Soldier a scenario in which he must perform an I&D of an abscess.

Performance Measures	**GO**	**NO-GO**
1. Solicited a patient history.	——	——
2. Gathered equipment.	——	——
3. Performed a patient care hand-wash.	——	——
4. Put on gloves.	——	——
5. Prepared the site.	——	——
6. Performed the incision, drainage, and packing.	——	——
7. Removed gloves.	——	——
8. Documented the procedure	——	——

Performance Measures   <u>GO</u>   <u>NO-</u>
                                                                       <u>GO</u>

9. Re-evaluated patient in 24-48 hours after the I&D procedure.   ___   ___

Evaluation Guidance: Score each Soldier according to the performance measures in the evaluation guide. Unless otherwise stated in the task summary, the Soldier must pass all performance measures to be scored GO. If the Soldier fails any step, show what was done wrong and how to do it correctly.

References
  Required   **Related**
  SF 600   ISBN 0-07-065351-9

TREAT HIGH ALTITUDE ILLNESS
081-833-0206

Conditions: You need to treat a patient for high altitude illness. You will need the patient's medical record, aspirin, humidified oxygen (O2) and O2 delivery equipment, hyperbaric bag, dexamethasone and acetazolamide. You are not in a CBRNE environment.

Standards: Treat high altitude illness without causing further injury to the patient.

Performance Steps

1. Recognize the signs and symptoms of various types of high-altitude illness.
    a. Acute mountain sickness (AMS).
NOTE: AMS is the most common form of altitude sickness and may develop at altitudes as low as 6,500 feet.
      (1) Headache.
      (2) Fatigue.
      (3) Nausea.
      (4) Dyspnea.
      (5) Sleep disturbances.
      (6) Symptoms are aggravated by exertion.
NOTE: AMS may evolve into high-altitude pulmonary edema (HAPE), high-altitude cerebral edema (HACE), or both.
    b. High-altitude cerebral edema (HACE).
NOTE: HACE is believed to be present to a mild degree in all forms of altitude sickness. HACE can occur within 3-5 days after arrival at 9,000 feet, but generally occurs at altitudes above 12,000 feet.
      (1) Gait ataxia is a reliable early warning sign.
      (2) Headache.
      (3) Mental confusion.
      (4) Hallucinations.
      (5) Irrational behavior progressing to coma.
WARNING: Coma and death may develop within a few hours of the first warning signs of HACE.
    c. HAPE.
NOTE: HAPE usually develops 24 to 96 hours after rapid ascent above 8,000 feet. HAPE accounts for the greatest number of fatalities.
      (1) HAPE is characterized by increasing dyspnea.
      (2) Irritative cough that produces frothy, often bloody sputum.
      (3) Weakness.
      (4) Cyanosis.
      (5) Low-grade fever.
      (6) Tachycardia.
      (7) Fine or coarse rales.
      (8) Coma.
WARNING: HAPE may worsen rapidly. Coma and death may occur within hours.

2. Initiate prophylactic measures to prevent high-altitude illnesses.
    a. Altitude sickness is best prevented by slow ascent.
    b. Most individuals can ascend to 5,000 feet in 1 day without symptoms.
    c. Above 5,000 feet, a rate of 1,500 feet per day is advisable.

Performance Steps
NOTE: All rates are variable. A climber should learn how fast he can ascend without developing symptoms. A climbing party should be paced to its slowest member.

      d. Although physical fitness enables greater exertion with lower oxygen (O2) consumption, it does not protect against any form of altitude sickness.

      e. Increased hydration with moderate salt restriction may prevent or diminish symptoms of AMS.

      f. Eating frequent, small meals that are high in easily digestible carbohydrates improves altitude tolerance.

      g. Acetazolimide 125-250 milligrams (mg) twice a day beginning on day 1 before ascent and continuing for 2 days at maximum altitude is an effective prophylactic for AMS.

      h. Dexamethasone 4 mg. Give 4 mg orally or IM every 6 hours and continuing for 2 days at maximum altitude.

NOTE: The combination of both drugs has been shown to be more effective than either drug alone.

   3. Manage high-altitude illnesses.
      a. AMS.
         (1) Descent, if mission allows, will provide the quickest resolution of symptoms.
         (2) Increase fluid intake.
         (3) Give aspirin or a nonsteroidal anti-inflammatory drug (NSAID) for altitude headache.
         (4) Antiemetic.
         (5) Light diet.
         (6) Avoid any additional ascent or exertion until symptoms resolve.
         (7) Monitor patient for symptom progression to HAPE/HACE.
      b. HACE.
         (1) HACE requires immediate descent.
         (2) Give acetazolimide 125 -250 mg twice a day.
         (3) Give dexamethasone 4 mg orally, intramuscularly, or intravenously every 6 hours.
      c. HAPE.
         (1) Bed rest and oxygen at high altitude may be tried with mild HAPE, depending on mission requirements.
         (2) If the condition worsens, immediate descent is essential.

NOTE: On the basis of mission parameters, a portable hyperbaric bag may be transported and used as a temporary substitute for descent. However, this will only buy time for mission completion, and descent must be accomplished at the earliest opportunity.

         (3) Once descent is accomplished, the patient should be continued on oxygen and managed as with other forms of pulmonary edema.
         (4) When promptly treated, patients usually recover from HAPE within 24 to 48 hours after descent.

   4. Record all episodes of altitude illness and treatment in the patient's medical record.

   5. Evacuate the patient, as required.

Evaluation Preparation: This task is best evaluated by verbalization of the steps. Give the Soldier a scenario in which he must manage high-altitude illness.

Performance Measures	<u>GO</u>	<u>NO-GO</u>
1. Recognized the signs and symptoms of various types of high-altitude illness.	——	——
2. Initiated prophylactic measures to prevent high-altitude illnesses.	——	——
3. Managed high-altitude illnesses.	——	——
4. Recorded all episodes of altitude illness and treatment in the patient's medical record.	——	——
5. Evacuated the patient, as required.	——	——

Evaluation Guidance: Score each Soldier according to the performance measures in the evaluation guide. Unless otherwise stated in the task summary, the Soldier must pass all performance measures to be scored GO. If the Soldier fails any step, show what was done wrong and how to do it correctly.

References

Required	**Related**
None	PHTLS

CONDUCT CRITICAL INCIDENT STRESS DEBRIEF
081-833-0215

Conditions: You are the leader of a critical event debriefing team assigned to conduct a critical incident stress debriefing to members of a unit following a traumatic event that resulted in the loss of several personnel. You will need a safe meeting place. You are not in a CBRNE environment.

Standards: Complete all the steps necessary to conduct a Critical Incident Stress Debriefing/Critical Event Debriefing.

Performance Steps

1. Define after-action debriefing (AAD): A briefing that adds to the usual after-action review (AAR), when team members "share" and "talk out" their emotional reactions to the events and clarify details of the events in order to understand actually what happened. This action is conducted within the resources of command, (mainly without facilitators). After the AAR and AAD, leaders can request the critical incident stress debriefing.

2. Define critical incident stress debriefing (CISD):
    a. A scheduled meeting (approved by leaders) for those units where a highly disruptive or traumatic event occurred.
    b. A facilitative technique similar to an after-action debrief, yet more dependent on the skills and experience of the facilitator.
    c. Not an operational critique, but a discussion to clarify what happened and help restore well-being and unit effectiveness.

3. Identify CISD guidelines:
    a. Conduct in an emotionally neutral location, relatively safe from enemy action or observations and other distractions.
    b. The process is confidential, and everyone involved is urged to maintain confidentiality regarding the session.
    c. All personnel are equal during the debriefing.
    d. Personnel are asked to speak for himself, and no one else.
    e. All personnel are encouraged to stay for the entire process.

4. Identify the purposes for a CISD:
    a. Quickly restore and enhance unit cohesion and effectiveness.
    b. Reduce short-term emotional and physical distress.
    c. Prevent long-term distress and burnout.
    d. Safeguard future effectiveness, happiness, unit and family well-being.

5. Identify the factors indicating the need for a CISD:
    a. Death of a unit member (by combat, accident, or suicide).
    b. Death or suffering of noncombatants (especially women and children).
    c. Having to handle dead bodies, other horrible sights, and smells.
    d. Friendly fire incidents, especially if it caused casualties.
    e. Situation involving a serious error, injustice or atrocity.
    f. Evident distress of many participants.
    g. Evident reluctance of unit members to talk through the event in the after-action debriefing under their own leadership.

Performance Steps

6. Identify the steps to conduct the CISD (must be done in sequence):
    a. Introductory phase.
        (1) Introduce the critical event debriefing (CED) team.
        (2) Explain the CED process.
        (3) No notes or recordings will be made.
        (4) No breaks are scheduled, but anyone may leave as needed and return as soon as possible.
        (5) No one is required to speak, but all are encouraged to express themselves.
NOTE: The debriefing team should notice those who keep apart, and check with them afterwards to see if they could use one-on-one debriefing or other assistance.
        (6) Each speaker speaks only for self, and not for others.
        (7) All persons are equal during the debriefing.
NOTE: All ranks should speak frankly (with proper courtesy) without fear of reprisal. Advise others that everyone, including leaders, should be ready to accept how others speak of their performance, and present their own viewpoint in turn.
        (8) Fact-finding, not fault-finding but "facts" include the team members' personal reactions to the event.
        (9) The team members are available after the debriefing.
NOTE: Ideally, team consists of a leader and an assistant, plus one assistant for every ten in a group of more than twenty.
    b. Fact phase.
        (1) To reconstruct the event in detail, in chronological order as an "unbroken" historical time-line.
        (2) View (facts) from all sides and perspectives.
        (3) Get participants to start at the beginning, before the critical event occurred, and work up to the event(s).
        (4) First person involved in the critical event is asked to tell how it started; what his role (duty position) was, and what he saw, heard, smelled and did, step by step.
        (5) Other participants are drawn in as the action (first person's story) reaches them; leader asks them to tell their observations and actions in detail.
NOTE: Expected duration of debriefing is 2 to 3 hours (depends upon number involved and the complexity of the critical event).
        (6) Leader asks those who don't come in on their own what they were doing (seeing, hearing, etc.); everyone is asked.
NOTE: Normally include only those directly involved in the event; may include close unit members who were absent by chance, or newly arrived replacements as listeners only; higher command personnel only if present at the event, and trusted support persons as listeners, such as chaplains and medics, even if they weren't present at the incident.
        (7) If there are disagreements about what happened, when, (the "what" is the focus, not "why"), leader tries to elicit observations from others, which could resolve the difference and reach consensus, (or at least clarify the difference in memories of the event).
        (8) Discussion may proceed to phases 3 and 4 before event reconstruction is complete and agreed upon. The leader should eventually bring the talk back to fill in the time-line.
NOTE: It is best to conduct debriefing 8 to 72 hours after the event (but better even weeks or months late than never).
    c. Thought phase.
        (1) To personalize the event.

Performance Steps

      (2) Shift focus from factual to emotional.

      (3) Participants are asked to share what "thoughts" they had.

      (4) Leader leads transition; allows participants to express themselves; "let it happen" (report thoughts that were in their mind as the event started, thoughts in their mind as they saw, smelled, or did something during the event, and thoughts in their mind when the event was over).

NOTE: No media or outsiders permitted.

    d. Reaction phase.

      (1) To identify feelings (emotions) raised by the event.

      (2) To allow ventilation of feelings (emotions) raised by the event.

      (3) Leader emphasizes that all emotional reactions deserve to be expressed, respected and listened to.

      (4) Participants are asked and encouraged to share "reactions" (i.e., worst thing about the event, how each reacted to what happened and how each feels about the event now).

      (5) Debriefing team listens for common themes, feelings, misperceptions (i.e. anger, blame, guilt).

      (6) Leader and group confirms the normality and commonality (universality) of the feelings (i.e. seeing that others had or have the same reactions, helping individuals "reframe" the meaning of the event(s).

      (7) Leader prevents "scape-goating" and personal verbal abuse.

NOTE: Leader must assure that distress and misunderstandings are clarified, not ignored or left to fester.

    e. Symptom phase.

      (1) To normalize and personalize physical stress responses.

      (2) Shift focus back from emotional to factual.

      (3) Participants are asked to describe how their bodies reacted physically before, during and since the event.

NOTE: Check for common symptoms such as: gastric-intestinal distress, urinary frequency, loss of bowel and/or bladder control, loss of breath, loss of sexual arousal or interest, heart-pounding, muscle, back, neck and headache, insomnia, bad dreams, intrusive memories, trembling, trouble concentrating, irritability, jumpiness and startle reactions (not all inclusive).

      (4) Participants are reassured; find other group members having same symptoms, often can find some humor in it.

    f. Teaching phase.

      (1) Reassurance by educating that feelings and stress symptoms are normal reactions to abnormal conditions.

      (2) Educate that symptoms are expected to resolve normally.

NOTE: Leader summarizes the thoughts, feelings, symptoms expressed by the group, reemphasizing normality, reducing feelings of uniqueness, weakness or injury.

      (3) Debriefing team may teach more about "stress" process; give stress management, anger management training, instruction on "grief" process or coping strategies (as needed).

      (4) Avoid prediction or glamorization of long-term disability.

    g. Reentry phase.

      (1) Complete the debriefing.

      (2) Leader gives final invitation for comments (makes summary statement to the group).

Performance Steps

(3) CED team distributes list of point of contacts (i.e., unit ministry team, medics, mental health section, and aid station personnel) who could follow-up with participants if needed.

(4) Group (participants) may help define the group self-support activities.

NOTE: It is important that participants are fully aware of follow up options (i.e. consolidated debriefing with other units, a second critical event debriefing and/or further professional assistance).

Evaluation Preparation: This task is best evaluated by verbalization of the steps. Give the Soldier a scenario in which he must conduct a critical incident stress debriefing.

Performance Measures	GO	NO-GO
1. Defined after-action debriefing.	——	——
2. Defined critical incident stress debriefing (CISD).	——	——
3. Identified CISD guidelines.	——	——
4. Identified the purposes for a CISD.	——	——
5. Identified the factors indicating a need for a CISD.	——	——
6. Identified the steps to conduct the CISD (must be done in sequence).	——	——

    a. Introductory phase.
    b. Fact phase.
    c. Thought phase.
    d. Reaction phase.
    e. Symptom phase.
    f. Teaching phase.
    g. Reentry phase.

Evaluation Guidance: Score each Soldier according to the performance measures in the evaluation guide. Unless otherwise stated in the task summary, the Soldier must pass all performance measures to be scored GO. If the Soldier fails any step, show what was done wrong and how to do it correctly.

References

Required	Related
None	ISBN 0-7637-4406-9

MAINTAIN A HUMAN PATIENT SIMULATOR
081-833-0219

Conditions: You have a situation in which the simulator is not functioning properly, or during PMCS of the simulator, you notice a problem that must be corrected for the simulator to function properly. You will need a SimMan and a standard SimMan tool kit.

Standards: Restore the simulator to mission capable status.

Performance Steps

   1. Install IV arm.

NOTE: The right arm is the only arm that is utilized for IVs.

      a. Expose the chest by removing the chest skin, and place the chest skin to the side.

        (1) Undo the four upper tabs, two on the back side of each shoulder.

        (2) Undo the four side tabs, two on each side of the chest.

      b. Raise the chest plate and place it to the left side of the head, ensuring that the wires are not placed under pressure.

      c. Expose the nut for the right arm.

      d. Remove the deltoid cushion to expose the bolt.

      e. Remove the nut, ensuring that the inside washer does not fall down into the mannequin. Place the bolt, spring, two washers, and the nut to the side.

      f. Repair the arm.

        (1) Cleaning the inside of the arm (without replacing the veins).

          (a) Take the two veins that are threaded through the hole in the upper arm, pull the arm skin out, and place the two veins between the skin and the arm.

          (b) Pour a water and dish soap mix between the skin and the arm.

          (c) Begin to roll the arm skin down to the hand.

          (d) If you need to remove the skin completely, pull it over the hand. The thumb is a part of the skin and will come completely away from the arm.

          (e) Clean the arm with water and dish soap. Do not use bleach to clean the arm; this will cause the arm to become soft and unusable.

          (f) When clean, replace the skin by placing over the hand, and pour a small amount of water and dish soap mix over the lower portion of the arm and roll the skin up.

          (g) Place the two veins back into the hole in the upper arm.

        (2) Replacing the veins and arm skin.

          (a) Cut the arm skin off of the arm; be careful not to cut the arm itself.

          (b) Remove the veins from the arm.

          (c) Clean the arm with water and dish soap. Do not use bleach to clean the arm; this will cause the arm to become soft and unusable.

          (d) Place the new veins on the IV arm; secure the veins by using a little bit of glue (super glue, contact cement, or a comparable fast drying glue).

          (e) Roll the arm skin down to the hand and place it over the hand of the IV arm.

          (f) Place a small amount of water and dish soap over the lower half of the IV arm and roll the new skin onto the arm.

          (g) With the skin all of the way on, place the two veins through the hole.

          (h) Place the bolt in the deltoid area, ensure that the bolt is set up in the following sequence bolt, spring, washer.

Performance Steps

       (i) Place the arm on the body by pushing the bolt through the hole in the shoulder and attaching the washer and nut and tighten.

       (j) Move the arm to ensure the arm has full range of motion. If it is too stiff to move, loosen the nut and check the range of motion again.

       (k) Replace the deltoid cushion on the IV arm.

       (l) Replace the chest plate and place the chest skin back and secure it in place.

2. Install chest needle decompression bladders.

   a. Ensure that the simulator has the breathing set to apnea (zero breaths).

   b. Expose the chest by removing the chest skin, and place the chest skin to the side.

     (1) Undo the four upper tabs, two on the back side of each shoulder.

     (2) Undo the four side tabs, two on each side of the chest.

   c. Lift the chest plate and disconnect the bladder(s) from the Y connection that needs to be replaced.

   d. Place the chest plate back on the simulator and pull the pneumothorax bladder(s) through the lower hole of the chest plate.

   e. Place the new bladder in place by feeding the tube end through the opening and pushing the bladder in the lower hole.

   f. Lift the chest plate and connect the new bladders to the Y connector.

   g. Hold off on replacing the chest skin at this time; perform a test of the bladders first.

   h. Put the mannequin in pneumothorax and observe the bladders to ensure they inflate. If the bladders do not inflate, ensure that the air tube is not kinked over in any fashion to block the air. If the bladders still do not inflate, go back to step 2a and troubleshoot the problem.

   i. Once the bladders inflate, replace the chest skin and secure it.

3. Install breathing bladder.

   a. Ensure that the simulator has the breathing set to apnea (zero breaths).

   b. Expose the chest by removing the chest skin, and place the chest skin to the side.

     (1) Undo the four upper tabs, two on the back side of each shoulder.

     (2) Undo the four side tabs, two on each side of the chest.

   c. Raise the chest plate and place it to the left side of the head, ensuring that the wires are not placed under pressure.

   d. Place the lung bags up towards the mannequin's head.

   e. Remove the CPR compression shaft.

   f. Remove the foam insert from inside the mannequin.

   g. Remove the breathing bladder by following the connector tube to the manifold and removing the tube.

   h. Attach the new bladder by attaching the tube to the manifold. Ensure that the bladder tube runs exactly the same way as the one that was removed.

   i. Turn breathing on (to any number) and observe to see if the bladder inflates with every breath of the mannequin. If not, check the bladder to ensure that it is securely connected. Once the bladder inflates, continue to reassemble the simulator.

   j. Place the mannequin back to apnea (zero breaths).

   k. Replace the foam cushion, CPR compression shaft, and lungs.

4. Install upper teeth (replace hard teeth with soft or vice/versa).

   a. If you are replacing hard teeth, ensure that you recover any of the broken teeth.

   b. Slightly retract the upper lip.

   c. With your other hand, grab the teeth and pull them out of the mouth.

   d. Spray a small amount of silicone lubricant on the new teeth.

Performance Steps

    e. Retract the upper lip.

    f. With the other hand place the new teeth in place.

5. Install new chest tube module.

    a. The left side is the only side of the mannequin that is used for chest tubes; the right side is for pneumothorax.

    b. Ensure that the simulator has the breathing set to apnea (zero breaths).

    c. Remove the left side of the chest skin and lift it onto the chest of the mannequin.

    d. Pull the old chest tube module out of the mannequin.

    e. Ensure that the new module is facing with the ribs out. You can feel ribs on one side and the other side feels flat.

    f. Insert the opening of the module towards the mannequin's head.

    g. Press into place until flush with the mannequin's side.

    h. Replace the chest skin.

6. Install blood pressure (BP) arm.

    a. Ensure that the mannequin is off.

        (1) Monitor off.

        (2) Link box off.

        (3) Compressor off.

        (4) Software not running.

    b. Expose the chest by removing the chest skin, and place the chest skin to the side.

        (1) Undo the four upper tabs, two on the back side of each shoulder.

        (2) Undo the four side tabs, two on each side of the chest.

    c. Raise the chest plate and place it to the left side of the head, ensuring that the wires are not placed under pressure.

    d. Expose the nut that holds the left arm in place.

    e. Remove the deltoid cushion.

    f. Remove the nut, ensuring that the inside washer does not fall down into the mannequin. Place the bolt, spring, two washers, and the nut to the side.

    g. Place the bolt, spring, and washer on the arm.

    h. Prior to attaching the arm to the body the wire connector must be placed through the hole in the shoulder below the bolt hole. This is a critical step that must be done to ensure that the wire connector is not damaged in any way. Fold the wire connector back onto the wires and slide it into the hole sideways. If any resistance is encountered, stop immediately.

    i. Place the arm on the body by pushing the bolt through the hole in the shoulder and attaching the washer and nut and tighten. Do not test the arm's range of motion with the wire harness connected to the mannequin. This could cause the harness to be broken.

    j. Connect the wire harness.

    k. Power up the mannequin and test the arm by taking a BP on the mannequin and checking the pulses at the radial and brachial sites. Now calibrate and check the BP.

    l. If all are functioning properly, place the chest plate on the mannequin and secure the chest skin to the mannequin.

7. Install chest skin

    a. Remove the old chest skin.

        (1) Undo the four upper tabs, two on the back side of each shoulder.

        (2) Undo the four side tabs, two on each side of the chest.

    b. With the old chest skin removed, discard of the old chest skin and get the new one.

Performance Steps
   c. Turn the chest skin upside down and place on a flat hard surface. Sprinkle some baby powder on the inside of the chest skin and rub it around. This lubricates the chest skin and keeps if from rubbing on the chest plate.
   d. Place the new chest skin on the chest plate and secure all of the tabs.

Evaluation Preparation:
Setup: None

Brief Soldier: Tell the Soldier to perform the appropriate steps to ensure the simulator is mission capable. You may specify which malfunctions are encountered.

Performance Measures	GO	NO-GO
1. Installed IV arm.	——	——
2. Installed chest needle decompression bladders.	——	——
3. Installed breathing bladder.	——	——
4. Installed upper teeth.	——	——
5. Installed new chest tube module.	——	——
6. Installed blood pressure arm.	——	——
7. Installed chest skin.	——	——

Evaluation Guidance: Score each Soldier according to the performance measures. Unless otherwise stated in the task summary, the Soldier must pass all performance measures to be scored GO. If the Soldier fails any steps, show what was done wrong and how to do it correctly.

References
   Required   **Related**
   None   None

PROVIDE INITIAL SCREENING FOR TRAUMATIC BRAIN INJURY
081-833-0234

Conditions: A casualty requires an assessment to determine whether a traumatic brain injury (TBI) has occurred. You will need a military acute concussion evaluation (MACE) screening tool. You are not in a CBRNE environment.

Standards: Assess the casualty for a TBI.

Performance Steps

  1. Obtain a history of the incident.
NOTE: A concussion is a mild TBI. The purpose of the MACE is used to evaluate a person in whom a concussion is suspected.

NOTE: Anyone who was dazed, confused, "saw stars" or lost consciousness as a result of an explosion/blast, fall, motor vehicle accident, or other event involving abrupt head movement, a direct blow to the head, or other head injury should be evaluated using the MACE tool.
  a. Description of the incident.
  (1) What happened?
  (2) Tell me what you can remember.
  (3) Were you dazed, confused, saw "stars"?
  (4) Did you hit your head?
  (5) Was a helmet worn?
  (6) Any history of amnesia before or after the incident? If yes, how long?
  (7) Any loss of consciousness? If yes, how long?
  b. Cause of Injury.
  (1) Explosion/blast.
  (2) Blunt object.
  (3) Motor vehicle crash.
  (4) Fragment.
  (5) Fall.
  (6) Gun shot wound.
  (7) Other.
  c. Symptoms.
  (1) Headache.
  (2) Memory problems.
  (3) Dizziness.
  (4) Balance problems.
  (5) Nausea or vomiting.
  (6) Difficulty concentrating.
  (7) Irritability.
  (8) Visual disturbances.
  (9) Ringing of the ears.
  (10) Other.

  2. Perform a physical examination on the casualty.
NOTE: There are 5 domains of neurological function: orientation, immediate memory, neurological screening, concentration and delayed recall.
  a. Orientation.

Performance Steps

    (1) Ask the casualty to tell you the present month, date, and day of the week, year, and time.
    (2) Award 1 point for each correct response with a maximum of 5 points.
    (3) Record the results.
  b. Immediate memory.
    (1) Ask the casualty to remember five words and repeat them in any order.
    (2) Repeat this two more times for a total of three times.
    (3) Award 1 point for each correct response for a maximum of 15 points.
  c. Neurological screening.
    (1) Check eyes for pupillary response and tracking.
    (2) Check for speech fluency and word finding.
    (3) Evaluate for motor responses, gait and coordination.
    (4) Do not award points for these, but record any abnormalities.
  d. Concentration.
    (1) Provide the casualty with four strings of numbers and ask him to read them to back to you in reverse order.
    (2) Ask the casualty to read the months of the year in reverse order.
    (3) Award 1 point for each string of numbers correctly repeated and 1 point for the correct reversed sequence of months for a maximum of 5 points.
    (4) If the casualty fails to repeat the first two strings of numbers, stop and move to delayed recall.
  e. Delayed recall.
    (1) Ask the casualty to recall the five words from the previous memory test.
    (2) Award 1 point for each correct response for a maximum of 5 points.

3. Compute the total score on the MACE screening tool and determine whether a concussion has occurred or not.

4. Refer casualty for further evaluation if score is below 25.
NOTE: Scores below 25 may represent clinically relevant neurocognitive impairment and require further evaluation for the possibility of a more serious brain injury.

Performance Measures   <u>GO</u>   <u>NO-GO</u>

1. Obtained a casualty history.   —— ——

2. Performed a physical examination.   —— ——

3. Computed the total score on the MACE screening tool and determined whether a concussion has occurred or not.   —— ——

4. Referred casualty for further evaluation if score was below 25.   —— ——

Evaluation Guidance: Score each Soldier according to the performance measures. Unless otherwise stated in the task summary, the Soldier must pass all performance measures to be scored GO. If the Soldier fails any steps, show what was done wrong and how to do it correctly.

References
  Required   **Related**
  None   MACE

PERFORM A NEUROLOGICAL EXAMINATION ON A PATIENT WITH SUSPECTED CENTRAL NERVOUS SYSTEM (CNS) INJURIES
081-833-3014

Conditions: You need to perform a neurological examination on a patient with a suspected CNS injury. You will need a flashlight or penlight, a pin or sharp object, and a rubber hammer. You are not in a CBRNE environment.

Standards: Perform a neurological examination without causing further injury to the patient.

Performance Steps

  1. Look for the cause(s) of the injury.
     a. Observe the patient's position.
     b. Observe the environmental conditions.
  NOTE: If the patient is unconscious, ask bystanders for information.

  2. Evaluate the patient's mental status.
     a. Determine the level of consciousness.
        (1) Alert--awake and responsive (verbal and motor). The patient responds immediately, fully, and appropriately to commands.
        (2) Lethargic--sleepy or drowsy. The patient can be aroused and responds appropriately, but will fall asleep again as soon as he is left alone.
        (3) Comatose--partial to complete unconsciousness. Use the Glasgow Coma Scale to determine the level of coma. (See task 081-835-3030.)
     b. Ask the patient to perform calculations (basic math) to assess cognition. For example, have the patient count backward from 100 by three or sevens.
     c. Observe the patient's verbal and nonverbal behavioral responses to evaluate affect (mood). For example:
        (1) Does the patient laugh inappropriately?
        (2) Does the patient display excessive or inappropriate anger, fear, anxiety, or confusion?
        (3) Does the patient respond to stimuli in a normal manner?
     d. Question the patient to evaluate long and short term memory.
        (1) Discuss the patient's past to evaluate remote recall (long term memory). Verify the patient's response with information on what company or unit he is assigned to and the company's mission or with the unit's members.
        (2) Discuss current events to evaluate recent recall (short term memory). For example, ask the patient what he was doing just before being injured, or what his unit was doing the previous day.
     e. Question the patient to evaluate his orientation to person, place, and time.
        (1) Ask the patient to spell his name, name family or unit members, and recite his home or unit address. (This determines whether the patient knows who he is and who others are.)
        (2) Ask the patient to identify his location, naming the city, state, or country. (This determines whether the patient knows where he is.)
        (3) Ask the patient to identify the day of the week, month, and year.

  3. Evaluate the patient's cerebellar functions.
     a. Test coordination and balance.

Performance Steps

  (1) Ask the patient to extend both arms, close the eyes, and alternately touch the index finger to the nose.

  (2) Ask the patient to slap the palms of the hands on his legs, and then the backs of the hands on the legs, alternating in a rapid motion.

  (3) Ask the patient to stand relaxed with the eyes open. Watch for movement.

  (4) Perform the "Romberg test".

   (a) Have the patient stand up and relax. Instruct the patient to close his eyes.

   (b) If the patient cannot maintain balance when the eyes are closed, the test is positive.

NOTE: The medic should stand close to the patient to support the patient if he starts to fall.

 b. Check the patient for normal gait and heel-toe-heel walking.

  (1) Ask the patient to walk a straight line both forward and backward.

  (2) Observe the patient for coordination, balance, and posture. Note inability to walk heel-toe-heel with one foot in front of the other.

4. Evaluate the patient's motor function.

 a. Check for mild weakness.

  (1) Have the patient stand with the arms outstretched, palms upward, and eyes closed for 20 to 30 seconds.

  (2) Observe the patient's arms for the "pronator sign" (the arm starts dropping and the hand turns over slightly).

 b. Test muscle tone.

  (1) Ask the patient to relax.

  (2) If the patient is ambulatory, have him sit on the edge of the examining table. Watch the freedom of movement of the legs. This indicates tone.

  (3) If the patient is in bed, lift the patient's arm, drop it, and observe the arm as it falls. Look for atrophy--loss of muscle tone or strength.

 c. Test muscle strength.

  (1) Ask the patient to walk on his heels.

  (2) Ask the patient to walk on his toes.

  (3) Extend your hands to the patient, and ask the patient to firmly grip and squeeze your hands. Note strength and equality of grip.

  (4) Ask the patient to alternately flex and extend the feet while providing resistance with your hands. Look for atrophy.

5. Evaluate the patient's cranial nerve function.

 a. Test pupillary reflexes.

  (1) Dim the lighting and shine a light into one of the patient's eyes.

  (2) Observe the pupillary response.

  (3) Repeat the procedure on the other eye.

  (4) If the pupils are unreactive or unequal, they are abnormal.

  (5) If the pupils are equal and reactive, record PERRLA (pupils equal, round, and reactive to light and accommodation).

 b. Test facial nerves.

  (1) Ask the patient to smile and raise his eyebrows.

  (2) Look for weakness or drooping on either side of the face when smiling.

  (3) Look to see if there is even movement of both eyebrows.

6. Evaluate the patient's sensory functions.

NOTE: When doing this test, ask the patient not to watch what you are doing.

 a. Allow the patient to assume a comfortable position with the eyes closed.

Performance Steps

    b. Test perception of pain by using a safety pin. Lightly touch the skin with the sharp and dull areas of the pin.

      (1) Ask the patient to identify the sensation felt (sharp or dull).

      (2) Ask the patient to identify where the sensations were felt.

    c. Test perception of touch by using a cotton ball to lightly brush the skin, asking the patient to tell you when and where he felt the sensation.

7. Check for the presence of a Babinski reflex.

    a. Grasp the ankle with your left hand.

    b. With a blunt point and moderate pressure, stroke the sole near its lateral border, from the heel toward the ball of the foot. The course of the stroke should curve to the middle to follow the bases of the toes.

    c. Normal reflex--toes curl. (Recorded as the absence of a Babinski reflex.)

    d. Abnormal reflex. (Recorded as the presence of a Babinski reflex.)

      (1) Dorsiflexion of the great toe.

      (2) Fanning of all the toes.

      (3) Dorsiflexion of the ankle.

      (4) Flexion of the knee and hip.

8. Evaluate the patient's deep tendon reflexes (DTRs).

    a. Biceps.

      (1) Position the elbow at about a 90 degree angle of flexion with the arm slightly pronated.

      (2) Grasp the elbow with your left hand so the fingers are behind it and your abducted thumb presses the biceps brachia tendon.

      (3) Strike your thumb a series of blows with the rubber hammer, varying your thumb pressure with each blow until the most satisfactory response is obtained.

      (4) A normal response will be elbow flexion.

    b. Triceps.

      (1) Grasp the patient's wrist with your left hand and pull the arm across the chest so the elbow is flexed about 90 degrees and the forearm is partially pronated.

      (2) Tap the triceps brachia tendon directly above the olecranon process.

      (3) A normal response is elbow extension.

    c. Knee.

      (1) Legs dangling.

        (a) Have the patient sit on a table, high bed, or litter to permit free swinging of the legs.

        (b) Tap the patellar tendon directly.

NOTE: The tendon is distal to the patella.

        (c) A normal response is extension of the knee.

      (2) Lying supine.

        (a) With your hand under the popliteal fossa, lift the knee from the table.

        (b) Tap the patellar tendon directly.

        (c) A normal response is extension.

    d. Ankle.

      (1) Legs dangling.

        (a) With your left hand, grasp the foot and pull it in dorsiflexion. Find the degree of stretching of the Achilles tendon that produces the optimal response.

        (b) Tap the Achilles tendon directly.

Performance Steps

        (c) A normal response is contraction of the gastrocnemius muscle and plantar flexion of the foot.

    (2) Lying supine.

        (a) Partially flex the hip and knee. Rotate the knee outward as far as comfort permits.

        (b) With your left hand, grasp the foot and pull it in dorsiflexion.

        (c) Tap the Achilles tendon directly.

        (d) A normal response is plantar flexion.

Performance Measures	GO	NO-GO
1. Looked for the cause(s) of the injury.	——	——
2. Evaluated the patient's mental status.	——	——
3. Evaluated the patient's cerebellar functions.	——	——
4. Evaluated the patient's motor function.	——	——
5. Evaluated the patient's cranial nerve functions.	——	——
6. Evaluated the patient's sensory functions.	——	——
7. Checked for the presence of a Babinski reflex.	——	——
8. Evaluated the patient's deep tendon reflexes.	——	——

Evaluation Guidance: Score each Soldier according to the performance measures. Unless otherwise stated in the task summary, the Soldier must pass all performance measures to be scored GO. If the Soldier fails any steps, show what was done wrong and how to do it correctly.

References

    Required   **Related**

    None   BTLS FOR PARAMEDICS

SUTURE A MINOR LACERATION
081-833-3208

Conditions: You have a casualty with a minor laceration requiring closure. The laceration does not involve the face, hands, feet, or genitalia. You will need sterile suture set, appropriate type and size of suture, staples, skin adhesive, steri strips, lidocaine 1% with and without epinephrine, saline irrigation solution, antiseptic solution, sterile gloves, antibiotic ointment, sterile dressing, and SF 600 (Medical Record - Chronological Record of Medical Care). You are not in a CBRNE environment.

Standards: Properly clean, anesthetize, and suture the laceration without causing further harm.

Performance Steps

1. Prepare the site.
    a. Put on gloves.
    b. Expose the area to be sutured.
    c. Gently scrub the site with an antiseptic solution using circular motions for a minimum of 5 minutes.
NOTE: Use ample pressure to remove dirt and microorganisms.
    d. Irrigate the wound with a copious amount of normal saline at a low pressure.
    e. Dry the site using sterile gauze pads.

2. Anesthetize the area.
    a. Cryoanesthesia.
        (1) Apply a moistened ice cube to the skin for about 5 minutes.
        (2) Spray the area with commercial refrigerants, as directed.
    b. Topical applications.
        (1) Apply the agent directly to the mucus membrane, serous surface, or onto the open wound.
        (2) Slightly saturate a gauze with the appropriate agent and place it on the wound for 5 to 10 minutes.
        (3) Check the area for tissue blanching which indicates adequate anesthesia.
NOTE: Often topical application is suboptimal for suture placement.
    c. Simple infiltration.
        (1) Ensure the casualty does not have an allergy to the agent.
        (2) Using a needle and syringe, draw up an adequate amount of 1% lidocaine.
NOTE: Lidocaine with epinephrine is never used on the tip of the nose, ears, fingers, toes, or genitalia due to vasoconstriction.
        (3) Enter directly into the dermis through the laceration.
        (4) Aspirate prior to injecting the solution to ensure the needle is not in a vessel. (If blood returns into the syringe, withdraw, change the needle, and try a new site.
        (5) Slowly inject solution beneath the skin surface, raising a wheal in the area to be anesthetized.
        (6) Repeat steps 2c(3) through 2c(5) depending on the size of the laceration.

3. Select the method of closure.
    a. Skin adhesive.
        (1) Hold the wound edges together and slightly everted with tissue forceps.
        (2) Apply adhesive with the applicator tip by lightly wiping along the long axis of the wound.

Performance Steps

NOTE: Three to four thin layers should be applied successively. Avoid droplets or a single thick layer.

   (3) Hold the wound edges together for approximately 1 minute.

   (4) Instruct the casualty not to apply ointment or dressing to the wound.

  b. Steri strips.

   (1) Apply benzoin to a 2 to 3 cm area beyond the wound edges. Do not allow benzoin to enter the wound.

   (2) Using forceps, attach the strip to the skin on one side and then pull it across the wound to close the wound edges.

   (3) Start in the center and progress toward each end. Leave some space between individual strips.

   (4) Instruct the casualty not to get the area wet.

  c. Staples.

   (1) Hold the wound edges together with tissue forceps.

   (2) Place the stapling device gently against the skin surface.

   (3) Slowly squeeze the trigger.

   (4) Evenly place only the necessary amount of staples to close the wound.

NOTE: There is little to no benefit to locally infiltrating an area for 1 to 2 staples to be placed. The anesthetic is more discomforting than the procedure.

  d. Suture.

   (1) Select the proper size and type of material.

   (2) Check for adequate anesthesia by grasping the wound edges with tissue forceps. Note if the casualty can feel pain.

   (3) Grasp the needle with the needle holder about 1/2 to 1/3 the distance from where the suture is attached.

   (4) Hold the needle holder in the palm, using the index finger for fine control.

   (5) Enter the skin at approximately a 90 degree angle on the far side of the wound and exit on the near side.

NOTE: You should enter and exit the skin about 2 mm from the edge. Entry and exit points should be directly across from each other.

   (6) Pull the suture through the wound until approximately a 2 cm tail remains on the far side of wound.

   (7) Hold the end of the suture attached to the needle in the nondominant hand.

   (8) Hold the needle holder in the dominant hand.

   (9) Loop the suture twice around the needle holder.

   (10) Grasp the free end of the suture with the blades of the needle holder.

   (11) Cross the hands so that the hand holding the swaged end is on the far side and the hand holding the needle holder and free end are on the near side of the wound.

   (12) Pull upward on the suture ends when clinching the first throw.

   (13) Adjust the tension of the first throw so that the wound edges come together snugly but not tightly.

   (14) For the second throw of the knot, the needle end is on the far side of the wound and the free end on the near side.

   (15) Hold the needle end of the suture in the nondominant hand and lay the needle holder on top.

   (16) Loop the suture only once around the needle holder.

   (17) Grasp the free ends with the blades of the holder.

   (18) Cross the hands so that the sutures smoothly intertwine.

   (19) Cinch down the throw.

Performance Steps

CAUTION: Take care not to cinch down too tightly on the second throw because the tightness will be transmitted to the wound.

      (20) Pull the knot to the side so that it will not directly overlie the laceration.

      (21) The pattern of looping the suture around the holder on alternate sides of the wound is repeated until the desired number of throws are completed.

      (22) Cut the ends of the suture material to approximately 3 to 5 cm length.

  4. Apply antibiotic ointment to the site.

  5. Apply a sterile dressing to the site.

  6. Remove gloves.

  7. Document the procedure.

Performance Measures

	GO	NO-GO
1. Prepared the site.	——	——
2. Anesthetized the area.	——	——
3. Selected the method of closure and sutured the laceration.	——	——
4. Applied antibiotic ointment.	——	——
5. Applied a sterile dressing.	——	——
6. Removed gloves.	——	——
7. Documented the procedure.	——	——

Evaluation Guidance: Score each Soldier according to the performance measures. Unless otherwise stated in the task summary, the Soldier must pass all performance measures to be scored GO. If the Soldier fails any steps, show what was done wrong and how to do it correctly.

References

    Required　　　　　　　　　**Related**
    SF 600　　　　　　　　　　　BASIC NURSING 7

ADMINISTER BLOOD
081-835-3000

Conditions: You have verified a medical officer's order requiring the administration of blood. You have identified the patient and explained the procedure. A patient care hand-wash has been performed. You will need a blood pack with SF 518 (Medical Record - Blood or Blood Component Transfusion), thermometer, blood pressure cuff, stethoscope, blood transfusion recipient set ("Y" type), intravenous (IV) stand, tourniquet, needle and syringe, IV catheter, tape, alcohol and betadine prep pads, gloves, a container of 0.9% normal saline for injection, and the patient's clinical records. You are not in a CBRNE environment.

Standards: Administer the blood IAW the medical officer's orders and without causing injury to the patient.

Performance Steps
1. Verify and inspect the blood pack received from the laboratory.
    a. Note the time the blood pack was received and record the time on the SF 518.
NOTE: Infusion of a blood pack should be initiated within 30 minutes of being issued.
    b. Two people must verify and match the information on the blood pack label with the data on the requisition form (SF 518).
NOTE: One of the verifiers must be a Registered Nurse when directed by local policy.
    c. Inspect the blood for abnormalities such as gas bubbles or black or gray colored sediment (indicative of bacterial growth).
NOTE: Return the blood pack to the blood bank if any abnormality is present or suspected.
    d. Match the blood pack with the patient's identification.
        (1) The same two people must compare the information on the blood unit with the data on the patient's wristband. Ensure the patient's name, blood type, and hospital number positively match the data on the blood pack.
        (2) Sign the SF 518 IAW local policy when all the data has been confirmed as a positive match.

2. Establish baseline data.
    a. Reconfirm data from the patient's history regarding allergies or previous reactions to blood or blood products.
    b. Measure and evaluate the vital signs.
    c. Record the vital signs on the SF 518 and in the nursing notes.

3. Prepare the blood and the blood recipient set.
NOTE: Use only tubing that is designed for the administration of blood products. It is equipped with a filter designed for the fine filtration required for blood products.
    a. Close all three clamps on the "Y" tubing.
    b. Aseptically insert one of the tubing spikes into the container of normal saline. Invert and hang this container about 3 feet above the level of the patient.
    c. Open the clamp on the normal saline line and prime the upper line and the blood filter.
    d. Open the clamp on the empty line on which you will eventually hang the blood. Normal saline will flow up the empty line to prime that portion of the tubing.
NOTE: Use only 0.9% normal saline for injection with blood. Other solutions are not compatible.
    e. Once the blood line is primed with saline, close the clamp on the blood line.
    f. Leave the clamp on the normal saline line open.
    g. Open the main roller clamp to prime the lower infusion tubing.

Performance Steps

    h. Close the main roller clamp.

    i. Aseptically expose the blood port on the blood pack.

    j. Aseptically insert the remaining spike into the blood port and hang the blood at the same level as the normal saline container.

NOTE: If "Y" type recipient tubing is not available, use regular infusion tubing for the normal saline and the available blood recipient tubing for the blood pack. Prime each set. Attach a sterile, large bore (16 or 18 gauge) needle to the end of the blood tubing and "piggyback" the blood into the normal saline line below the level of the roller clamp. Hang the blood pack at least 6 inches higher than the normal saline.

  4. Perform the venipuncture. (See task 081-833-0033.)

NOTE: Insert a large gauge IV catheter (14, 16, or 18) for administering blood to an adult patient. This will enhance the flow of blood and prevent hemolysis of the cells.

  5. Begin the infusion of blood.

    a. Attach the primed infusion set to the catheter, tape it securely, and open the main roller clamp.

NOTE: If a preexisting catheter is being used, run in 50 cc of normal saline to flush out any incompatible solution. If a new catheter was inserted, this step is not required.

    b. Close the roller clamp to the normal saline and open the roller clamp to the blood.

    c. Adjust the flow rate with the main roller clamp.

      (1) Set the flow rate to deliver approximately 10 to 25 cc of blood over the first 15 minutes.

NOTE: When delivering blood by piggyback, begin the infusion by opening the roller clamp on the normal saline line and setting it to keep open (TKO) rate. Adjust the roller clamp on the blood line to deliver 10 to 25 cc of blood over the first 15 minutes.

Performance Steps

(2) Monitor the vital signs closely for the first 15 minutes and observe for indications of an adverse reaction to the blood. Refer to Figure 3-74.

PATIENT'S SYMPTOMS	REACTION TIME	TYPE OF REACTION
* Nausea * Severe chills * Rapid elevation of temperature * Pain in the lumbar region * Flushed appearance * Tachycardia * Hypotension	* After only 25 cc of blood have been transfused	HEMOLYTIC * Extremely serious * Can be fatal * Transfusion of incompatible RBCs * RBC destruction
* Mild to severe chills * Normal to elevated temperature * Headache * Flushed appearance * Anxiety * Hypotension	* During the transfusion or 30 to 60 minutes after the transfusion is completed.	PYROGENIC * Serious * Contaminants may have been introduced into the blood or the IV equipment
* Flushed appearance * Edema of the face and lips * Dyspnea (from laryngeal edema) * Wheezing * Anxiety * Itching * Hives * Anaphylaxis	* During the transfusion or 1 to 2 hours after the transfusion in completed	ALLERGIC * Serious * Allergic response (hypersensitivity) by the recipient to substances in the donor blood.

Figure 3-74. Monitor the vital signs chart

CAUTION: Any time an adverse reaction is suspected, immediately stop the blood and infuse normal saline. Notify the charge nurse and medical officer immediately.

(3) Set the main roller clamp to deliver the prescribed flow rate if, after the first 15 minutes, no adverse reaction is suspected and the vital signs are stable.

NOTE: Use the correct formula to calculate flow rate.

6. Monitor and evaluate the patient throughout the procedure.
   a. Monitor vital signs every hour or more frequently IAW local SOP.
   b. Compare the vital signs with previous and baseline vital signs.
   c. Observe for changes that indicate an adverse reaction to the blood.
   d. Stop the blood, infuse normal saline, and notify the charge nurse and medical officer if a reaction is suspected.

CAUTION: When a transfusion reaction occurs or is suspected, the unused blood and recipient tubing must be sent to the laboratory along with a 10 ml specimen of the patient's venous blood and a post transfusion urine specimen.

7. Discontinue the infusion of blood.
   a. When the blood pack has emptied, close the clamp to the blood and open the clamp to the normal saline.
   b. Flush the tubing and filter with approximately 50 cc of normal saline to deliver the residual blood.
   c. After the residual blood has been delivered, run the normal saline at a TKO rate or hang another solution, if one has been prescribed.

Performance Steps

d. Take and record the vital signs at the completion of the transfusion and 1 hour after completion.

NOTE: As a rule, a unit of blood should be infused within 2 to 4 hours unless contraindicated by risk of circulatory overload. If the prescribed flow rate will deliver the blood within a shorter or longer period of time, verify the order with the charge nurse or prescribing medical officer.

8. Dispose of the used blood pack IAW local SOP.
   a. Return it to the laboratory blood bank with a copy of SF 518.
   b. Discard it in a container for contaminated waste.

9. Document the procedure and significant nursing observations on the appropriate forms IAW local SOP.
   a. Complete the SF 518.
      (1) Return one copy to the laboratory blood bank.
      (2) Place one copy in the patient's chart.
   b. Record the procedure and the patient's response in a nursing note entry.

Performance Measures	GO	NO-GO
1. Verified and inspected the blood pack.	——	——
2. Established baseline data.	——	——
3. Prepared the blood and blood recipient sets.	——	——
4. Performed the venipuncture.	——	——
5. Began the infusion of blood.	——	——
6. Monitored and evaluated the patient.	——	——
7. Discontinued the infusion of blood.	——	——
8. Disposed of the used blood pack IAW local SOP.	——	——
9. Documented the procedure on the appropriate forms IAW local SOP.	——	——

Evaluation Guidance: Score each Soldier according to the performance measures. Unless otherwise stated in the task summary, the Soldier must pass all performance measures to be scored GO. If the Soldier fails any steps, show what was done wrong and how to do it correctly.

References

Required	Related
SF 518	BASIC NURSING 7
	KOZIER & ERB

ADMINISTER MEDICATIONS BY IV PIGGYBACK
081-835-3002

Conditions: You have a medical officer's orders requiring the administration of a medication by the intravenous (IV) piggyback route. You must prepare the piggyback unit. A patient care hand-wash has been performed. You will need medication, diluent, needle, syringe, alcohol (or other antiseptic) prep pads, label, container of IV solution, IV administration tubing, tape, DA Form 4678 (Therapeutic Documentation Care Plan (Medication)), and the patient's clinical record. You are not in a CBRNE environment.

Standards: Prepare the IV piggyback unit without contamination and administer it to the patient without complications.

Performance Steps

1. Identify the patient, explain the procedure, and ask about allergies.

2. Check the DA Form 4678 against the medical officer's order.
   a. Name of the medication.
   b. Amount (dose) of medication.
   c. Route of administration.
   d. Time to be administered.

3. Select the medication.
   a. Check the medication label three times to ensure that the correct medication is being prepared for administration.
   b. Check the expiration date of the medication.
   c. Handle only one medication at a time.
NOTE: If unfamiliar with a medication, look it up to determine contraindications, precautions, and side effects.

4. Prepare the medication.
   a. Calculate the amount of medication required to equal the prescribed dose.
NOTE: If the medication is in powdered form, prepare it for use by adding the diluent specified on the drug information instructions.
   b. Draw the prescribed amount of the prepared medication into a syringe.
   c. Check the medication and calculations again to ensure that the correct medication and correct dose have been prepared.

5. Prepare the piggyback unit.
NOTE: Refer to the drug manufacturer's instructions to determine the type and amount of solution to be used as the piggyback unit.
   a. Use an alcohol prep pad to swab the injection port on the container of IV solution to be used as the piggyback unit.
   b. Inject the prepared medication into the container of IV solution.
   c. Mix the solution and medication into the container of IV solution.
   d. Label the piggyback unit with the name of the medication, the amount added, the time added, the date added, and the initials of the person who prepared the piggyback unit.
   e. Dispose of the needle and syringe IAW local SOP.

6. Prime the piggyback infusion tubing.
   a. Close the clamp on the piggyback tubing.

Performance Steps

    b. Aseptically insert the spike on the piggyback tubing into the solution port on the piggyback unit.

    c. Squeeze the drip chamber to fill it half full.

    d. Open the clamp on the piggyback tubing, allowing the solution to prime the tubing.

    e. Close the clamp on the piggyback tubing when the solution reaches the end of the tubing.

NOTE: Attach a sterile needle to the end of the piggyback tubing if one is not provided by the manufacturer.

CAUTION: Take care not to waste any medicated IV solution while priming the tubing.

  7. Connect the piggyback unit to the primary tubing.

    a. Swab the injection port on the primary tubing with an alcohol prep pad.

    b. Insert the needle into the injection port of the primary tubing.

    c. Secure the connection with tape.

NOTE: Attach the piggyback tubing to the primary tubing below the level of the roller clamp. This will allow the piggyback unit to flow at its set rate without adjusting the flow rate of the primary solution.

  8. Hang the piggyback unit on the IV pole, ensuring that the piggyback unit is at least 6 inches higher than the primary container.

  9. Ensure patency of the primary IV.

  10. Begin the secondary (piggyback) infusion.

    a. Calculate the flow rate in accordance with the medical officer's order.

NOTE: If the medical officer does not specify a flow rate, set the flow rate IAW the drug manufacturer's instructions.

    b. Adjust the roller clamp on the piggyback tubing to regulate the flow rate of the piggyback solution.

CAUTION: Do not adjust the flow rate of the primary container.

NOTE: When fluid from the secondary line enters the primary tubing, the primary infusion is automatically interrupted. When all the solution in the piggyback unit has been delivered, the primary infusion will resume flow at the set rate.

  11. Label the piggyback infusion tubing with the time and date the medication was initiated.

  12. Observe the patient for signs of infusion complications or reaction to the medicine. (See task 081-833-0034.)

  13. Document the procedure and significant nursing observations on the appropriate forms IAW local SOP.

Performance Measures

	GO	NO-GO
1. Identified the patient, explained the procedure, and asked about allergies.	——	——
2. Checked the DA Form 4678 against the medical officer's order.	——	——
3. Selected the medication.	——	——
4. Prepared the medication.	——	——

Performance Measures	<u>GO</u>	<u>NO-GO</u>
5. Prepared the piggyback unit.	——	——
6. Primed the piggyback infusion tubing.	——	——
7. Connected the piggyback unit to the primary tubing.	——	——
8. Hung the piggyback unit on the IV pole, ensuring that the piggyback unit was at least 6 inches higher than the primary container.	——	——
9. Ensured patency of the primary IV.	——	——
10. Began the secondary (piggyback) infusion.	——	——
11. Labeled the piggyback infusion tubing with the time and date the medication was initiated.	——	——
12. Observed the patient for signs of infusion complications or reaction to the medicine.	——	——
13. Documented the procedure and significant nursing observations on the appropriate forms IAW local SOP.	——	——

Evaluation Guidance: Score each Soldier according to the performance measures. Unless otherwise stated in the task summary, the Soldier must pass all performance measures to be scored GO. If the Soldier fails any steps, show what was done wrong and how to do it correctly.

References

Required	**Related**
DA FORM 4678	BASIC NURSING 7

ADMINISTER BLOOD PRODUCTS
081-835-3054

Conditions: Your assigned patient has an order requiring the administration of a specific blood product. You will need the prescribed blood component, appropriate administration set, clean gloves, the patient's medical record, and SF 518 (Medical Record - Blood or Blood Component Transfusion). You are not in a CBRNE environment.

Standards: Safely and correctly administer a prescribed blood component.

Performance Steps

1. Verify the medical officer's order and ensure the patient has a current signed informed consent for transfusion.

2. Review the procedure for administering a blood transfusion. (See task 081-835-3000.)
CAUTION: The administration of blood and blood products is generally outside the scope of practice of the 68WM6 Practical Nurse. Administration of blood and blood products is restricted to the domain of a medical officer. However, this task is an expected skill of the 68WM6 Practical Nurse in a combat field environment.

3. Obtain the blood component from the blood bank.

4. Obtain the appropriate administration set.

5. Review the directions and facility protocol for administration of the blood product.
  a. Rate of administration.
  b. Route.
  c. Risk factors.
  d. Possible complications.

6. Obtain a positive identification of the patient and provide privacy.

7. Wash your hands and follow standard precautions.

8. Explain the procedure to the patient/family.

9. Administer the prescribed blood product.
NOTE: When using an infusion control device (pump), review the manufacturer's instructions for use to ensure you are using the correct pump, configured correctly, with the proper tubing and filter.
  a. Fresh plasma - rapid administration with straight line administration set.
  b. Platelets - administer at a rate of 10 minutes per unit, using the platelet infusion set and filter.
  c. Granulocytes - slow administration over 2 to 4 hours with Y-type blood tubing and normal saline. Do not use a microaggregate filter.
  d. Serum albumin - slow administration at 1 ml/min using the special tubing supplied with the solution.
  e. Gamma globulin - given IM 0.25 to 0.5 ml.
  f. Coagulation factors - administer with standard syringe or component drip set.
    (1) Factor VIII - 1unit/5min
    (2) Factor IX - reconstitute 10 to 20 ml of diluent.

Performance Steps

10. Monitor the patient for unexpected outcomes and report abnormal findings to the medical officer.

11. Conduct patient education based on specific patient needs.

12. Document the nursing activity.
    a. Type and amount of blood product administered.
    b. Volume of saline or diluent used.
    c. Patient's response to the procedure.
    d. Unexpected outcomes and interventions.
    e. Patient education provided.

Performance Measures	GO	NO-GO
1. Verified the medical officer's order and the presence of a current signed informed consent.	——	——
2. Reviewed the procedure for administering a blood transfusion.	——	——
3. Obtained the prescribed blood component from the blood bank.	——	——
4. Obtained the correct administration set.	——	——
5. Reviewed the directions and facility protocol for administration of the blood product.	——	——
6. Obtained positive patient identification and provided privacy.	——	——
7. Washed hands and followed standard precautions.	——	——
8. Explained the procedure to the patient/family.	——	——
9. Correctly administered the prescribed blood product.	——	——
10. Monitored the patient for unexpected outcomes and reported abnormal findings to the medical officer.	——	——
11. Provided patient education based on individual needs.	——	——
12. Documented the nursing activity.	——	——

Evaluation Guidance: Score each Soldier according to the performance measures in the evaluation guide. Unless otherwise stated in the task summary, the Soldier must pass all performance measures to be scored GO. If the Soldier fails any step, show what was done wrong and how to do it correctly.

References

Required
SF 518

Related
CARLSON & LYNN-MCHALE
DUELL, MARTIN, & SMITH

INSPECT BASE CAMPS
081-850-0049

Conditions: You must inspect a base camp site to ensure compliance with general sanitation requirements. You will need a clipboard, a site survey checklist, and FM 4-02.17.

Standards: Inspected the base camp to ensure compliance with general sanitation IAW FM-4-02.17

Performance Steps

1. Identify the person-in-charge of the base camp to be inspected, notify him or her of the reasons for the inspection, and request an escort as required.

2. Conduct an inspection of the entire base camp area that includes all aspects of the unit's preventive medicine status.
   a. Check that individual preventive medicine measures (PMM) are conducted by individuals and provided for by the command.
      (1) Showering devices.
      (2) Handwashing devices.
      (3) Soakage pits located under handwashing and showering devices.
      (4) Laundry facilities.
   b. Inspect the base camp water supply.
      (1) Quantity of water required for soldiers is available.
      (2) Inspect quartermaster water distribution points within the base camp.
      (3) Inspect water sources supporting base camp operations.
      (4) Inspect water containers.
   c. Inspect food service sanitation.
      (1) Transportation of food.
      (2) Food storage operations.
      (3) Food preparation and serving operations.
      (4) Field dishwashing facilities.
   d. Inspect base camp waste disposal operations.
      (1) Inspect disposal of human waste.
      (2) Inspect solid waste disposal and any temporary storage.
      (3) Inspect handling of regulated medical waste and hazardous materials/waste that may occur within the base camp.
   e. Inspect base camp arthropod control operations.
   f. Inspect base camp rodent control operations.
   g. Inspect preventive measures against heat and cold injuries.
   h. Inspect preventive measures against industrial chemical hazards.
   i. Inspect preventive measures against noise hazards.
   j. Inspect any unit field sanitation teams within the base camp.

3. Annotate noted deficiencies and determine corrective measures for each.

Performance Measures	GO	NO-GO

1. Identified the person-in-charge of the base camp to be inspected, notified him of reasons for the inspection, and requested an escort as required. —— ——

2. Conducted an inspection of the entire base camp area that included all aspects of the unit's preventive medicine status. —— ——
   a. Individual PMMs.
   b. Water supply.
   c. Food service sanitation.
   d. Waste disposal operations.
   e. Arthropod control operations.
   f. Rodent control operations.
   g. Preventive measures against heat and cold injuries.
   h. Preventive measures against industrial chemical hazards.
   i. Preventive measures against noise hazards.
   j. Unit field sanitation teams.

3. Annotated noted deficiencies and determined corrective measures for each. —— ——

Evaluation Guidance: Score the Soldier GO if passes all of the performance measures. Score the Soldier NO-GO if fails any performance measure. If the Soldier fails any performance measure, demonstrate how to perform it correctly. The Soldier is expected to review the performance steps and practice the task until can perform it correctly.

References
Required	Related
FM 4-02.17 (FM 4-02.17)	

This page intentionally left blank.

APPENDIX A -

DRUG DOSAGE CALCULATIONS

Calculate Intravenous Drip Rates

1. To calculate the drip rate per minute (flow rate) of intravenous (IV) fluids, first obtain the following information:

    a. Delivery rate (drops per cc) of the IV tubing set being used. This is also referred to as the "tubing factor." (The IV tubing package will state the rate of delivery for that particular IV set; for example, 10 drops per cc for standard drip tubing or 60 drops per cc for mini drip tubing.)

    b. Volume of fluid (in cc) to be infused. (This can be expressed in an hourly amount or in a total volume; for example, "100 cc/hour" or "2 liters over 6 hours.")

    c. Amount of time (in minutes) the fluid is to be infused. (This can be expressed in an hourly rate or total time; for example, "150 cc per hour" or "infuse 1 liter over 4 hours.")

2. Calculate the flow rate in drops per minute using the following formula:

$$\text{gtt/min} = \frac{\text{volume to be infused} \times \text{gtt/cc of administration set}}{\text{infusion time in minutes}}$$

1. To convert grams (Gm) to milligrams (mg), multiply Gm by 1000 and move the decimal point three places to the right; for example, 0.075 Gm = 75 mg and 0.25 Gm = 250 mg.

2. To convert milligrams to grams, divide milligrams by 1000 and move the decimal point three places to the left; for example, 1000 mg = 1 Gm and 500 mg = .5 Gm.

Calculation of Doses from Drugs in Solution

1. Some drugs are dispensed as solutions. The strength of the solution is written on the label of the drug container; for example, "10 mg per ml." The problem is to determine what quantity of solution will contain the required dose of the drug. The method of solving the problem is by ratio and proportion. The formula is as follows:

Required amount of drug	:	Unknown amount of solution	::	ratio of strength of solution on hand
_____ is to		_____	as	_____ is to _____

2. EXAMPLE: The physician has ordered Benadryl Elixir, 25 mg p.o. The Benadryl Elixir on hand contains 10 mg per ml. How many ml (cc) of the Elixir must be administered to achieve the required dose?

    a. Write out the formula.

        25 mg : x ml :: 10 mg : 1 ml

    b. Multiply the inner values.

        x times 10 (10x)

    c. Multiply the outer values.

        25 times 1 (25)

    d. The multiplied inner values equal the multiplied outer values, so:

        10x = 25

    e. Divide 25 by 10 to find x.

        $x = \dfrac{25}{10}$ or 2.5

    f. 2.5 ml of Benadryl Elixir must be administered to achieve the required dose of 25 mg.

Convert from Apothecary to Metric

1. To convert grains to milligrams, multiply grains by 60 to obtain milligrams; for example, 1/4 grain = 15 milligrams.

2. To convert milligrams to grains, divide milligrams by 60 to obtain grains; for example, 30 mg = 1/2 grain.

Liquid Measure		Liquid Measure	
Metric	**Approximate Apothecary Equivalents**	**Metric**	**Approximate Apothecary Equivalents**
1,000 ml	1 quart	3 ml	45 minims
750 ml	1 1/2 pints	2 ml	30 minims
500 ml	1 pint	1 ml	15 minims
250 ml	8 fluid ounces	0.75 ml	12 minims
200 ml	7 fluid ounces	0.6 ml	10 minims
100 ml	3 1/2 fluid ounces	0.5 ml	8 minims
50 ml	1 3/4 fluid ounces	0.3 ml	5 minims
30 ml	1 fluid ounce	0.25 ml	4 minims
15 ml	4 fluid drams	0.2 ml	3 minims
10 ml	2 1/2 fluid drams	0.l ml	1 1/2 minims
8 ml	2 fluid drams	0.06 ml	1 minim
5 ml	1 1/4 fluid drams	0.05 ml	3/4 minim
4 ml	1 fluid dram	0.03 ml	1/2 minim
Weight		**Weight**	
Metric	**Approximate Apothecary Equivalents**	**Metric**	**Approximate Apothecary Equivalents**
30 Gm	1 ounce	30 mg	1/2 grain
15 Gm	4 drams	25 mg	3/8 grain
10 Gm	2 1/2 drams	20 mg	1/3 grain
7.5 Gm	2 drams	15 mg	1/4 grain
6 Gm	90 grains	12 mg	1/5 grain
5 Gm	75 grains	10 mg	1/6 grain
4 Gm	60 grains/1 dram	8 mg	1/8 grain
3 Gm	45 grains	6 mg	1/10 grain
2 Gm	30 grains/1/2 dram	5 mg	1/12 grain
1.5 Gm	22 grains	4 mg	1/15 grain
1 Gm	15 grains	3 mg	1/20 grain
0.75 Gm	12 grains	2 mg	1/30 grain
0.6 Gm	10 grains	1.5 mg	1/40 grain
0.5 Gm	7 1/2 grains	1.2 mg	1/50 grain
0.4 Gm	6 grains	1 mg	1/60 grain
0.3 Gm	5 grains	0.8 mg	1/80 grain
0.25 Gm	4 grains	0.6 mg	1/100 grain
0.2 Gm	3 grains	0.5 mg	1/120 grain
0.15 Gm	2 1/2 grains	0.4 mg	1/150 grain
0.12 Gm	2 grains	0.3 mg	1/200 grain
0.1 Gm	1 1/2 grains	0.25 mg	1/250 grain
75 mg	1 1/4 grains	0.2 mg	1/300 grain
60 mg	1 grain	0.15 mg	1/400 grain
50 mg	3/4 grain	0.12 mg	1/500 grain
40 mg	2/3 grain	0.1 mg	1/600 grain

ABBREVIATIONS

ac	before meals
ad lib	as much as desired
bid	twice a day
c	with
cc	cubic centimeter
caps	capsule
Gm	gram
gr	grain
gtt	drop
h	hour
hs	bedtime (hour of sleep)
kg	kilogram
l	liter
mg	milligram
ml	milliliter
od	right eye (oculo dextro)
os	left eye (oculo sinistro)
ou	both eyes (oculus uterque)
pc	after meals
po	by mouth
prn	when needed/as necessary
qd	every day (daily)
qid	four times daily
qod	every other day
qs	in sufficient quantity
q2h	every 2 hours
q4h	every 4 hours
q6h	every 6 hours
q8h	every 8 hours
s	without
stat	at once/immediately
sq or sc	subcutaneously
ss	one half
tab	tablet
tsp	teaspoon
tbsp	tablespoon
tid	three times daily

GLOSSARY

1SG
first sergeant

AAR
after-action review; after-action report

ACCP
The Army Correspondence Course Program

AED
automatic external defibrillator

AIPD
Army Institute for Professional Development

ALC
Advanced Leader Course

AMEDD
Army Medical Department

ARNG
Army National Guard

ARNGUS
Army National Guard of the United States

ATNAA
antidote treatment, nerve agent, autoinjector

AVPU
alertness, responsiveness to vocal stimuli, responsiveness to painful stimuli, unresponsiveness

BAS
battalion aid station

Battle focus.
A process to guide the planning, execution, and assessment of the organization's training program to ensure they train as they are going to fight.

BDO
battle dress overgarment

BDU
battle dress uniform

BSA
> battalion aid station

BSA(MED)
> body surface area

BSI
> body substance isolation

BVM
> bag-valve-mask

CAM
> chemical agent monitor

CANA
> convulsant antidote for nerve agent

CASEVAC
> casualty evacuation

C-A-T
> combat application tourniquet

CBRN
> Caribbean Basin Radar Network; chemical, biological, radiological, or nuclear

CBRN*
> chemical, biological, radiological, or nuclear

CBRNE
> chemical, biological, radiological, nuclear, and high-yield explosive

CC
> Chemical Corps; cubic centimeter

cc*
> cubic centimeter

cc/hr
> cubic centimeters of fluid per hour

CCP*
> casualty collection point

cGy
> centigray

CL
> class/combat lifesaver (depends on use)

cm
centimeter(s); Chemical

cm H2O
centimeters of water

CMS
central material service

CNS
central nervous system

Collective training.
Training, either in institutions or units, that prepares cohesive teams and units to accomplish their combined arms and service missions on the battlefield.

Common task.
A critical task that is performed by every Soldier in a specific skill level regardless of MOS.

cont
continued; continuous

COPD
chronic obstructive pulmonary disease

CPR
cardiopulmonary resuscitation

Critical Task
See "Task," "Critical collective task," and "Critical individual task."

Cross training.
The systematic training of a Soldier on tasks related to another duty position within the same military occupational specialty or tasks related to a secondary military occupational specialty at the same skill level.

CSF
cerebrospinal fluid

CSM
command sergeant major

CTC
combat training center or combined training center

DCAP-BTLS
deformities, contusions, abrasions, punctures or penetration, burns, tenderness, lacerations, and swelling

DCS(MED)
  division clearing station

DTR
  deep tendon reflex

EKG
  electrocardiogram/electrocardiograph

EPW
  enemy prisoner of war

ET*
  endotracheal

ETD*
  emergency trauma dressing

ETS
  expiration time of service

F
  Fahrenheit; fail; failed; frequency; full

FAS*
  forward aid station

FM
  field manual; frequency modulated/modulation

FMC*
  Field Medical Card

FROPVD
  flow-restricted oxygen-powered ventilation device

gtts
  drops

HD*
  hemostatic dressing

HEPA
  high effeciency particulate air filters

HTH
  high test hypochlorite

I&O
  intake and output

IAW
   in accordance with

ICS*
   intercostal space

ID
   identification; infantry division

IET
   Initial-entry training

IM
   intramuscular

IMSA
   installation medical supply activity

Individual training.
   Training which prepares the Soldier to perform specified duties or tasks related to the assigned duty position or subsequent duty positions and skill levels.

Integrated training
   Training of a critical task in a formal course of instruction by integrating or consolidating the proponent-provided TSP material into an existing lesson. The task MAY be one in which the performer has received prior training, i.e., it is best used to sustain/refine previously acquired skills. The training must be applicable to the block of instruction in which it is integrated; trains the task to standard; and evaluates task performance during instruction under conditions prescribed in the TSP.

IV
   intravenous

JSLIST
   joint service lightweight integrated suit technology

JVD
   jugular vein distention

KED
   Kendrick Extrication Device

kg
   kilogram(s)

KVO
   keep the vein open

L/min
   liters per minute

LCE
load-carrying equipment

LCV
load-carrying vest

LPM/lpm
liters per minute

LZ
landing zone

MAS**
main aid station

MCL*
midclavicular line

MD
medical doctor

MDI
metered dose inhaler

MEDCEN
medical center

MEDCOM
medical command

MEDDAC
medical department activity

MEDEVAC
medical evacuation

Merger training.
Training that prepares noncommissioned officers to supervise one or more different military occupational specialties at lower skill levels when they advance to a higher level in their career management field.

MES
medical equipment set(s)

METL
mission essential task list

METT-TC
mission, enemy, terrain and weather, troops and support available, time available, civil considerations

mg**
milligram(s)

min(s)
minute(s)

Mission essential task list.
A compilation of collective mission essential tasks which must be successfully performed if an organization is to accomplish its wartime mission(s).

ml or mL
milliliter

mm Hg
millimeters of mercury

mm/sec
millimeters per second

MOI*
mechanism of injury

MOOTW
military operations other than war (joint only)

MOPP
mission-oriented protective posture

MOS
military occupational specialty

MOSC
military occupational specialty code

MRE
meal, ready to eat

MSO
medical supply officer/medical supply office

MTF
medical treatment facility

MTP
mission training plan; MOS training plan

MVA
> motor vehicle accident

MTP*
> MOS training plan

NAAK
> nerve agent antidote kit

NATO
> North Atlantic Treaty Organization

NBC
> nuclear, biological, and chemical

NCO
> noncommissioned officer

NPA*
> nasopharyngeal airway

NPO
> nothing by mouth

NRB
> nonrebreather

NSN*
> national stock number

OPA*
> oropharyngeal airway

OSHA
> Occupational Safety and Health Administration

P
> needs practice; pass; passed; barometric pressure; mean radius of curvature

PA*
> physician assistant

PAC
> Personnel Administration Center / program activity code / Pacific (depends on use)

PASG
> pneumatic anti-shock garments

PCS
> permanent change of station

PEA
pulseless electrical activity

PMM
preventive medicine measures

PMS*
pulse, motor function, and sensation

pnt
patient

PPE
protective posture equipment; personal protective equipment

ppm
parts per million

prn
as necessary

psi
pounds per square inch

PVC
premature ventricular contraction

RC
Reserve Component

RDIC
resuscitation device, individual chemical

RDL
Reimer Digital Library

RN
registered nurse

RTO
radio/telephone operator

RYE
retirement year ending

SC or SQ
subcutaneous

SELF-DEVELOPMENT
    Self-development is a planned, progressive, and sequential program followed by leaders to enhance and sustain their military competencies. Self-development consists of individual study, research, professional reading, practice, and self-assessment.

SL
    squad leader; skill level

SLC
    Senior Leader Course

SM
    Soldier's manual

SMCT
    Soldier's manual of common tasks

SOAP
    subjective, objective, analysis, plan

SOI
    signal operating instructions

SOP
    standing operating procedures

SSN
    social security number

STAT
    immediately

STB
    supertropical bleach

STP
    Soldier training publication

SURG
    surgeon

Sustainment training.
    The provision of training to maintain the minimum acceptable level of proficiency required to accomplish a critical task.

TADSS
    training aids, devices, simulators, and simulations

TBSA
    total body surface area

TG
trainer's guide

TIC*
tenderness, instability, or crepitus

TKO
to keep open

TRADOC
United States Army Training and Doctrine Command

Training-up
The process of increasing the skills and knowledge of an individual to a higher skill level in the appropriate MOS. It may involve certification.

TRD
tenderness, rigidity, and distension

TSC
Theater Support Command/Training Support Center (depends on use)

TSOP
tactical standing operating procedure(s)

TTP
tactics, techniques, and procedures

Unit training.
Training (individual, collective, and joint or combined) conducted in a unit.

US
United States

USAR
United States Army Reserve

WP
white phosphorus

This page intentionally left blank.

REFERENCES

New reference material is being published all the time. Present references, as listed below may become obsolete. To keep up-to-date, see DA Pam 25-30. Many of these publications and forms are available in electronic format from the sites listed below:

Army Publishing Directorate
 Administrative Departmental Publications and Forms
  (ARs, Cirs, Pams, OFs, SFs, DD & DA Forms)
Soldier's Training Homepage – RDL Services
 Army Doctrinal and Training Publications
  (FMs, PBs, TCs, STPs)

Department of Army Forms
DA FORM 3949	*Controlled Substances Record.*
DA FORM 4678	*Therapeutic Documentation Care Plan (Medication).*

Field Manuals
FM 4-02.17)	*Preventive Medicine Services.* 28 August 2000.
FM 8-10-6	*Medical Evacuation in a Theater of Operations Tactics, Techniques, and Procedures.* 14 April 2000.

Other Product Types
DD FORM 792	*Nursing Service - Twenty-Four Hour Patient Intake and Output Worksheet.*
DD FORM 1380	*U.S. Field Medical Card.*
OF 520	*Electrocardiograph Record.*
SF 511	*Vital Signs Record.*
SF 518	*Medical Record - Blood or Blood Component Transfusion.*
SF 600	*Medical Record - Chronological Record of Medical Care.*
SF 601	*Immunization Record.*

Special Texts
BATES, B.	*A Guide to Physical Examination and History Taking*, 6th Edition; J. B. Lippincott Co 1 January 1995

Related Publications

Related publications are sources of additional information. They are not required in order to understand this publication.

Army Correspondence Course Program Subcourses
IS0 871	*Combat Lifesaver Course: Student Self-Study.* 1 September 2006
IS0 873	*Combat Lifesaver Course: Instructor Guide.* 1 September 2006
IS0 875	*Combat Lifesaver Course: Examinations.* 1 September 2006
MED 751	*Outpatient Medical Records Branch.*

Army Regulations
AR 40-66	*Medical Record Administration and Health Care Documentation.* 17 June 2008.
AR 40-562	*Immunizations and Chemoprophylaxis.* 29 September 2006.
AR 350-1	*Army Training and Leader Development.* 3 August 2007.
AR 600-63	*Army Health Promotion.* 7 May 2007.

Department of Army Forms
DA FORM 2028	*Recommended Changes to Publications and Blank Forms*

Department of Army Pamphlets
DA PAM 600-24	*Suicide Prevention and Psychological Autopsy.* 30 September 1988.
DA PAM 600-70	*U.S. Army Guide to the Prevention of Suicide and Self-Destructive Behavior.* 1 November 1985.

Field Manuals
FM 3-21.8 (FM 7-8)	*The Infantry Platoon and Squad.* 28 March 2007.
FM 3-21.38	*Pathfinder Operations.* 25 April 2006.
FM 4-02 (FM 8-10)	*Force Health Protection in a Global Environment.* 13 February 2003.
FM 4-02.2	*Medical Evacuation.* 8 May 2007.
FM 4-02.4 (FM 8-10-4)	*Medical Platoon Leaders' Handbook Tactics, Techniques, and Procedures.* 24 August 2001.
FM 4-02.6 (FM 8-10-1)	*The Medical Company, Tactics, Techniques, and Procedures.* 1 August 2002.
FM 4-02.285 (FM 8-285)	*Multiservice Tactics, Techniques, and Procedures for Treatment of Chemical Agent Casualties and Conventional Military Chemical Injuries.* 18 September 2007.
FM 4-25.11 (FM 21-11)	*First Aid.* December 2002.
FM 4-25.12	*Unit Field Sanitation Team.* 25 January 2002.
FM 7-0	*Training for Full Spectrum Operations.* 12 December 2008.
FM 7-1	*Battle Focused Training.* 15 September 2003.
FM 8-10-6	*Medical Evacuation in a Theater of Operations Tactics, Techniques, and Procedures.* 14 April 2000.

FM 8-38   *Centralized Materiel Service/Section.* 28 February 1979.

FM 21-60   *Visual Signals.* 30 September 1987.

Other Product Types

EWS NATO HANDBOOK   *Emergency War Surgery* (Third United States Revision, Emergency War Surgery NATO Handbook, 2004))

Soldier Training Publications

STP 21-1-SMCT   *Soldier's Manual of Common Tasks Skills Level 1.* 14 December 2007.

STP 21-24-SMCT   *Soldier's Manual of Common Tasks (SMCT) Warrior Leader Skill Levels 2, 3, and 4.* 9 September 2008

Special Texts

0-7637-4406-9   AAOS, *Emergency Care and Transportation of the Sick and Injured,* 9th Edition, Jones & Bartlett Publishers, 8 June 2006

BASIC NURSING 7   Rosdahl, *Textbook of Basic Nursing,* 7th Edition, Lippincott, Williams, & Wilkins (ISBN: 0-7817-1636-5), 1 October 1999

BERHOW, R (16)   *The Merck Manual of Diagnosis and Therapy,* 16th Edition; Merck Company, 1 January 1992

BTLS FOR PARAMEDICS   Campbell, *Basic Trauma Life Support for Paramedics and Other Advanced Providers,* 4th Edition, Prentice Hall, 1 August 1999

CARLSON & LYNN-MCHALE   *AACN Procedure Manual for Critical Care,* 4th Edition, W B Saunders (ISBN: 0721682685), 15 January 2001

DUELL, MARTIN, & SMITH   *Clinical Nursing Skills: Basic to Advanced Skills,* 5th Edition, Prentice Hall (ISBN: 0838515665), 15 January 2000

EMERG CARE AND TRANS 9   AAOS, *Emergency Care and Transportation of the Sick and Injured,* 9th Edition, Jones & Bartlett Publishers (ISBN: 0-7637-4738-6), 1 March 2005

EMERGENCY CARE   O'Keefe (Editor), *Brady Emergency Care,* 8th Edition, Prentice Hall (ISBN: 0835950735), 1 July 1997

FUNDAMENTALS RESP CARE   Wilkins/Scanlan/Stoller, *Egan's Fundamentals of Respiratory Care,* 8th Edition, C. V. Mosby (ISBN: 0323018130), 2 June 2003

HABIF, T. B.   *Clinical Dermatology,* 3rd Edition; C. V. Mosby, 1 January 1996

ISBN 0-07-065351-8   *Emergency Medicine,* 5th Edition, Tintinalli, McGraw Hill, 1 January 2000

ISBN 0-13-084584-1   *Basic Trauma Life Support for Paramedics and Other Advanced Providers,* Fourth Edition, Campbell, Brady, 1 January 2000

ISBN 0-316-12891-0   *Emergency Care in the Streets,* Fifth Edition, Caroline, Little, Brown and Company, 1 January 1995

ISBN 072166055X   *Clinical Procedures in Emergency Medicine,* 3rd Edition; W. B. Saunders, 1 January 1998

ISBN 0-7637-3901-4   *Emergency Care and Transportation of the Sick and Injured,* 9th Edition (AAOS, Textbook, Hard Copy), 1 January 2006

ISBN 0-8151-80002-9   *EMT-Intermediate Textbook. Mosby,* 1 January 1997

ISBN 08359-5073-5   *Emergency Care,* 8th Edition, O'Keefe, Brady, 1 January 1998

ISBN 08359-5089-1 (PBK)   *Emergency Care,* 8th Edition, O'Keefe, Brady, 1 January 1998

ISBN 0803606621	Barnes/Schreuder/Israel, *Respiratory Care Principles: A Programmed Guide to Entry Level Practice*, 3rd Edition, FA Davis (ISBN: 0803606621), 1 March 1991
ISBN 0-07-065351-9	*Emergency Medicine, A Comprehensive Study Guide*, 6th Edition, Tintinalli, McGraw Hill, 2004. 1 June 2004
ISBN13:978-0-7637-4406-9	AAOS, *Emergency Care and Transportation of the Sick and Injured*, 9th Edition, Jones and Bartlett, 1 January 2005
KOZIER & ERB	*Techniques in Clinical Nursing: Basic to Intermediate Skills*, 5th Edition, Prentice Hall (ISBN: 0130281573), 10 September 2001
PHTLS	NAEMT, PHTLS: *Basic And Advanced Prehospital Trauma Life Support* (Military Version), 5th Edition, Mosby-Year Book (ISBN: 0-32303-271-0), 27 August 2004
TACTICAL EMERGENCY CARE	De Lorenzo/Porter, *Tactical Emergency Care*, Brady, Prentice Hall (ISBN: 0-8359-5325-4), 1 January 1998
USAMRIID	*Medical Management of Biological Casualties Handbook*, 4th Edition, 1 March 1996

Technical Manuals

| TM 9-2355-311-10-1-1 | *Operator's Manual Volume 1 of 2 Common Items for Stryker Family of Vehicles Infantry Carrier Vehicle, M1126, (NSN 2355-01-481-8575) Mortar Carrier Vehicle B, M1129E1, (2355-01-505-0871) Commander's Vehicle, M1130, (2355-01-481-8573) Recon/Scout Vehicle, M1127, (2355-01-481-8572) Engineer Squad Vehicle, M1132, (2355-01-481-8570) Antitank Guided Missile Vehicle, M1134, (2355-01-481-8576) Medical Evacuation Vehicle, M1133, (2355-01-481-8580) Fire Support Vehicle, M1131, (2355-01-481-8574) Fire Support Sensor System, M1131E1, (2355-01-528-1274) Stryker.* 16 November 2006. |
| TM 9-2355-311-10-1-2 | *Operator's Manual Volume 2 of 2 Common Items for Stryker Family of Vehicles Infantry Carrier Vehicle, M1126, (NSN 2355-01-481-8575) Mortar Carrier Vehicle B, M1129E1, (2355-01-505-0871) Commander's Vehicle, M1130, (2355-01-481-8573) Recon/Scout Vehicle, M1127, (2355-01-481-8572) Engineer Squad Vehicle, M1132, (2355-01-481-8570) Antitank Guided Missile Vehicle, M1134, (2355-01-481-8576) Medical Evacuation Vehicle, M1133, (2355-01-481-8580) Fire Support Vehicle, M1131, (2355-01-481-8574) Fire Support Sensor System, M1131E1, (2355-01-528-1274) Stryker.* 16 November 2006. |

Training Aids

| MACE | Defense and Beterans Brain Injury Center, http://www.dvbic.org/ |

Training Circulars

Training Support Packages

| 081-MEV1 | Stryker Medical Evacuation Vehicle (MEV) Crew Training |

By Order of the Secretary of the Army:

GEORGE W. CASEY, JR.
General, United States Army
Chief of Staff

Official:

JOYCE E. MORROW
Administrative Assistant to the
Secretary of the Army
0909105

DISTRIBUTION:

Active Army, Army National Guard, and US Army Reserve: Not to be distributed. Electronic Media Only.

www.ingramcontent.com/pod-product-compliance
Lightning Source LLC
Chambersburg PA
CBHW040752220326
41597CB00029BA/4726